ALLEN PARK
PUBLIC LIBRARY
8100 Allen Rd
Allen Park, MI 48101-1708
(313) 381-2425
W9-BVU-311
ALLEN PARK PUBLIC LIBRARY #4
ALLEN PARK 48101
313-381-2425
DISCARD

Robert Markley

Encyclopedia of ROSES

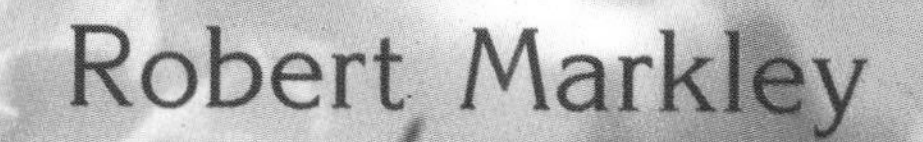

Encyclopedia of ROSES

History

Botany

Characteristics

Design Examples

Planting and Care

The Best Species and Varieties

Translated from the German by Elizabeth D. Crawford

Contents

Culture and History

Of all the plants cultivated by humankind, the rose is the richest in tradition. Whether in antiquity, the Middle Ages, or modern times, whether in music, literature, painting, or architecture—the rose has played and continues to play an outstanding role. Information from many sources allows us to reconstruct the exceptional career of the most spectacular flowering shrub in our gardens.

The Rose in History

To stand in front of a rose bed with the awareness that one is looking at one of the oldest and most tradition-rich cultivated plants of humankind is not without a certain fascination. The rose family tree unarguably extends back for many thousands of years. It may even trace its very earliest beginnings back for millions of years. Thus the rose family tree came into existence before the beginning of human existence. However, those ancient Methuselahs of roses have little or nothing to do with the modern garden rose. The modern varieties we see today in home gardens, in public displays, or on terraces are hardly more than a hundred years old. In fact, the repeat-flowering climbing and shrub roses are not really even seventy years old. However, the roses of the dim past were far less spectacular in their appearance and of a rather simpler beauty.

In the history of the rose, fact and legend have been closely interwoven, occasionally inseparably, since the beginning. No other plant has so stirred the imagination of people, and also that of scholars, as has the rose. Thus, as early as the sixteenth century, gardener, physician, and scholar Hieronymus Bock was recommending that people not become too involved in the history of the rose "because so much of it has become so badly filled with lies." That need not be the case in this chapter, for the facts of the rose's place in history exist in sufficient quantity.

Roses in Antiquity

It is accepted that roses have existed on the earth for twenty-five to thirty million years. In layers of rock from the Tertiary, paleobotanists—that is, scientists who study the flora of long-gone times—have found some leaves, spines, and twig fragments that they attribute to the rose. A flower, which would have removed all doubt, has not been found to date.

Still, of the twenty-five fossil rose species hitherto recognized, three are regarded with great probability as being roses.

Rose omnipresence: Indian miniature with a pair of pheasants (1633–1642).

China—Origin in the Middle Kingdom: With the beginning of Chinese garden cultivation about 2700 B.C., the first roses were probably planted in gardens for decoration. Two thousand years later, the Chinese philosopher Confucius reported extensive rose plantings in the imperial gardens in Peking. Yet the standing of the rose was modest compared with the importance of the rose of Asia, the peony. Also, the chrysanthemum was (still) ahead in playing the queen of the flowers.

All the same, it was probably in the period after the birth of Christ that roses in China were selected and crossed. (The group of China roses descended from the species *Rosa chinensis* revolutionized European rose breeding at the end of the eighteenth century. The highly bred China roses introduced repeated flowering and lower growth height into the gene pool with catapulting results. In eighteenth-century Europe, only a few dozen varieties were known. By 1815, 250 were already known, by 1828, over 2,500.)

Persia: The art of the production of rose oil and rose water was already known in ancient Persia. Rose culture has a long tradition in all of Persia but particularly in the region of northern Iran. The Persian words for *rose* and *flower* are the same. It is thought that the rose plants reached Asia Minor, Greece, Mesopotamia, Syria, and Palestine from Persia.

Greece: The oldest certain representation of a rose was found in Crete. The palace at Knossos houses the approximately 3,500-year-old *Fresco with Blue Bird.* It was discovered at the end of the last century by archeologist Sir Arthur Evans. In the second volume of his work *The Palace of Minos,* he writes, "From the rocks spring wild peas or vetches—the pods shown simultaneously with spiky flowers—clumps of what seem to be dwarf Cretan irises, blue fringed with orange, and—for variety's sake—rose edged with deep purplish green. To the left, for the first time in Ancient Art, appears a wild rose bush, partly against a deep red and partly against a white background, while other coiling sprays of the same plant hang down from a rock-work arch above. The flowers are of a golden rose colour with orange centres dotted with deep red. The artist has given the flower six petals instead of five, and has reduced the leaves to groups of three like those of the strawberry." That the fresco depicts a rose is considered certain, but it is still undetermined to this day which rose species the artist meant. The fresco has been improperly restored several times, and many fine details have been crudely changed. Nevertheless, most probably, stylized roses depicted in the original parts of the fresco are a form of *Rosa gallica.*

After the decline of the Minoan-Mycenaean culture, we look for a long time in vain for information about the rose. Not until Homer, who lived in the ninth century before Christ, does a clue reappear. In the *Iliad,* he describes how, after the death of Achilles during the seige of Troy, Achilles' body is embalmed by Aphrodite and rubbed with rose oil. Barely two hundred years later, the poet Sappho celebrated the rose as the queen of the flowers. Five hundred years before the birth of Christ, the musician and poet Pindar reported that in Athens, wreaths of roses were being worn—an indication of the importance that the cultivation of roses must have had.

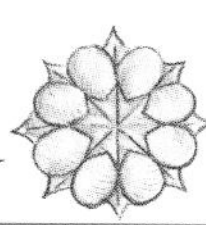

Later, Herodotus described the "rose with sixty petals" of King Midas, which was supposed to be more fragrant than all other roses. One can conjecture that this rose was a doubled form of *Rosa gallica* or an offspring of *Rosa* x *alba,* which already existed at that time.

The father of botany, Theophrastus, has provided many details about gardening around 300 B.C. He differentiated between *rhodon,* the double rose, and *kynosbaton,* the dog rose (*Rosa canina*). His contemporary, Epicurus, had a large rose garden installed in Athens and always had fresh flowers at his disposal. The island of Rhodes was a rose island, and coins minted there display a rose flower.

Souvenir of a lavish rose bouquet: Roman mosaic of a basket of flowers.

Rome: From Greece, the rose also spread to Rome to continue its triumphal procession. The Roman high society took to the queen of the flowers with the same intensity it did all other means of pleasure. Contemporary witnesses have passed down a picture of the excesses and orgies on the basis of which the rose acquired a dubious reputation that lasted into the Middle Ages. For example, Nero, the emperor renowned for his extravagant orgies during his rule from A.D. 37 to 68, spared neither cost nor effort to lend his feasts a special note with unbelievable quantities of roses. For a single banquet he sometimes spent four million sesterces—which has a current value of about a $750,000. For centuries, succeeding emperors also had the rooms of their banquets and feasts ornamented with lavish rose decorations. Thus, a feast at the beginning of the third century A.D. attained notorious fame when the guests of the Emperor Heliogabalus were showered with such a quantity of roses that some of them suffocated.

No wonder that the rose became debased to a kind of mass-produced article. Even before the birth of Christ, people had honored successful battle leaders with rose crowns because the rose was seen as something uncommonly costly. That was over now. With the brawling rose feasts and the drastic increase in rose growing that resulted from them, the rose's social standing also altered. To gain sufficient quantities of roses, the growing areas were so expanded that even the raising of vital grain crops was endangered. Space became scarce, and more intensive growing methods were sought. This period saw the beginning of rose forcing. Blooming was hastened with the help of warm water. Thus, the harvest was increased.

The center of Roman rose growing was Paestum. Lying 25 miles (40 km) south of Naples were huge flat areas used for the growing of roses. Another growing region was in Praeneste, today's Palestrina, a city lying about 19 miles (30 km) southeast of Rome. The growing fields, mainly planted with forms of *Rosa gallica,* were supposed to have extended right up to Rome. However, the enormous demand could still not be filled. Additional roses had to be imported from Egypt—whose very advanced civilization interestingly remained almost untouched by the rose intoxication of the era. How the fragile goods withstood the more than six-day voyage to Rome unharmed has, to this day, remained one of the unsolved riddles of history.

Roses in the Middle Ages

The end of the Roman Empire in the fifth century brought an important twist in the career of the rose. No one paid court any longer to the star of earlier feasts. In the centuries that followed, the queen of the flower became a wallflower. It survived the unrosy time only in some cloister gardens as a medicinal plant.

Charlemagne was the first to free the rose from its Sleeping Beauty existence. He had to ordain its rediscovery by law in 794. *Capitulare de villis imperialibus,* the imperial ordinance about country estates, listed the medicinal, vegetable, fruit, and decorative plants that were to be grown in the free imperial towns. The rose headed the list.

However, the rose's comeback moved along only hesitantly. The greatest botanist of his time, Albertus Magnus, described the few rose species that were being planted in central Europe in the thirteenth century: *Rosa* x *alba, Rosa rubiginosa, Rosa arvensis,* and forms of *Rosa canina*—really not much more than the Romans had already had in cultivation.

Redouté color print of *Rosa gallica* 'Versicolor' ('Rosa Mundi'), often confused with *Rosa* x *damascena* 'Versicolor' ("York-and-Lancaster rose"). The former has distinctly striped petals. The "York-and-Lancaster rose" either has flowers with a pink and a white half or has individual pink and white flowers on one bush, but the individual petals are never striped.

An enlargement of the variety was a present from the **crusaders.** Through them, *Rosa* x *damascena* found its way from Syria, which can be translated as something like *land of the roses,* into the West. The rose's appellation goes back to the name of the Syrian capital city, Damascus. *Rosa* x *damascena* was acknowledged to be purely and simply the most fragrant rose of its time. In the Orient, people had already been obtaining rose oil from it for centuries.

Thibaut IV of Champagne, King of Navarra, brought back a double-flowered offspring of *Rosa gallica* from the Holy Land. Very probably this produced *Rosa gallica* 'Officinalis', the Apothecary Rose. The semidouble, crimson rose, which became the most important medicinal and cosmetic rose of its age, was cultivated on a large scale beginning in the thirteenth century in Provins near Paris. On the main street of the town were countless druggists and apothecaries, which shipped all over the world the medicines produced from the roses.

In the previous century, **Hildegard von Bingen** (1098–1179) had already recom-mended the rose in her books of medicines. Along with the cult of the Virgin Mary, which had been developing since the eleventh century, this assisted the queen of the flowers to a new flowering. The representation of Mary as a "blooming rose branch," "a stem bearing rose blossoms," and, above all, as a "rose without thorns" underlines the position of the rose. In Christian symbolism, the thornless, white rose before the fall is the symbol of innocence, while the red rose symbolizes the blood of Christ, and the thorns are said to be the emblems of sin.

The prayer form of the **rosary,** which has its origin in a saint's legend, has persisted until the modern day. Edmund Lauert writes of it, "According to legend, the Archangel Gabriel had three garlands plaited from 150 roses to honor Mary, a golden 'glorious' one, a white 'joyous' one, and a red 'grievous' one. Since the Middle Ages the prayer of the faithful has been compared with an unfolding rose: A succession of prayers became the rosary. In its smaller form it consists of thirty-three small beads, each year in the life of the Savior, and of five large ones, each representing one of his wounds. Early rosaries were made of beads that consisted of rose petals kneaded with a binding agent."

Thousand-Year Rosebush of Hildesheim: In the late Middle Ages, probably the oldest rose in the world, the so-called thousand-year rose, must have been planted at the cathedral in Hildesheim for it was already being referred to as the old rose in 1573. Over the course of time, it repeatedly suffered severe damage (last in World War II). Each time, though, it recovered, and it is still blooming today.

■ Roses and War—The Wars of the Roses

Roses rather seldomly play a role in connection with military conflicts, even though the jousting fields on which the knights held their combats were known in the Middle Ages as "the rose garden."

During the Crusades in the Holy Land, Saladin, the sultan of Egypt and Syria, succeeded in reconquering Jerusalem from the Christians in 1187. Since the rose was holy to the Muslims, huge quantities of rose water were brought in on the backs of five hundred camels, with which the Omar Mosque was "cleansed" of Christian beliefs. About three hundred years later, after the conquest of Constantinople (the

From Shakespeare's King Henry VI.

Plantagenet: . . . Let him that is a true-born gentleman
And stands upon the honour of his birth,
If he suppose that I have pleaded truth,
From off this brier pluck a white rose with me.

Somerset: Let him that is no coward nor no flatterer,
But dare maintain the party of the truth,
Pluck a red rose from off this thorn with me.

. . .

The adherents of the two houses choose their rose as well.

. . .

Plantagenet: . . . Now, Somerset, where is your argument?

Somerset: Here, in my scabbard; meditating that
Shall dye your white rose in a bloody red.

Plantagenet: Meantime your cheeks do counterfeit our roses;
For pale they look with fear, as witnessing
The truth on our side.

Somerset: No, Plantagenet,
'Tis not for fear; but anger that thy cheeeks
Blush for pure shame to counterfeit our roses,
And yet thy tongue will not confess thy error.

modern Istanbul), Mohammed II also had a mosque purified with rose water.

The widely known term *Wars of the Roses* refers to a conflict between the English noble houses of York and Lancaster. At the center, however, stood not a rose but nothing less than the throne of England.

Nevertheless, this struggle became known in history as the Wars of the Roses because the emblem of the house of York was a white rose, probably a semidouble *Rosa* x *alba* descendant. On the other hand, the house of Lancaster bore the red Apothecary Rose, *Rosa gallica* 'Officinalis' on its coat of arms. (It is also sold today as the red rose of Lancaster.) According to legend, the Wars of the Roses were triggered by an argument between the two Plantagenet lines of the English royal house.

In his play *King Henry VI*—part I, act II, scene 4 (see box at left), Shakespeare set the crucial scene between the dukes of York and Lancaster. This was supposed to have taken place in the the gardens of the temple in London, 150 years after the event.

Evidence shows that generations before this verbal falling out, the two houses were already quarreling terribly. However, the actual fighting began in 1455 with bloody battles in St. Albans, Towton, Hexham, and Tewkesbury. Edward IV mounted the throne in 1460; the York line held the upper hand for ten years. Then Henry VI conquered the throne for a short time, again to be deposed by Edward. Edward's brother Richard III carried on the succession but was killed on August 14, 1485 in the battle of Bosworth Field. After thirty bloody years, the wars of succession were ended. Henry VII, a Tudor and linked to both houses through rather murky connections, married Edward's daughter Elizabeth of York and, as an outsider, came to the throne. His son Henry VIII, later famous for his divorces unsolemnized by the Roman Catholic Church (among other things, by means of the scaffold), included in his coat of arms the red-and-white Tudor rose as the symbol of the uniting of both houses. *Rosa* x *damascena* 'Versicolor' (not to be confused with 'Rosa Mundi', also called *Rosa gallica* 'Versicolor'), at whose bush the quarrel in the temple garden was said to have had its warlike beginning, was supposed to recall this history-laden event as the "York-and-Lancaster rose." (However, this story is probably folklore since the "York-and-Lancaster rose" was first described in 1551 in a book by Monardes more than seventy years after the end of the Wars of the Roses.) The Tudor rose, which is stylized as a small rose on top of a larger red rose, still decorates the coat of arms of the English royal house today.

Roses That Never Bloom: Roses in Art

The rose's reputation since antiquity as the flower of the gods of pleasure (Bacchus) and love (Venus, Aphrodite) made it appear to the early Christians as a depraved flower. Only in the Middle Ages did the rose become a Christian symbol. In a long process, it changed to the mystic rose, the enigmatic *rosa mystica.* There now existed a rose to

Italian illuminated manuscript (*Handbook of Cerruti*).

symbolize Mary that had never really existed in nature. However, the worldly, earthly side of the rose remained, parallel to the symbolism of purity (see "The Rose as Erotic Symbol," page 18). For example, in the fifteenth century, Botticelli always presented the rose as the flower of Venus. It also stood as the allegory of spring at the center of his work of the same name (see picture on page 16).

■ . . . In Painting

The rose as the symbol of purity but also of the pain of Mary held special sway over the fresco painters in cathedrals and churches. Taddeo Gaddi, at the beginning of the fourteenth century, depicted a rosebush on a fresco. He was also one of the first to paint a rose in a vase.

When Martin Schongauer had completed his picture *Maria im Rosenhag* (Mary in the Rose Garden) in Colmar in 1473, the manner of representing the rose had changed from a purely stylized to an entirely realistic presentation. (The *rosa mystica* nevertheless remained true to its name despite the precisely detailed representation. The Schongauer rose with its oval leaves, fiery red flowers, slender buds, and long sepals has raised countless questions for botanists until the present day: Which actual rose of this period dominated by wild roses like *Rosa gallica* could Schongauer have meant?) The rose garden as a whole serves as a metaphor for paradise and links the face of Mary with the purity and beauty of the rose.

From the late sixteenth century on, the rose is found as the central theme of numerous still lifes. The Flemish and Dutch painters, above all, brought this art form of representing unmoving or lifeless objects, usually flowers, fruit, and utensils, to full flower. These artists include Ambrosius Bosschaert, Jan Breughel (many still lifes of roses), Jan Davidsz de Heem, Jacob Marell, Rachel Ruysch, and Gerrit van Spaëndonck (the teacher of Pierre-Joseph Redouté), to name some examples. Today, without having special previous knowledge, the symbolism of these pictures is no longer so easily reconstructible as at the time they were painted. However, the artists' contemporaries knew that, for example, fully opened roses in fragile glass vases symbolized the transitoriness of beauty and of life.

Such fiery red roses did not yet actually exist: Martin Schongauer's famous painting *Maria im Rosenhag* **(detail).**

The important painters of that period represented the rose so precisely in some instances that identifying them today is still easy. The striped *Rosa gallica* 'Versicolor' decorates countless works, as does *Rosa foetida* after the middle of the seventeenth century.

Redouté—Raphael of the Rose

Pierre-Joseph Redouté is acknowledged as a legendary painter of roses. He left behind a monumental work in the area of botanical illustration.

Redouté was born on July 10, 1759 in Saint Hubert in the Belgian Ardennes as the youngest of six children. His father was a professional painter although without fortune and with modest income. However, from the profession of the father came a decisive impetus for the son.

His talent for painting was recognized early. By the time he was fifteen years old, he was drawing with great ability. As a young man in Paris, at first he assisted his brother, who worked as a painter of large pieces of theatrical scenery. In his free time, Redouté sketched the plants in the royal gardens. Through the sale of some of these works he came to the notice of the amateur botanist and professional judge L'Heritier. This rich patron introduced Redouté to the nature of plants, granted the talented young man free entry to his library, and provided him with commissions. The career of the Raphael of flowers was underway.

His work enabled Redouté, already quite well-known, to meet with various personalities of his day. He also came to know the court painter Gerrit van Spaëndonck. Redouté relieved him of the burden of the tiresome, poorly paying botanical painting to which Spaëndonck, as professor of plant painting in the Royal Gardens in Paris, was obligated. Spaëndonck concentrated instead on the lucrative commissioned works for rich burghers. He thereby prepared the way for Redouté ultimately to become official court painter—first under Queen Marie Antoinette, later under the Empress Joséphine. However, Redouté, who survived all the turmoil of the French Revolution entirely unscathed, first worked at the court as designer for the imperial knitting. He was not yet the famous painter of roses he is known as today.

Among the botanists in the royal service was E. P. Ventenat, who asked Redouté to color his plant arrangements. The empress supported Redouté in this undertaking and thus commenced his most productive creative period. It resulted in his greatest work until that time, *Les Liliacées*, which he dedicated to Joséphine.

Only then did he begin work on *Les Roses*. It is often written that Joséphine got him to do it—unfortunately a legend. After all, when the first of the three volumes of *Les Roses* appeared, the empress had already been dead for three years. Also, the models for the illustrations were not the roses of the royal gardens at Malmaison but mainly ones from the gardens in Paris and Versailles. Redouté and botanist Claude Antoine Thory, who wrote the text, specifically acknowledged the cooperation of famous tree and rose nurseries of their time—Bosc, Dupont, Vilmorin, and Noisette. One must thus conclude from this that Redouté in his famous work—counter to all stories—did not picture the roses of the Empress Joséphine.

Redouté enjoyed life and spent money faster than he earned it. Despite an annual income of 18,000 francs with which Joséphine had provided him, his debts grew rapidly. Maintaining his expensive Paris home and his estate in Fleury-sous-Meudon contributed decisively to this circumstance. When he was able to sell the original drawings of *Les Roses* in 1828 for 30,000 francs, it produced only a drop in the sea of creditors' demands.

On the death of the eighty-year-old Redouté, Gerd Krüssmann wrote in *Rosen, Rosen, Rosen* (*Roses, Roses, Roses*), "Like

Typical color-stippled drawing of a Gallica hybrid from Redouté's mammoth work *Les Roses*.

Empress Joséphine was born on June 23, 1763. At the age of sixteen she came to Paris, where she married the Vicomte de Beauharnais. The couple had two children. Beauharnais died at the guillotine in 1794, but his wife retained his fortune. Napoleon met Marie-Josèph-Rose Tascher de la Pagerie, her actual name, at a ball. The two were married in 1796. Two years later, Napoleon acquired for Joséphine, as she was called, the palace of Malmaison.

Empress Joséphine was passionately interested in gardens and began to create an incredible plant collection modeled on the English garden. Green treasures were brought to her from all over the world; the most famous botanists of the period were in her service. One of her specialists, the Irish gardener John Kennedy, even possessed during the war years a blank pass that allowed him to pass the frontiers unmolested in order to pursue unhindered the search for and purchase of plants.

Moreover, Joséphine benefited from her husband's wars. Napoleon's warships brought many magnificent exotic plants to France and he enjoyed the respect of the British enemy for it. After their victory in the battle of Trafalgar, the British guaranteed free escort to Malmaison for the plants and seeds found on the captured ships.

The most famous rose collector of her day: Empress Joséphine.

The empress was especially fond of roses. André Dupont laid out a rose garden for her that contained all the rose varieties known at that time. With it, Joséphine was able to call her garden the largest rose collection of her day.

so many artists earlier and today he (Redouté) lived for the day and did not think of the future. So he was without means in his old age and had to keep painting in order to live. . . . On June 19, 1840, he was just about to paint a lily a young student had brought him when his heart suddenly stopped. He dropped his head, closed his eyes, never to open them again."

The success and fame of Redouté's drawings continue to this day. Countless objects, from napkins to wrapping paper to calendars, are decorated with his timeless illustrations.

Anne Marie Trechslin: Since Redouté, artists have made numerous drawings and watercolors of roses. It is impossible to describe them all here. However, mention must be made of the fantastic rose watercolors by the Swiss artist Anne Marie Trechslin that have been appearing since the 1960s.

Botticelli's *Spring* (around 1477–1478) shows the rose as an allegory of this season.

■ . . . In Music

"Music itself has no content. It is, like the art of a period, merely structure. Thus, in addition, it cannot 'express' the rose. In relation to verbal elements, music is only the 'handmaiden'. It cannot paint a rose, cannot represent its fragrance, or interpret its symbolism. But in harmony with the text it can deepen the sensual effect of speech." This is the matter-of-fact, sensible analysis by the music historian Nowottny.

In the musical form of one of the best-known German poems, the "Heidenröslein" ("Heath Rose"), composed by Goethe in 1771, poetry and music blend and reinforce one another: "Sah ein Knab ein Röslein stehn, Röslein auf der Heiden; war so jung und morgenschön, lief er schnell es anzusehn . . ." ("A young boy saw a little rose growing, a little rose on the heath, new and lovely as the morn, he quickly ran to look at it . . .").

Roses in music often go along with veneration of the fair sex. In his operetta *Gasparone,* Austrian Karl Millöcker has the tenor sing, "What my heart feels, I cannot say, dark-red roses mean tenderness. Deep-hidden feelings lie in flowers; were it not for the language of flowers, how would lovers manage. If it is hard to speak, there must be flowers, for what one dares not say, is said through flowers. What my heart feels, you know exactly, dark-red roses I bring you, beautiful lady . . . ."

The Irish poet Thomas Moore describes "the last rose of summer." It turns up again in Friedrich von Flotow's opera *Martha,* "Last rose, how do you like blooming here so lonely, your dear sisters are already long gone, long gone away . . . ."

In the United States, or more precisely in South Texas, a folk heroine is celebrated in "The Yellow Rose of Texas": "She's the sweetest little rosebud that Texas ever knew, her eyes they shine like diamonds, they sparkle like the dew. You may talk about your Clementine, and sing of Rosalie, but the Yellow Rose of Texas is the only girl for me . . . ." From the botanical point of view, this is supposed to be about *Rosa* x *harisonii* ('Harison's Yellow'), a hybrid species with bright yellow, semidouble flowers.

The symbolic power of the rose extends to modern rock music. An example is the ambitious song called "Róisín Dubh" ("Black Rose") by the rock group Thin Lizzy. Gary Moore wrote the music. The text comes from the Irish rock musician Philip Lynott, who took his inspiration from the Gaelic poem by lyricist James Clarence Mangan (1803–1849), "My Dark Rosaleen." Like Mangan, Lynott used the dark rose, the very "My Dark Rosaleen," as a symbol for Ireland and dipped into the myths and legends of his native country. The cover of the long-playing record "Black Rose" makes an allusion to the sorrowful history of Ireland. It shows a symbolic drawing by artist Jim Fitzpatrick of a dark violet rose, from whose center beads of blood drip over the petals to the outside.

■ . . . In Literature

"'Have you not heard anything in our garden that was particularly pleasing to you?' he asked.

'The birdsong,' I said suddenly.

'You have observed well,' he replied.

'The birds in this garden are our agents against caterpillars and damaging insects. It is they that clean the trees, shrubs, the small plants, and of course also the roses far better than human hands or whatever other contrivance is supposed to be capable of it. Since these pleasant workers have been helping us this year, we have never seen any caterpillar damage in our garden that might have have been, at the least, noticeable.'"

Adalbert Stifter—renowned romantic of the last century for whom "only the

natural things are right"—revealed himself in his novel *Nachsommer* (*Indian Summer*) to be an ecogardener par excellence almost 150 years ago. In *Nachsommer* as in other Stifter stories, the rose appears as an important leitmotiv. In an expressive, painterly style, Stifter rarely describes the rose with concrete details. He often does not recount species or varieties. However, like many of his fellow poets before him, Stifter uses the rose as a metaphor. All follow Goethe's motto, "Roses are for poetry, apples are for eating."

Certainly, the rose is used most often as the eternal flower in poetic art. The previously mentioned Greek poet Sappho, who lived in the seventh century before Christ, wrote, "If Zeus intended to give the flowers a queen, the rose must bear this crown." The foundation stone for terming the rose *queen of the flowers* was thus laid.

The most popular book in medieval France was the allegorical *Roman de la Rose.* It was begun around 1230 by Guillaume de Lorris and finished some years later by Jean de Meun. In not less than 22,817 verses, it encompassed a quasi encyclopedia of medieval knowledge, decisively influencing the thoughts and poetry of that time. In the *Roman de la Rose,* which turns on a dream by the twenty-year-old poet Guillaume, allegory plays an important role. The young poet dreams that he comes as an accomplished knight to a garden of love, which is surrounded by a high wall. While in the garden, he sees during the spring the reflection of a magnificent rose—emblem of a lady love. Listed in detail in verse form are, besides many other things, the ways he has to approach the beloved. He must do so with fidelity, respect, and generosity. Simultaneously, personified "Danger," "Slander," "Shame," and "Fear" seek in vain to drive these virtues out of him. Also "Reason," appearing in the form of a proud lady, cannot talk him out of pursuing the beloved. After further battles, as described in thousands of verses full of philosphy and erudition, love triumphs, and the young man is allowed to pluck the rose.

In the material modern times of the nineteenth and twentieth centuries, not very much remains of the myth of the rose. Kitsch and commerce have almost destroyed its spell. The main character in Antoine de Saint-Exupéry's *The Little Prince* also regrets this.

"The men where you live," said the little prince, "raise five thousand roses in the same garden—and they do not find in it what they are looking for."
"They do not find it," I replied.
"And yet what they are looking for could be found in one single rose, or in a little water."
"Yes, that is true," I said.
And the little prince added:
"But the eyes are blind. One must look with the heart . . ."

■ . . . In Churches

During the eleventh century, consummate masters in the art of stained glass developed, produced large stylized rose windows in the gothic cathedrals. Primarily, the tradition of window rosettes began in France and spread throughout Europe: from Amiens, Chartres, Laône, Reims, and Notre-Dame de Paris to Strasbourg, Freiburg, Cologne to England (Exeter, Canterbury, York, and Westminster). They adorned the front end of the nave. The origins of these rose windows are ascribed to the crusaders, who saw the rose windows in the Mosque of Ibn Tulun in Cairo and got the idea of placing these magnificent jewels in Christian churches too.

However, the rose is also found in the stone representations of plants in the medieval cathedrals. In her work *Die Planzenwelt der mittelalterlichen Kathedralen* (*The Plant World of the Medieval Cathedral*) (Cologne, 1964, cited by Krüssmann in *Rosen, Rosen, Rosen*), Lottlisa Behling described many instances in detail.

◆ France

Amiens, Cathedral: Rose branch (central west portal, between the columns of the lateral arcade on the right side)
Paris, Cathedral: Gable rose, roses with leaves (south transept); rose and seven other plant species (west facade, north side portal)
Reims, Cathedral: Climbing rose over the head of Saint Nicasius (north portal of the west facade, left wall)

◆ Switzerland

Basel Münster: Rose arbor (west facade, center portal, left capital area)

◆ Germany

Freiburg Münster: Rose branch (west portal, upper door frame)
Marburger Elisabethkirche: Rose garden (field of arch of west portal, right half)

Even in modern times, the rose has lost nothing of its strength as a Christian religious symbol. It also plays an important role in contemporary church architecture. An example of this is the rose window in the Cathedral of Mary built in 1969 at Neviges/Velbert. Especially on sunny days, the fiery red of the rose window draws every visitor into its spell.

Rose window in the west facade of the gothic cathedral at Chartres.

A Symbol Rich in Meanings: Let Roses Speak

Etymologically, the term *symbol* goes back to the Greek word *symballein*, which means *to throw together.* In the symbol, two things are thrown together in such a way that they cannot be separated again: visible and invisible, real and unreal.

Like no other plant, the rose, its flower shape, its petals, its fragrance, and its thorns, has always animated humans to use it as a symbol for all sorts of things. The history of the rose as a symbol for love and beauty is almost as old as human cultural history itself.

The power of its symbolism, rooted in countless peoples and cultures, is unique. For example, every child knows what a single red rose presented as a gift means.

The Language of Roses: Out of the symbolism of the rose has arisen the language of roses. This symbolic language originated in the Orient. Later, in Europe, the French and English took over and refined the custom of sending messages with the help of roses and other plants. This fashion reached a climax during the reign of Queen Victoria. At that time, countless books were published about the subject, most of which went back to the work *Le Langage des Fleurs* by Madame de la Tour.

As might be expected, love is at the center of all these messages. Each part of the rose was assigned a very particular symbolism and value. Thorns presaged danger, leaves promised hope. Thornless but leafy rose canes stood for a love full of hope and confidence. However, if these canes were turned upside down, it signaled the opposite.

Not only were all kinds of signals concealed in the rose itself, the mode and manner in which the rose was offered and received were also heavy with symbolism. The rose bent to the right meant *I*, to the left meant *thou* or *you.* If the right hand of the person opposite accepted the rose, the giver could give a sigh of relief for the answer meant *yes.* However, the left hand amounted to a curt *no.*

A woman who carried a single rose fastened to her dress over her heart was considered engaged, a rose stuck in the hair warned caution, in the décolleté it symbolized friendship. Red roses meant fiery love. Pink roses stood symbolically for youth or beauty, white roses for yearning passion or impatience. Yellow roses were ascribed to jealous affection, sometimes even to envy or unfaithfulness.

In addition, certain rose species and varieties possessed very precisely defined symbolism. The following box lists the symbolic meaning for ten different roses.

- Damask roses (*Rosa* x *damascena*) = enduring beauty
- Chinese rose (*Rosa chinensis*) = eternal beauty
- Austrian briar (*Rosa foetida*) = Everything about you is charming
- Provence rose (*Rosa centifolia*) = Messenger of love, my heart is in flames, modesty
- Dog rose (*Rosa canina*) = Modesty, joy and pain in one
- Bud of the moss rose = Confession of love (e.g., *Rosa centifolia* 'Muscosa')
- Rosa multiflora = Dignity
- Musk rose (*Rosa moschata*) = Charming, ill-tempered beauty
- *Rosa gallica* 'Versicolor' = Diversity
- *Rosa* x *alba* 'Maiden's Blush' = When you love me, you will know it

Sub Rosa Dictum—The Rose as Emblem of Secrecy: Whether in the confessional, in city halls (as in Bremen for example), or as a sign for the secret societies in the seventeenth and eighteenth centuries, the rose often appears as a symbol of secrecy. As early as 1742, it was written about in Zedler's *Universallexicon,* "In times past it was the custom to hang a rose over the table, so that anyone, as soon as he saw it, would be mindful that what he heard in secret should be kept secret . . . . *Sub rosa dictum,* that is, 'said under the rose.'"

A relic of this custom is the plaster rose that even today still ornaments many ceilings. As originally, this plaster rose is always located over the center of the table

La vie en rose: Curtains in the only rose museum in the world, in Bad Nauheim-Steinfurth, Germany.

at which confidential discussion takes place. People also amuse themselves sub rosa: In Paris in 1780, the Duke of Chartres founded an Order of Roses. His "ordre de chevaliers et nymphes de la rose" was an association that indulged in secret, dissolute pleasures.

The Rose as Erotic Symbol: Not by chance the word *eros* contains the same letters as the word *rose.* With its fragrance and its color, the rose has offered itself from time immemorial as a symbol of femininity. Dr. Aigremont showed this in his book *Volkserotik und Pflanzenwelt* (*Folk Erotica and the Plant World*), published in 1907: "On one ground, in particular, the rose has become the sexual symbol for women: because of its blood-red color. The color red has held sexual and erotic connections in all ages . . . . Among primitive peoples the sexual organs as well as the gods of love are depicted in red . . . . The love deities of Tibet are painted red. With us also, the rose and blood have stood in the closest relationship since ancient times . . . . In the Middle Ages women's red menstrual flow was called 'the rose,' also 'monthly rose,' 'female rose,' 'woman's rose' . . . . Besides the color there is the fragrance, the enchanting, intoxicating scent of the flower, which may perhaps be sexually arousing, when the woman, as she has done from ancient times, perfumes herself with rose scent . . . . Because of its color, scent, and form, the rose became queen of the flowers, it became the symbol of the woman . . . . Thus the rose garden in the Middle Ages became the embodiment of all pleasure and delight . . . . The Romans took over from the Greeks the erotic significance of the rose. In ancient Rome,

Durable symbolism: In the Steinfurth Rose Museum one strolls on a rose parquet floor.

roses were carried in the sensual, obscene Feast of the Flowers . . . . "

In folk sayings, roses and young girls are often compared to one another. "The purest rose, which falls into the thorns, tears her petals" means association with low company is harmful to even the purest girl. "The most beautiful rose will finally turn to hips" says that every girl who is so beautiful now will some day become fat and ugly.

A warning to parents not to put off the marriage of their children excessively is the meaning of the saying, "One must not let the rose wither on the stem," or "One must pluck the roses when they bloom." These mean that sexually mature girls should be married.

Other rose circumlocutions for slightly disreputable things are *rose garden, rose plan* (or *design, map, scheme, intention*), *rose alley,* and *rose corner.* These are city districts in medieval cities, and sometimes also in modern ones, in which ladies of pleasure or public prostitutes live. Very graphic in Ehrfurt's notorious red light district is a house named *Zur grossen Rose,* which is a clear reference to genitalia. In earlier times, the prostitutes in Frankfurt am Main even wore a rose as a badge.

Rose petals are used as an oracle of love. They are allowed to float on the water like little ships. If the little ships of a boy and a girl float toward each other, the boy and girl will fall in love and/or remain true to one another. If the ships float away from each other, the boy and girl will separate.

Bedded on roses: Rose armchair in the Steinfurth Rose Museum.

Rose 2000—Roses Today

Rose breeding could be described as the perpetual search for the perfect rose. Many people were and are caught by the fascination of this search. The actual process of artful crossing proves to be not very spectacular and is easily understandable (see "Rose Breeding," page 178). However, the most successful professional rose breeders often come from a long family tradition, with often only the second or third generation receiving the reward for their own pains and those of their forebears. The chief reason for this is that the development of first-class breeding parents for promising crosses and acquiring knowledge of their genetic characteristics require many years or even decades of patient work. Rose breeding is no occupation for the impatient. The selection of promising seedlings is successful only with the help of a practiced eye and years of experience.

Notes on Some Famous Rose Breeders

Besides experience, diligence, and persistance, a certain amount of luck, too, is needed to hit the bull's-eye—and numerous hardworking colleagues who carry out the breeder's Sisyphean toils of thousands upon thousands of crossings.

In Germany, the rose firms of W. Kordes' Söhne, of Klein Offenseth-Sparrieshoop, Strobel & Co., of Pinneburg (as German representatives of the French firm of Meilland), and of Rosen Tantau, of Uetersen have a long tradition as substantial breeders of new varieties. The cooperative association of the rose union in Steinfurth, in Hesse, also plays a considerable role in introducing new varieties from well-known foreign breeders. "Newcomers" include breeders like Karl Hetzel, who has been very successful with, among other things, 'Super Dorothy' and 'Super Excelsa', and also Werner Noack, the shooting star from Gütersloh.

W. Kordes' Söhne

On October 1, 1887, Wilhelm Kordes I founded a "horticultural nursery and market garden," which after very few years concentrated on the breeding of roses. His eldest son, Wilhelm Kordes II, with his brother, Hermann Kordes II, led the firm to world fame as Wilhelm Kordes' Sons. Without exaggeration, Wilhelm Kordes II can be characterized as one of the most successful rose breeders of this century. Varieties like 'Crimson Glory', 'Dortmund', 'Flammentanz', 'Kordes Sondermeldung', and 'Raubritter', to name only a few, are all due to him. From about 1955, Reimer Kordes, Wilhelm's son, took over the breeding. Being no less successful, he also created countless classics. Some shrub rose varieties include 'Bischofsstadt Paderborn', 'Lichtkönigin Lucia', 'Schneewittchen' (also called 'Iceberg'), and 'Westerland', which dominate to this day and have made the name of Kordes a synonym, as it were, for the shrub rose group. Wilhelm Kordes III, Reimer Kordes' son, joined the firm in 1977 and carried on the work of breeding. Numerous national and international awards have crowned his work. Wilhelm Kordes III is one of the most profound rose authorities of his generation in Germany.

Rosen Tantau

Mathias Tantau Sr. founded a nursery on January 6, 1906. Like Wilhelm Kordes I, Tantau developed a pronounced preference for roses and their breeding. As a result, from about 1918, the queen of the flowers stood absolutely at the center of his work. 'Garnette', 'Märchenland', 'Tantaus Überraschung', and many more turned out to be breeding triumphs. In 1953, Mathias

'Montana'®—Time-tested bedding rose classic from the breeding workshop of Rosen Tantau.

'Schneewittchen'® ('Iceberg')—One of the many great successes of Reimer Kordes. For millions of rose lovers it will be a lasting memorial to the rose breeder, who died in the spring of 1997.

Tantau Jr. took over the rose breeding and achieved spectacular success. For instance 'Super Star' became a sensation in the rose world because of its unique, hitherto unknown, salmon orange color. Other Tantau hits are 'Pariser Charme', 'Fragrant Cloud', 'Blue Moon'®, 'Whisky', and 'Montana'. Countless outstandingly fragrant roses are inseparably linked to the name of Tantau. In 1985, the firm's fortunes and rose-breeding operations passed into the hands of long-time colleague Hans Jürgen Evers. Successful new introductions like 'Diadem' and 'Monica', which helped bring the moderately profitable field-growing of cutting roses in Germany to a renaissance in the 1980s, are his works.

Strobel & Co./Meilland

For almost half a century, the firm of Strobel & Co. has functioned as the German representative of the French rose-breeding dynasty of Meilland-Richardier. Therefore, the histories of both firms are closely linked with one another. At the end of the 1940s, Gustav Strobel established the connnection with Francis Meilland, the breeder of such famous roses as 'Peace' and 'Baccara'. However Francis Meilland is remembered not only for creating these rose classics but also because of his service in the struggle to protect the originator of new rose varieties. He actively participated in the fight for the proper legal regulations—to the benefit of all rose breeders. He offered the apt comparison between rose breeders and authors. He had never heard of an author who sold a single copy of his book and then allowed the entire world to reprint it for nothing.

Francis Meilland died on June 15, 1958 of incurable cancer at the age of 46. In his last days, recognizing the very short time he had left to him, he arranged his remaining personal and business affairs. In particular, though under great physical stress, he instructed his then only seventeen-year-old son, Alain, in rose breeding. He was probably also confident that his wife Marie Louise—called Louisette for short and the most successful female rose breeder—would provide decisive support to Alain Meilland in breeding roses in the years to come because of her all-embracing knowledge. The worldwide successes of the young rose breeder in the following years—producing among others 'Sonia', 'Carina', 'Papa Meilland', 'Starina', and later 'Bonica'—secured the future of the rose-breeding family.

Francis Meilland had built up a worldwide net of foreign representatives early. He understood that although ideal rose-breeding conditions exist on the Côte d'Azur because of the light conditions, the selection of varieties for use in a particular country must take place locally. The nursery of Strobel & Co. in Pinneberg took over this task. Since the death of Gustav Strobel in 1979, the one who has the final say as to which variety is introduced to the German market and when is Klaus-Jürgen Strobel, born in 1931. He is a highly valued consultant in expert circles. (As a modest and humanely loyal rose expert, Strobel does not like to hear it said that he is the leader of the chorus in his field—but he must be mentioned.)

Werner Noack

Werner Noack, born in 1927 and located in Gütersloh since 1948, is the most successful "newcomer" among the German rose breeders. Since the 1980s, he has created a furor as a clean sweeper in the ADR rose trials (see page 163). His numerous ADR rose awards are the result of more than forty

'Flower Carpet'®—Werner Noack's great shot has been one of the most successful garden roses all over the world in recent years.

'Abraham Darby'®—One of the most beautiful English roses of breeder David Austin.

years of patient breeding work. The international breakthrough came for him in 1988 with the rose 'Flower Carpet'.

Noack has a special partiality for the area, bedding, and shrub roses. In spite of the immense success of the firms Tantau, Kordes, Rosen-Union, and Strobel/Meilland in this field, Werner Noack has aimed for and achieved different varieties that are "just a little bit better." Possibly, this has come about because Noack has, from the beginning, concentrated his breeding efforts on robust garden roses. By always seeking an easy-care rose, he has not treated his selection fields with a wholesale application of fungicides for decades.

Son Reinhard, current proprietor of the firm, will certainly be pulling more healthy Noack roses out of the breeder's treasure chest of his father in the upcoming years and presenting them to the public. Specialists and rose lovers are intently awaiting further developments.

'Bonica'®—A Meilland cross with the best frost hardiness.

Other Rose Breeders

Rose breeders are active all over the world. The following is a selection of other famous rose breeders and some of their successes:
David Austin (Great Britain):
'Heritage', 'Graham Thomas' (see also the section "English Roses," page 122)
Alexander Cocker (Great Britain):
'Alec's Red', 'Silver Jubilee'
Georges Delbard (France):
'Centenaire de Lourdes'
Gijsbert de Ruiter (The Netherlands):
'Minimo'—pot rose series
Jack Harkness (Great Britain):
'Yesterday', 'Marjorie Fair'
Sam McGredy (New Zealand):
'Picadilly' , 'Händel'
Ralph Moore (USA):
'Yellow Dagmar Hastrup', countless miniature roses
Toru Onodera (Japan):
'Nozomi'
Niels Dines Poulsen (Denmark):
'Royal Dane', 'Shalom'

Jackson & Perkins—The Largest Rose Nursery in the World

With annual rose graftings of 15 million, the American nursery Jackson & Perkins, called J & P for short, today represents the superlative in the worldwide rose market. The history of the firm goes back to 1872 when Charles H. Perkins established a modest farm in Newark, New York with his father-in-law Albert E. Jackson. The focus was fruit production. Then Perkins switched to running a tree nursery. This would not have been worth mentioning today had not Perkins hired Alvin Miller, who began breeding roses.

Miller showed himself to have a lucky touch. His most spectacular cross turned out to be 'Dorothy Perkins', a pink climbing rose that was named for Perkins's granddaughter. The enormous demand gave the nursery a powerful boost; the name J & P was now on everyone's tongue.

By 1920, 250,000 roses of many varieties were being offered. With the entry of Eugene Boerner and Charles H. Perkins, nephew and namesake of the founder, the great period of J & P began.

"Charlie" Perkins was a born salesman. Boerner—after first working in sales—took over the breeding department. Among the marketing actions of the dream team of that period, for example, was the establishment of a rose garden in Newark. Year by year in the 1940s and 1950s, it developed into the scene of a two-week rose festival and attracted more than 250,000 visitors annually. By the end of the 1950s, J & P was producing 20 million rose plants per year.

However, in the 1960s, the firm's star began to sink. Perkins and Boerner had failed to establish a new leadership generation in time. When Perkins died, Boerner had to take over the operation of the business. He was not lucky. In 1966, the Harry & David fruit company took over the world-famous rose firm.

Since then, J & P has been bought by the Yamanouchi Pharmaceutical Co. Ltd., of Japan, and is again on the road to success. The management is located in Medford, Oregon. The rose breeding takes place in sunny Somis, near Los Angeles, under the direction of Dr. Keith Zary. One hundred thousand crossings are made annually, using 350 parent varieties.

One more superlative must be mentioned: According to company statistics, J & P sends out 29 million catalogues annually; this number is also without parallel worldwide.

■ Rose Gardens and Rosariums—Living Rose Museums

Anyone who wants to experience the rose in the summertime fully should make a pilgrimage to some of the countless rosariums, large and small. A rosarium is, as a rule, a collection of the many existing species and varieties whose plants are maintained with great care and are usually very well labeled. In addition, rosariums are important for their inclusion of a wealth of rose varieties, some of them not available commercially for decades. In this regard, rosariums have a role as an important gene bank for future breeding. With their help, we can reach back to old but long-missing breed lines.

To rose lovers, rosariums offer the chance to learn about important characteristics of individual species and particular varieties, about flower color (which cannot always be accurately pictured or described in catalogs), and about scent and growth form of a rose.

What rosariums do not provide is an opportunity to determine the vigor of a specific variety. Because of the enormous number of varieties and the closeness of the plantings, roses in a rosarium are treated several times a year with pesticides. A basic rule is, the larger the area one plants only with roses, the more robust the varieties planted should be. However, since the rosarium is a collection that is supposed to contain as many varieties as possible, less-vigorous varieties must be planted as well.

In Dortmund, for instance, they have embarked on a new course. They are trying, through species-rich plantings of shrubs, perennials, and annuals, to develop easier-maintenance roses. However, even in the future, entirely ruling out the use of pesticides in a rosarium will not be possible. Anyone who falls in love with a particular variety on a visit to a rosarium should first look at a planting in a catalog or in a rose nursery or specifically ask the breeder whether the variety is suitable for being planted in his or her own garden with a justifiable expenditure of care. The following lists a choice of gardens worth seeing throughout Europe.

◆ Germany

Baden-Baden: Two rose gardens are in this spa city. The Gönneranlage (Gönner gardens) (named for the former mayor of Baden-Baden) in Lichtentaner Allee contains, among other things, a great many climbing roses. The Rosen-Neuheitengarten (Rose novelty/newness garden) on the Beutig (Moltkestrasse) includes an area for the international rose trials. Roses are judged there several times a year. A large prize jury composed of rose experts gives awards to the most beautiful varieties. The rose trials were initiated by Walter Rieger. Currently, they are directed by Bernd Weigel, director of the Gardens and Park Department and president of the Vereins Deutscher Rosenfreunde (VDR) (German Rose Fanciers' Society). Many rose plantings are also in the city.

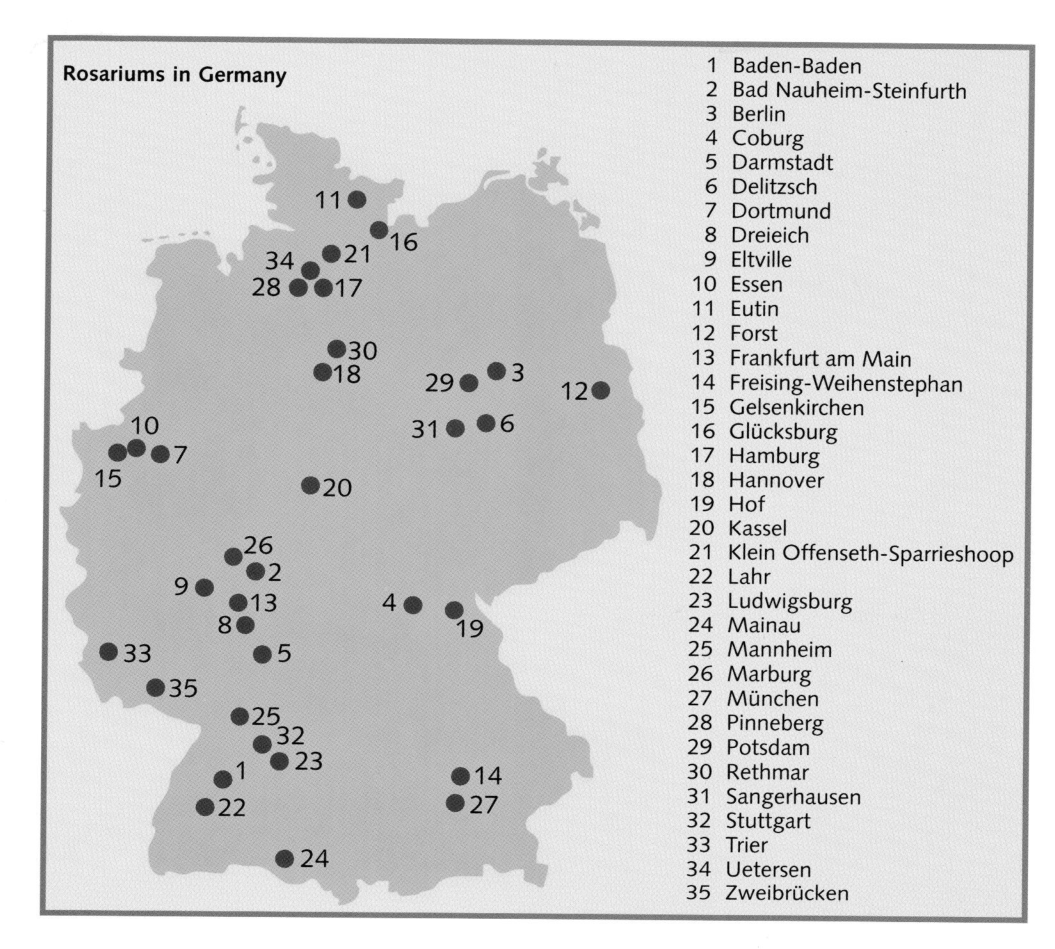

Bad Nauheim-Steinfurth: Steinfurth is the heart of the tradition-rich rose-raising district in the Hessian Wetterau. As one enters the little rose town when coming from Bad Nauheim, on the left side is the free-entry display rose garden of the Rose Union, a cooperative association of numerous Steinfurth rose firms. After that are visits, obligatory for rose lovers, to the display gardens of the nurseries of Gönewein and Heinrich, Schultheis, the oldest German rose nursery. Numerous other firms, among them Germany's only organic rose nursery, Michel & Ruf, offer roses and articles about roses. A visit to the **rose museum,** the only museum of this sort in the world, rounds out the expedition.

Berlin: Worth seeing, among others, are the rose gardens in the Grossen Tiergarten (big zoo), in Volkspark (people's park) Mariendorf, and in the Britzer Garten (Britzer garden)—the location of the national garden show in 1985.

Coburg: A small but fine, still relatively young, rose garden in Coburg is open to the rose lover.

Darmstadt: A lovely rose garden at the so-called *Rosenhöhe.*

Delitzsch: The rose garden at Wallgraben in Delitzsch, located north of Leipzig, is not too large and should be visited during the summer.

Dortmund: The Deutsche Rosarium Dortmund on Kaiserhain in Westfalia Park was founded in 1969 by the VDR (Vereins Deutscher Rosenfreunde). Under the direction of the world-renowned dendrologist and rosarian Gerd Krüssmann, an imposing rose garden has arisen in a very short time. Under the aegis of Dr. Otto Bünemann in the new rose garden on Kaiserhain, the combination of roses with other outdoor plants has been installed in a large area. The Kaiserhain is, for any fancier of variety-rich roses, a source of inspiration and a fund of design ideas. Also worth seeing is the *Rosenweg* (rose path) in Westfalia Park.

Dreieich: Burg Hayn in Dreieich, laid out by Lore Wirth, has herb and rose gardens in front of the charming backdrop of a castle.

Eltville: An expedition into the Rheingau must include a visit to the rose gardens in the castle moat with its imposing wall of climbing roses. Anyone who in June marvels at the 26-foot (8-m) high climbing rose 'Tausendschön' or would like to see the rambler 'Bobbie James' ensnaring and climbing a larch Methusaleh should not miss Eltville.

Essen: The rosarium in spacious Grugapark offers a Sunday excursion. In the rose gardens, the hybrid tea and bedding roses in classic geometric arrangements predominate.

Eutlin: Inviting rose gardens, countless roses in home gardens and city green spaces can be found.

Forst: The Ostdeutsche Rosengarten in Lausitz was established in 1913 by Alfred Boese. For many decades, the rosarian and rose book author Werner Gottschalk directed the garden with its extensive variety.

Frankfurt am Main: The rose garden in the Palmengarten (palm garden) rewards a journey to Frankfurt. The reconstruction at the end of the 1980s followed the geometric principles of the rose parterre. Paths divide the area into rectangular and triangular beds; standard roses and pergolas emphasize the formal elements. The beds are partially planted with fragrant waves of lavender. A visit at the peak blooming period in June or July is recommended; a rose show with the awarding of prizes takes place annually. In addition, there is a well with numerous English Roses from breeder David Austin planted around it.

Freising-Weihenstephan: This is the rose garden of the test gardens. It contains numerous rose combinations with perennials and ornamental shrubs. The garden is definitely worth seeing for rose lovers who also like perennials.

Gelsenkirchen: Rose gardens of the national garden show of 1997 contain many new introductions.

Glücksburg: A masterwork of modern landscape architecture is the show garden of the nursery of Ingwer Jensen, directly beside the famed Glücksburger Wasserschloss. Since its establishment in 1991, the garden shows many old and English Roses in spacious arrangements combined with numerous perennials and

'American Pillar' in Baden-Baden—A journey back to grandmother's rose romanticism.

In the Lahr rose garden one can stroll under the lus
arches of the climbing rose 'Paul's Scarlet Climber'

other shrubs. This abundance of species was masterfully mounted by Hamburg garden designer Günther Schulze. The best time to visit is between June 20 and the end of July, when the numerous once-blooming varieties reach the peak of their performance. This is probably one of the most enchanting rose gardens in Germany and a private tip for connoisseurs.

Hamburg: The park Planten und Blomen has a new rose installation; roses are in the new botanical garden.

Hannover: Schloss Herrenhausen has formal gardens.

Hof: Roses are in the botanical garden.

Kassel: A historic rose collection appears in the park of Schloss Wilhelmshöhe, whose origins are documented back to monastic times in the eleventh century. Visitors are offered a planted history of the shrub rose as well as a rose collection renewed since 1978 by Kassel rose lovers. Many wild roses, old varieties, and modern shrub roses are under the direction of Dr. Wernt and Hedi Grimm.

Klein Offenseth-Sparrieshoop Near Elmshorn: The nursery of W. Kordes' Söhne has display gardens. Over 200 different varieties in small beds allow a direct comparison of varieties. All are carefully labeled, and an overall map facilitates the search. The best time to visit is at peak blooming in July and at second flowering at the end of August. Besides, the rose fields of the Kordes nursery can be visited at any time. A current map showing the fields is obtainable at the information stand in the rose garden. This is a must for any rose fan.

Lahr: The city park has a rose garden.

Ludwigsburg: This is flowering baroque with a fantastic mixture of baroque architecture and a true rose garden ambiance. A visit is a journey into a long-past, elegant time.

Mainau Island: The original rose garden, laid out in 1871 by Grand Duke Friedrich von Baden, offers countless ideas for any rose garden design. It includes an imposing street of wild roses and rose gardens. Numerous inspired rose ideas of the former garden director and VDR president Jose Raff found their expression and annually attract numerous visitors.

Mannheim: Rose promenades occur in Luisenpark, Herzogenriedpark.

The rose parterre in the Palmengarten in Frankfurt
experience for any garden lover from the middle of

Undiluted roses are presented to the visitor to Mainau Island in summer.

Marburg: A rose garden is in Schlosspark Marburg.

Munich: Rosengarten Westpark was established for the international garden show in 1983. (The shrub rose 'IGA München '83' is a memento of this imposing exhibition.)

Pinneberg: Rosengarten im Fahlt was opened in 1934. Embedded in the city woods of the city of Pinneberg lies a rose garden that is regrettably little known. It contains very well-cared-for installations, which display many shrubs as well as roses.

Potsdam: Rose gardens are in the Schlosspark of Sanssouci.

Rethmar: Rose test gardens of the federal district office (Bundesortenamts) are in the area of Hannover. Comparison plantings of many currently marketed varieties appear. These gardens may be visited on request.

Sangerhausen: Europa-Rosarium Sangerhausen, founded in 1903, can be found here. It contains the largest rose collection worldwide. In a total area of more than 31 acres (12.5 h), 6,500 species and varieties of roses of all classes are represented. The rose paths lead to picturesque pavilions beside ponds and monuments. The wild rose demonstration path "Helmstal" is next to an immense shrub collection. Plan at least one full day for a visit.

Stuttgart: Höhenpark Killesberg was created for the international garden show in 1993. Worth seeing, it is the valley of the roses.

Trier: The rose garden in Nellspark was laid out in 1958–1959. It shows, among other things, many varieties of Trier rose breeds by, for example, breeders like of Peter Lambert.

Uetersen: Rosarium Uetersen was reorganized in 1932 by the firms Kordes, Tantau, and other Holstein firms. Today more than 30,000 roses of more than 800 varieties appear on extensive grounds with a café, restaurant, and hotel. The diversified park grounds with old stands of trees and a large pond are worth a visit.

Zweibrücken: Extensive rose gardens exist, and the wild rose garden separates them from the pheasantry. In addition, numerous plantings are in the municipal parks.

◆ Austria

① ***Baden bei Wein:*** An Austrian rosarium with a broad variety of old and new varieties, since 1967.
② ***Linz:*** A rose garden in the Botanical Garden, since 1967.
③ ***Vienna:*** A 5-acre (2-h) rosarium in Donaupark, since 1964.

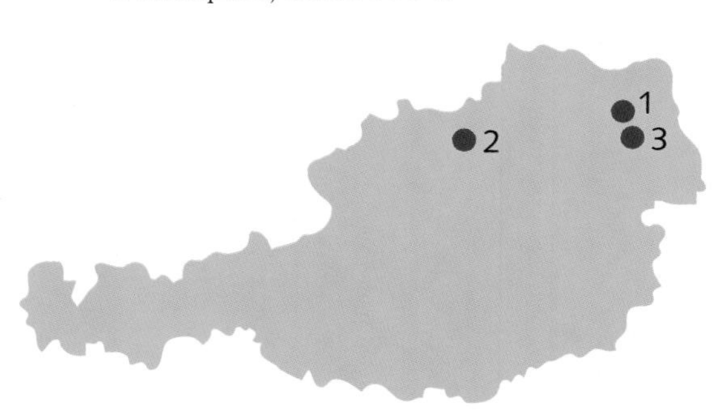

Some rose gardens in Austria.

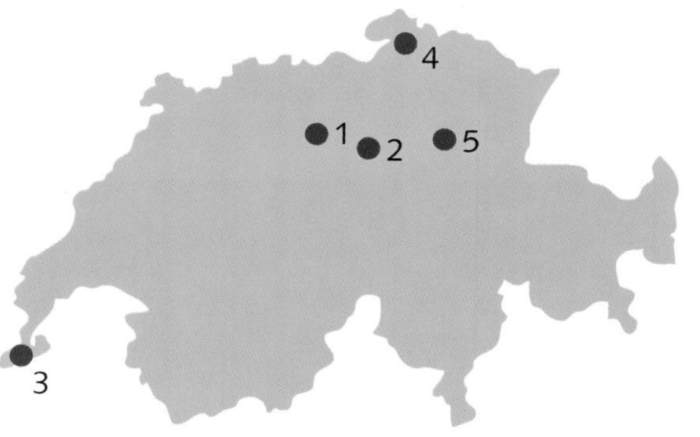

Some rose gardens in Switzerland.

◆ Switzerland

① ***Dottikon-Rothenbül:*** The show garden of the Huber nursery contains many historic roses.
② ***Gelfingen bei Luzern:*** A baroque garden of Schloss Heidegg, with a magnificent outlay in a small space.
③ ***Geneva:*** Parc de la Grange is the location of international rose competitions.
④ ***Neuhausen:*** The rose gardens at the Schloss Charlottenfels, established in 1938 by Dietrich Woessner, are very highly recommended.
⑤ ***Rapperswil:*** The rose garden, established in 1965 by Dietrich Woessner, is especially noteworthy for the blind rose garden; numerous roses grow in the municipal parks.

◆ The Netherlands

Amsterdam: Rosarium in Amstel Park.
Arcen bei Venlo: Schlossgarten Arcen.
Den Haag: Rose garden in Westbroekpark.
Winschoten: Rosarium.

◆ France

Chalon-sur-Sâone: Parc St. Nicolas.
Lyon: Parc de la Tête d'Or.
Orléans: Parc Floral.
Paris: L'Hay-les-Roses, Parc de Bagatelle in Bois de Boulogne, Parc de Malmaison.

◆ Italy

Cabriglia d'Arezzo: Rose garden (private collection of Professor Fineschi).
Genoa: Villa Nervi.
Monza: Rosarium Villa Reale (rose garden of the Italian Rose Society).
Rome: Roseto di Roma.
For other rosariums in Europe see the Appendix on page 233.

■ Rose Gardens in Rose Villages, Rose Cities, and Rose Clubs

Rose Villages:
Nöggenschwiel: Nöggenschwiel in the southern Black Forest is always worth a trip thanks to its countless roses in public and private gardens.
Schmitshausen: Schmitshausen at Zweibrücken offers the visitor numerous roses in public gardens.
Seppenrade: Seppenrade in the vicinity of Lüdinghausen has been a rose village since 1972. Besides many old and new varieties, the Seppenrader rose gardens contain the crosses of amateur breeder Ewald Scholle.
Steinfurth (See "Rosariums": Bad Nauheim-Steinfurth).
Rose Cities (See under "Rosariums": Baden-Baden, Dortmund, Eltville, Uetersen, Zweibrücken).
Rose Routes:
Neunkirchen has its rose gardens. Others are in Illinging-Hüttigweiler, Wemmetsweiler, and Welschbach.

■Fanciers' Associations

American Rose Society

The American Rose Society (ARS) was founded in 1892 to promote the growth and appreciation of the rose, which is the official flower of the United States. The organization has now evolved into the largest rose organization in the United States for amateur rose fanciers and growers. It currently has a total national membership of approximately 22,000 with more than 385 local rose societies throughout the country.

Membership benefits include receiving *The American Rose* magazine and *American Rose Annual,* aimed at both the amateur and professional rose grower. The ARS also publishes the annual *Handbook for Selecting Roses,* which lists hundreds of roses and their ARS rating. The ARS holds national conventions each year. It offers a national network of officially approved consulting rosarians who can give free advice on rose growing and selection. For more information on joining the ARS, write or call: The American Rose Society, P.O. Box 30000, Shreveport, LA 71130. Tel. (318) 938-5402. FAX: (318) 938-5405.

■ 'Peace'—The Most Famous Rose in the World

By the middle of this century, the death knell had just about sounded for the hybrid tea roses. The newly arriving bedding roses were appreciably more vigorous and disease resistant than their "older" colleagues, which could survive only with the help of enormous and constant spraying. Of course, many rose lovers took these factors into consideration and bought hybrid tea roses anyway. However, the broad mass of gardeners declined to follow and selected more vigorous alternatives—insofar as they existed.

This continued until 1945, when a miracle occurred in the rose world—Francis Meilland established a new hybrid tea standard with a new introduction. 'Mme. A. Meilland' was vigorous, healthy, and a repeat-flowering, yellow-red hybrid tea rose. Subsequently, a true renaissance of the hybrid tea rose began.

A while ago, on the fiftieth anniversary of this superrose, the firm of Meilland published the old personal notes of Francis Meilland on the development of 'Peace', as the variety is called in the United States. Some excerpts can be seen on page 27, which give some plain and simple insights into rose breeding. 'Peace' was not only Francis Meilland's best garden rose, it was also his most successful. With over 100 million propagated plants, it is the most planted garden rose of all time.

On the 50th Birthday of 'Peace'

"On leafing through our notebooks, we found under the date of June 15, 1935, the note that indicated the first pollination procedures for the creation of the Peace rose, a variety that is probably the best we have ever bred. It is listed under the number 3-35-40. These numbers mean that it was the third crossing that we undertook in 1935 and that it was the fortieth plant that was selected from this crossing for a test propagation of some eyes a year later . . . .

"It was, then, in the summer of 1936 that some eyes were grafted for the first time. This grafting must certainly have been undertaken very early, for when my father and I first checked on the recently grafted roses on October 10, 1936, we found the first shoots with glossy foliage and the first buds, which were just about to open. Under the very favorable weather conditions that fall, the buds developed into magnificent flowers with a slightly green overcast on the yellow petals with the intense crimson colored edges.

"In June 1939 No. 3-35-40 was the great big surprise in our rose fields and the variety that visitors usually noticed. That summer eyes were sent to Germany, Italy, and the United States . . . . With brutal suddenness the second World War broke out on September 3, 1939. After the invasion of France in 1940, Italy and England were also countries assailed by war. It thus resulted that the German firm introduced this rose No. 3-35-40 to the trade under the name Gloria Dei, the Italians sold it under the name Gioia. For France, my father and I decided that we would name this rose in memory of my late mother, Mme. A. Meilland . . . . It was only in June 1945 that we had any idea of what had become of our rose in the United States. We then first learned of the successful attempts to grow it . . . and that it had been agreed to call this rose Peace. Simply for the reason that this rose could be given to the public the moment the dreadful war was over.

". . . [A]fter that the founding meeting of the United Nations took place in San Francisco. The leaders of all forty-nine delegations found a vase with one Peace rose in their hotel rooms . . . .

"If circumstances have led to this rose being known under various names throughout the world, it is nevertheless right that every name has its significance for the people. For all people of good will who love flowers and the rose in particular, it gives the opportunity to praise God with Gloria Dei, to conquer life with a smile—with Gioia, to wish for peace with Peace, and for us it is an everlasting remembrance of Mme. A. Meilland."

Family Tree of 'Peace'

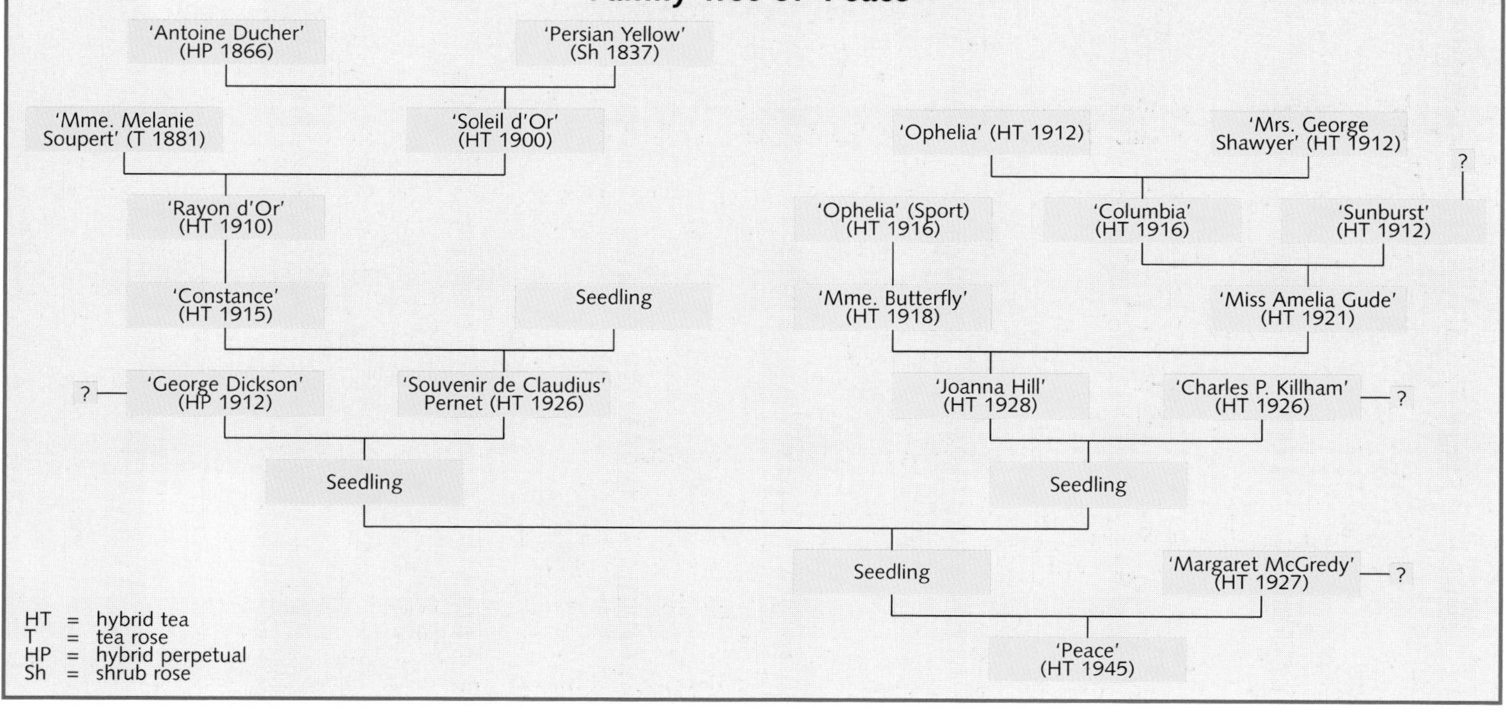

BOTANY

The rose has the greatest diversity of forms of any flowering shrub in our gardens. From the single flower of the wild roses to the lushly doubled Centifolias to differing foliage forms and fantastically spiny canes—the multiplication of these characteristics by the nuance-rich spectrum of colors of red, pink, yellow, and white flowers provides the basis for a wealth of varieties beyond compare.

Roses—Members of a Large Family

The rose is a member of the group of shrublike woody plants. This means that their canes become woody and live through the winter without leaves. Together with many other woody plants but also with some perennials like lady's-mantle and avens, the rose forms the rose family (Rosaceae)—a large family with more than 3,000 relatives (species). Among these are all the important fruit trees such as apples, pears, cherries, plums, peaches, and berry fruits such as strawberries, raspberries, and blackberries. This family also includes landscape-shaping wild woody shrubs such as blackthorn, bird cherry, and hawthorn and ornamental shrubs like Japanese quince, spiraea, cotoneaster, pyracantha, and kerria—they all belong to the family Rosaceae.

The rules for naming plants are complex. Scientists categorize every rose—in fact, every plant and every animal—by using a classiciation system. From most general to most specific, the divisions in this system are kingdom, phylum, class, order, family, genus, and species. When referring to a specific rose, botanists use just the genus and species—a system called binomial nomenclature. The genus is the first name, the species is the second. The genus of the rose bears the name *Rosa.* An x between the names stands for a cross, a hybrid. The rose species are almost innumerable; in just the wild roses alone are more than a hundred species, of which the dog rose (*Rosa canina*) is the best known.

Even a sun child like the rose does not live by light alone. Like all plants, its existence depends on warmth, water, and nutrients, which it receives in the right amounts, at the proper time, and in precisely measured proportional quantities. To enable a rose to take up this elixir of life and utilize it, to enable it to increase and maintain its species, it possesses different organs. Each fulfills a specific function in the life of a rose.

All important fruit species like pears . . .

. . . and strawberries are also members of the rose family.

The Precise Characteristics of a Rose

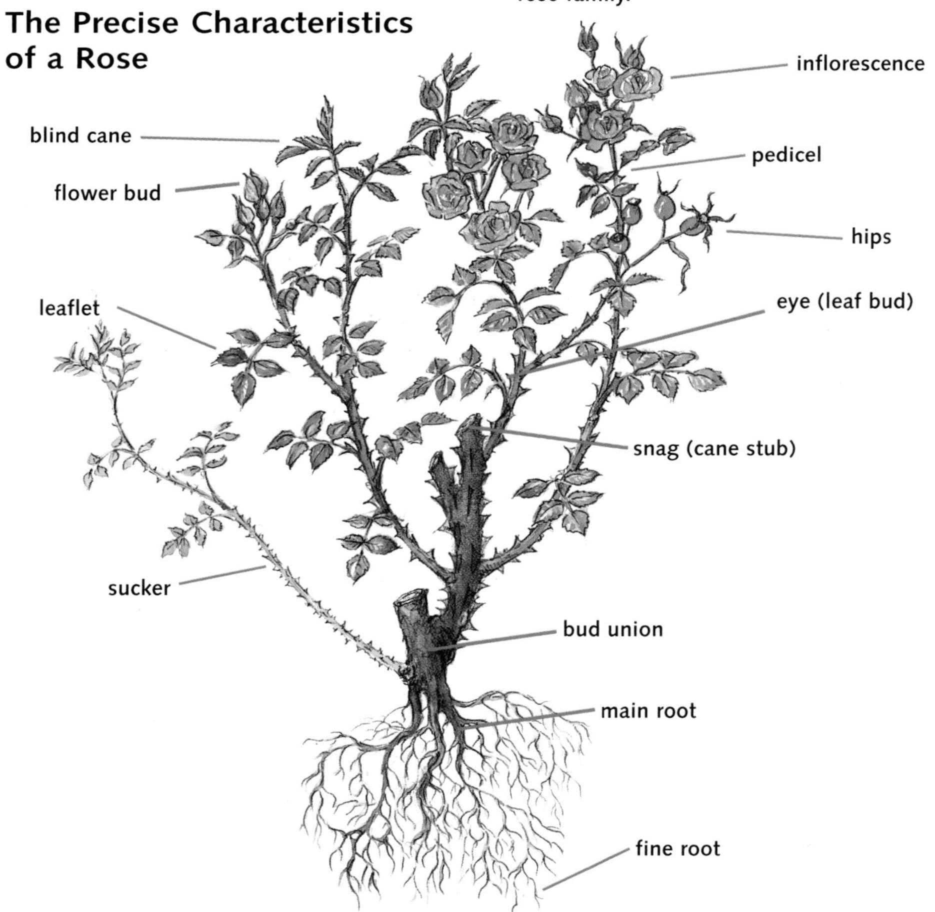

Parts of the Rose

The figure to the left shows all the characteristic parts of a grafted garden rose around an idealized drawing of the plant structure. The figure also names all these parts: the flower buds, which can be from pointed to rounded in form; an inflorescence with flowers, which can be single or double; the hips, the fruits of the rose; the pinnate leaves with leaflets, the solar cells of photosynthesis; the canes, which bear the flowers, leaves, prickles, and leaf buds (the so-called eyes); the flowerless canes, the so-called blind canes; the suckers, which arise below the bud union; the snag, the residue of poor pruning technique; the dormant buds, in effect the eyes on call; the old wood of several-years-old canes; and, last but not least, the roots with their main and fine roots.

■ The Flower—The Magic Five

Anyone involved with the structure of a rose flower repeatedly comes up against the number five. All wild roses have five petals and five modified leaves called sepals. However, even here this rule has its exception: *Rosa sericea* f. *pteracantha* has only four flower petals and four sepals.

Structure of the Rose Flower—Flower Fullness

single flower (5 to 9 petals)

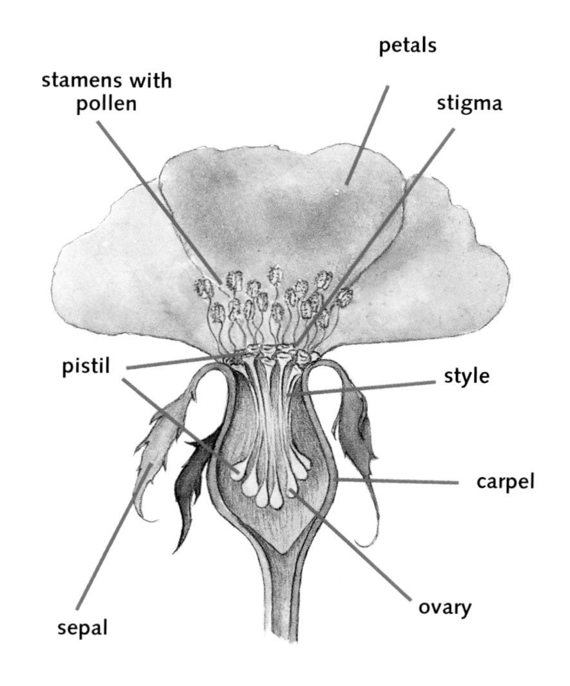

double flower (20 to 39 petals)

semidouble flower (10 to 19 petals)

very double flower (40 or more petals)

The flowers of all rose species are bisexual. Each flower contains both the male flower organs, anthers and stamens, as well as the female organ, the pistil, which consists of the stigma, style, and ovary. After fertilization, the hips with the seeds develop from the ovary.

Flower Fullness: A rose may develop up to a hundred stamens per flower. Through natural fertilization and/or human crossing, these stamens can gradually, over generations, be transformed into petals. This explains the differing flower fullnesses among roses, which extend from the single to the very double flower structure. These differing fullnesses make the rose the flowering shrub richest in variety of forms.

The classic single rose flower has five petals, with a maximum up to nine. At ten petals, we speak of semidouble roses, at twenty of double roses. Very double varieties display forty or more petals. Thus the conceptual definition of flower doubling in roses is—at least in theory—based on a multiple of five.

In practice, however, the number of petals can vary, sometimes markedly, even within an individual double variety. Therefore, a certain band width for the variety descriptions has been agreed upon.

In contrast with the number of petals, the number of sepals does not vary—it always remains five. When looked at carefully, one clearly sees that sepals form differently. An old verse stemming from the Latin describes this phenomenon very vividly, “We are five brothers born at the same time; two of us have beards, on two the beard is cut, one of us five has lost half of it.” With the dog rose, *Rosa canina*, the dissimilar brothers are especially pronounced and can be identified by merely looking closely.

Additionally, in many other rose species and varieties, one finds the simple, undivided sepals, the beardless, along with the pinnatifid, the bearded. The sepals of the repeat-flowering varieties are often also especially large on the flowers at the first flowering and decrease to clearly smaller ones at the second and—occasionally—third flowering.

Considered especially charming are the sepals of the moss roses. The sepals, pedicels, and ovaries of the most beautiful and powerfully fragrant kitchen garden rose, *Rosa centifolia* ‘Muscosa’, possess mosslike glands that make it a delight to the eye.

Flower Forms: The picture on the following page shows the various flower forms of the rose: flat (wild roses, many area roses), pointed (hybrid teas, many bedding, climbing, and shrub roses), quartered (usually in four zones with petals arranged like roof tiles and flower in raised form, e.g., ‘Souvenir de la Malmaison’ or English Roses like ‘Heritage’), rosette-shaped (either flat or balloon-shaped, e.g., English Roses and old and romantic roses but also climbing roses like ‘Rosarium Uetersen’), and rounded (Biedermeier form, numerous bedding and dwarf roses).

Inflorescences: Most wild rose species and many rose varieties develop their flowers in clusters arranged on inflorescences. Real single-stemmed roses are found only among the hybrid teas. When regarded strictly botanically, several different inflorescence types can be distinguished. Gerd Krüssmann, in his standard work *Roses, Roses, Roses,* names eight variants (see illustration page 33): the typical umbel (1), the pseudoumbel of the hybrid perpetual roses, an important connecting link between old and modern roses (2), the double umbel of *Rosa setipoda* (3), the compound pseudoumbel of the dog rose, *Rosa canina* (4), the pseudoumbel of *Rosa rugosa* (5), the narrowly compressed umbrella raceme (6), the irregularly branching, compound raceme with bent main axis (7), and the compound raceme (8).

The transitions between the individual types of inflorescences are fluid so that a clear distinction is not always very simple. However, whether an inflorescence is classified botanically as an umbel, raceme, or as neither of these—for the gardener knowing, for example, how the numerous bedding and ground cover roses with distinctive inflorescences can be used in designing is more important and more meaningful (see "Designing with Roses" beginning on page 47).

Flower Colors: The name says it all. The original color of the genus *Rosa* is rose. In the more than one hundred wild rose species, the forebears of all roses, many shades of pink dominate. Only a few exceptions exist: the yellow flowers of *Rosa foetida* and *Rosa hugonis* and the cream-colored flowers of the Scotch Briar rose, *Rosa pimpinellifolia.*

Thanks to the intensive rose crossing of the last two hundred years, today roses are available in red, pink, yellow, and white, between which are countless shadings. Only a pure blue is lacking.

The chemical elements of plant pigments have been thoroughly researched. Using a light microscope is sufficient to trace the genetics of flower colors. The chromoplasts, in which the pigments form, can be clearly distinguished. Among them are, for example, carotene and xanthophyll, which are responsible for the yellow flower colors among roses, and also anthocyan, which produces the red flower color.

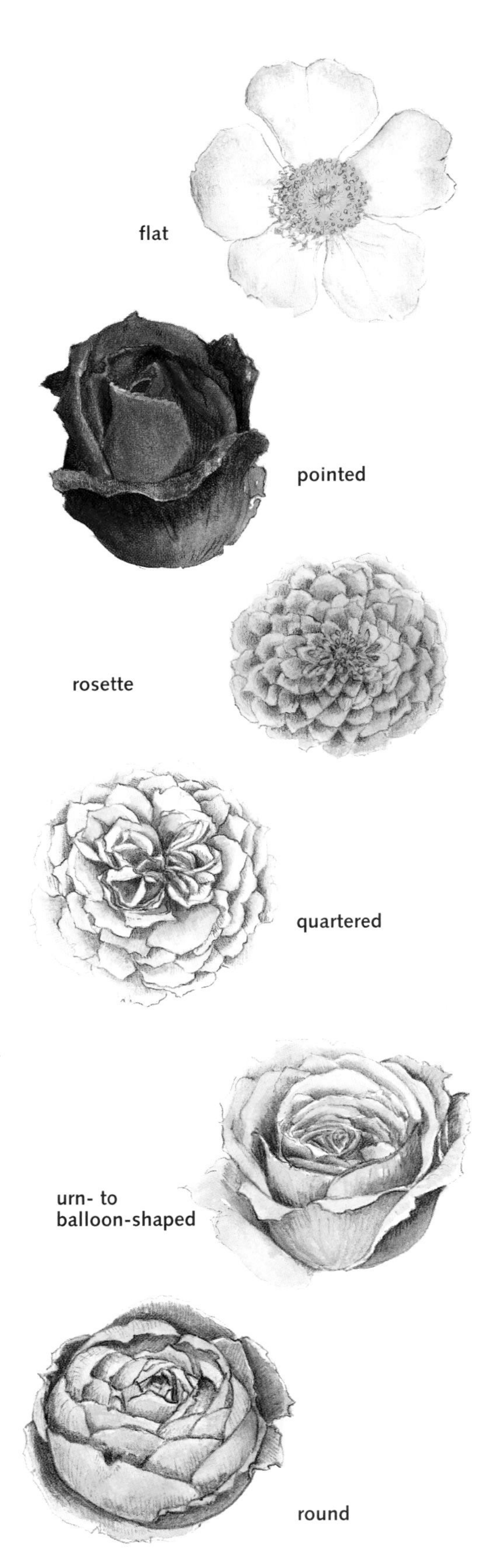

The syntheses of pigments in roses are very sensitive chemical processes. They can be disturbed by the lack of even one of the required enzymes, as the following examples show. The synthesis of a particular pigment is controlled by one or several genes. The striped petals of *Rosa gallica* 'Versicolor', a very old flower mutation, are thought to be produced by an inhibiting gene that prevents the synthesis of anthocyan in certain parts of the petal so that these places remain white or yellow. This is the way the particolored rose varieties in yellow or pale red have developed. A particular gene of another mutation, *Rosa foetida* 'Bicolor', is thought to be responsible for those roses in which the outer petals are colored differently from the inner ones.

If the synthesis of anthocyan takes place only in the outermost areas of the individual petals, flowers occur with a yellow or white eye, as is seen in 'Red Meidiland', for example. Once in a while, anthocyan forms only where light strikes the flower bud. As a result, bicolored flowers sometimes develop.

UV light influences the coloration of the petals. It blocks the color gene so that bud areas struck by UV light remain white. Thus, a multicolored effect may again appear when the rose blossoms.

During periods of cool weather, rose colors become more intense. Hot weather makes colors grow paler.

In addition, the color of a rose may change during the chronological course of flowering. Older red rose varieties like 'Paul's Scarlet Climber' take on a blue cast as they fade—they "go blue." Some pink or yellow roses lighten as they wither until they are almost white. In other yellow varieties, anthocyan synthesis begins only with the opening of the flower, which then turns yellow-red. Some white rose varieties owe their color to anthocyan blockade. However, where raindrops strike the petals, these areas lose their blockade—the white blossoms then appear sprinkled with red.

The problem of ambiguous color description has existed since rose books and catalogs have been available. Anyone who studies many catalogs has noticed that the descriptions of a particular rose can really differ between catalogs. A yellow

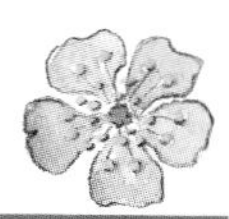

Different Inflorescences (according to v. Rathlef/W. Kordes/G. Krüssmann)

rose illuminated by the sinking sun looks much more intense than in the bright light of midday, in which it appears almost white. A rose that appears bluish looking when outdoors inside turns a fascinating red under artificial light because the high yellow content of the lightbulb obscures the blue to our eyes.

■ Leaf

Not only is the rose foliage the place where photosynthesis and respiration occur, the lungs of the plant as it were, but because of its manifold forms, the foliage also has decorative value. Unfortunately, the charm of the rose leaves is almost obscured, the glory of the flowers being so dominant. Therefore, taking a closer look at the green adornment of the rose is worthwhile.

Foliage Age: Among the rose species are **summer greens** and **evergreens.** The deciduous roses predominate in the continental United States. Unfortunately, the greenhouse cut roses are cultivated under glass year-round at constant summer temperatures. By using this unnatural environment on greenhouse cut roses, the flowers are foced to keep their leaves throughout the year. Even though these roses are actually summer greens, they appear to be evergreen. True evergreen roses, like *Rosa banksiae,* lack the necessary winterhardiness so that they occur only in decidedly mild regions. However, one can almost regard some modern varieties like 'New Dawn', 'Sommermorgen', or 'Flower Carpet' as evergreen for they hang onto their glossy foliage, bursting with health, far into the winter and sometimes even into the spring.

Foliage Form: The leaf of the rose consists of several composite parts, is odd-pinnate or compound off—pinnate, as the botanists say. What may at first seem to be independent leaves because of the very strong pinnation actually form a leaf unit, which is regarded as a whole. The rose leaflets can number three, five, seven, or more. Only *Rosa persica* displays simple leaves. Some examples of the countless variations of leaflets should be mentioned. Three to five leaflets appear on *Rosa gallica,* five to seven leaflets occur on *Rosa eglanteria. Rosa nitida* has seven to nine leaflets, while *Rosa sweginzowii* grows seven to eleven leaflets. Seven to thirteen leaflets appear on *Rosa moyesii,* and thirteen to fifteen leaflets grow on *Rosa sericea* f. *pteracantha. Rosa canina,* the native dog rose, usually has just seven leaflets.

The more wild rose blood that flows in a rose, the greater the number of its leaflets. As a rule, varieties that have resulted from repeated hybridization characteristically possess fewer leaflets. In addition, their individual leaflets are larger. The number of leaflets also decreases with their spatial nearness to the flower. For example, in hybrid tea and bedding roses, one finds only three-leafleted leaves directly under the flower. **Suckers,** wild canes that originate below the bud union from the grafting stock, can thus be easily recognized by their seven-leafleted leaves. Gardeners can use this visual cue and remove the suckers.

Modern area roses like 'Flower Carpet' or 'Alba Meidiland' indicate their wild ancestry by displaying strongly pinnated leaves.

Besides the leaflets, many rose species and varieties have undivided bracts present on the inflorescence. In their axes, these undivided bracts bear the flowers or other inflorescences.

The stipules are situated at the origin of the leaf stem. They are little leaflike structures that surround the leaf stem on both sides of the leaf base.

Foliage Color: Some rose species, like *Rosa glauca,* have quite unusual leaf color. *Rosa glauca,* which is also called *Rosa rubrifolia,* the red-leaved rose, starts to display frosted,

Leaf Form of the Rose

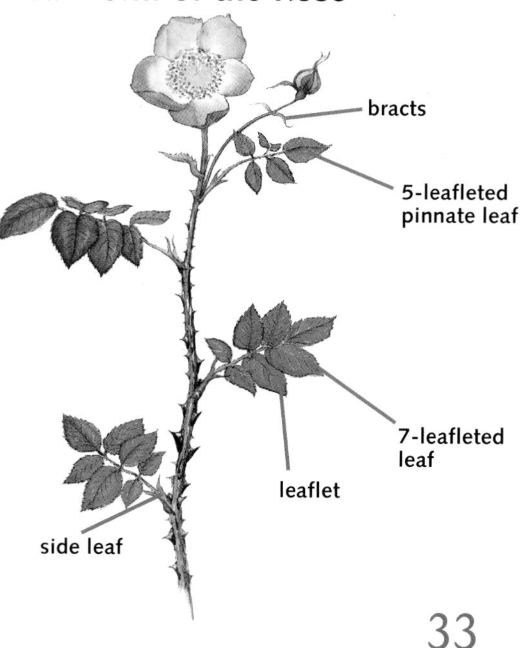

Rosa roxburghii **(Chestnut rose)**

Rosa virginiana

Rosa moyesii

Rosa pimpinellifolia **(The Scotch Briar rose)**

bluish red shimmering leaves in June. In addition, many hybrid tea roses first put out deep-red leaves and canes. These then gradually change in color to green.

'Flower Carpet' has extraordinarily attractive, shining foliage. The leaves of 'New Dawn' are also beautiful; they display a lacquered effect.

When using the various rose species and varieties to design a garden, considering the foliage structure is a good idea. One trait to pay attention to is the relative smoothness of the leaves of various rose species. For example, the wrinkled leaves of the *Rosa rugosa* hybrids do not go particularly well with the smooth, glossy leaves of the more modern varieties.

Foliage Blossoms: The light-green "petals" of the green rose, *Rosa chinensis* 'Viridiflora' are—from the botanical point of view—leaves.

Foliage Scent: Leaves can be carriers of aromatic essential oils. In damp weather, the leaves of the Sweet Briar rose, *Rosa eglanteria,* a native wild and very interesting hedge rose, exude an intense apple fragrance. The leaves of *Rosa primula* have the myrtlelike scent of incense.

Fall Coloring: The fall leaf drop of roses provides an aesthetic display in some species and varieties. Some roses display reddish fall coloring. Others display yellowish coloring during the fall. Those that have reddish coloring can be divided into four categories: rambler, climber, wild rose, and shrub rose. 'Bobbie James', 'Super Dorothy', and 'Super Excelsa' are examples of these ramblers. The 'Ilse Krohn Superior' is a climber that exhibits reddish fall coloring. *Rosa nitida* is a wild rose and *Rosa sweginzowii* 'Macrocarpa' is a shrub rose; both display this reddish coloring during the fall. Two classes of rose have yellowish fall coloring: *rugosa* hybrids and area roses. Examples of these *rugosa* hybrids include 'Dagmar Hastrup', 'Foxi', 'Yellow Dagmar Hastrup', 'Pierette', 'Polareis', 'Polarsonne', and 'Schnee-Eule'. Area roses that have yellowish coloring during the fall are 'Alba Meidiland' and 'The Fairy'. Page 68 describes roses with striking fall coloring.

■ Fruit

That the fruit of the rose has its very own name underlines its importance from ancient times to the present. Hips are the seed fruits, which develop from the flower. The real seeds of the rose, the nuts, ripen in the hips. The number of seeds varies greatly. *Rosa hugonis* produces only a few, sometimes even only one seed, while *Rosa moyesii* produces four to six large ones, and *Rosa rugosa* up to one hundred moderately large ones in one hip. *Rosa carolina* and *Rosa virginiana* have very small seeds. Numerous area, shrub, and wild roses are lavish producers of hips. Dwarf, bedding, hybrid tea, and climbing roses—differing among varieties—also fruit. However, many double varieties in these rose classes are self-sterile. They therefore do not develop hips.

The form and color of the hips are extremely diverse. In fall, they glow yellow, orange, or red but also greenish or brown to black. The forms vary from spherical to flattened spherical to egg-, pear-, or bottle-shaped. The fruits are either bare or bristled and sometimes so covered by dense prickles that they resemble chestnuts (as with *Rosa roxburghii,* the Chestnut rose). Most of the hips are fleshy.

> **Note** What is in hips for humans and animals and what delicious recipes can be prepared with them can be found in the chapter "Vitamins in Roses" beginning on page 134.

■ Prickles (Thorns)

Roses have, botanically speaking, prickles and not thorns, even if Joseph von Scheffel did write, "It is unpleasantly ordained in life, that even on the rose there are thorns, and though the poor heart yearn and write, yet at the end comes parting." When considered botanically, prickles are extrusions developed from epidermal cells of the canes, leaf nerves, or sepals. They are pointed in shape. Prickles can always be easily removed from the canes. To confuse

Spines of the Rose

Spine form in 'Feuerzauber'

'Königin der Rose' . . .

'Red Nelly' and

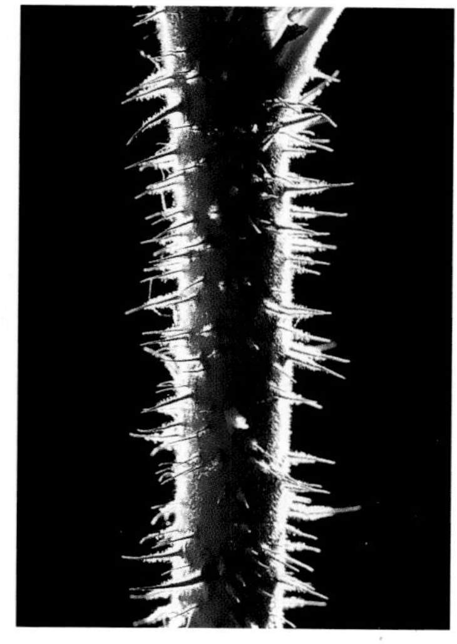

Rosa rugosa.

the reader and gardener totally, gooseberries have thorns (transformed shoots), which almost always grow attached to the cane and thus cannot be easily removed. Rose prickles can be shaped in all different kinds of ways. For example, in *Rosa rugosa* and its hybrids, they are bristley to needlelike. Those of *Rosa sericea* f. *pteracantha* are triangular, flat, and pointed. In *Rosa eglanteria,* they are hooked, and in *Rosa multiflora* they are clawlike.

Sometimes, too, different prickle shapes occur on one and the same cane. The color of the prickles varies also. In *Rosa rugosa* they are gray. In *Rosa eglanteria* and *Rosa pimpinellifolia,* they are rust brown. In hybrid tea and bedding roses, the spines are mostly green, and in *Rosa sericea* f. *pteracantha* they are a brilliant fiery red.

■ Root

In the truest sense of the word, the roots of the rose carry on a hidden existence. In the relevant technical books, there is almost never any attention paid to their treatment and maintenance. Roots form the foundation for the plant. Yet there are many rose fanciers for whom they are out of sight, out of mind. Thus, it is worth taking a look at rose roots. Knowledge of their growth forms and the requirements that derive from them are of primary importance for later success with rose planting in the garden or in pots.

Every form of root system results from the natural battle for survival of the particular plant in question. Every plant root utilizes the surrounding soil space differently in order—if possible—to avoid competition from other plants in procuring nutrients and water.

Knowing how the roots of a particular rose species are constituted allows planning of optimal communities of roses and other shrubs and perennials. As a rule, garden roses are grafted onto a wilding understock, usually *Rosa laxa.* In contrast with the wild roses, this grafting forms deep, wide-spreading roots, like underground extensions of the above-ground canes.

Grafted roses therefore draw on deeper layers of water and nutrients—provided that no impenetrable layers of soil block their way to the lower levels. Scientific research has shown that in grafted greenhouse cut roses, the intensive **root zone** lies between 35 1/2 and 47 1/4 in. (0.90 to 1.20 m).

To develop, the rose root needs aerated soil above all. Experiments have determined that the plants breathe above the roots almost as much as above the leaves. Oxygen deficiency in the ground, lost through compaction or sogginess, for example, stops root growth. As a consequence, the plant experiences poor growth and severe pest attacks.

ROSY VARIETY

The colors and scent of rose blossoms have been described many times. However, two special characteristics of the rose, namely its flowering patterns and its growth forms, have been given only little attention until now. They are quickly dealt with by using imprecise terms like *everblooming* **or** *beautiful growth habit.* **This occurs although these two criteria underline the diversity of the rose and substantially determine the success of the design of a rose planting.**

Characteristics of the Rose

The special position that the rose holds among the garden shrubs can be easily recognized if we consider its many attributes. Because of the innumerable flower colors and marked differences in flowering patterns, in heights of growth, and in flower and growth forms of the individual species and varieties, roses offer the garden lover an unparalleled design resource on which to draw.

So that each of the particular rose varieties or species can display its specific talent and tendencies to optimal advantage in the garden, it is wise to investigate certain rose character traits thoroughly before selecting and planting. If these character traits are taken into consideration when designing with roses, the gardener spares himself or herself some mistakes and some bad experiences. It is thoroughly annoying when, after years of carefully growing rose plants, their blooming period or their growth form does not harmonize with the accompanying shrubs and perennials.

Naturally, no normal roses exist, so their inclusion in particular groups is always a somewhat uncertain affair. In addition, the behavior of roses can differ greatly from year to year. In dry, very hot summers, for instance, the flowering rhythm of the repeat-flowering roses may become interrupted. Many varieties remain crouched on the starting line and undergo a pause in flowering. Then, with soil dampness suddenly present after a thorough rainfall, they start off again headlong. The beginning of new growth after a long, hard winter can be deferred for weeks. In contrast, during the course of a mild winter, the bud dormancy of roses is very shallow. This occurs because bud sprouting is very easily stimulated by increasing temperatures. As a result, the canes are already putting out the first leaves in February.

Nevertheless, in spite of weather factors and regional climatic differences, year-round observations permit the classification of roses into certain **characteristic groups.** They should be a help to the garden lover when making the first, rough choice of the appropriate rose varieties for the desired purpose. At the same time, knowledge of these characteristic groups sharpens the awareness of the rose as garden construction material. For the individual groups of rose varieties, examples are given that indicate the particular characteristic of the group that are especially strikingly. By using the further details for all varieties as given in "The Rose Atlas" beginning on page 185, other varieties can be classified into the characteristic groups as desired.

Flowering Habits

Along with their nuance-rich flower variety, the flowering pattern of roses is the primary foundation for their reputation as queens of flowers. No other shrub makes bringing color into the home garden from May until the first frost in the fall possible. Naturally, a single rose variety does not do this all by itself. However, with the help of different groups of roses with varying flowering patterns, one can design a half year of rose flowering that constantly provides new high points.

Basically, we distinguish between once- and repeat-flowering varieties. The transitions between them are very fluid.

The Once-Flowering Rose

The current rose market is dominated by repeat-flowering varieties. Little attention is paid to the once-flowering varieties. This occurs without good reason. To this group belongs a list of the most interesting varieties of the shrub and climbing roses (ramblers). Once-flowering roses generally trump the repeat-flowering varieties chronologically. Once-flowering roses are enchanting throughout their blooming period, which may last for as long as five weeks. Many once-flowering shrub roses are outstandingly frost hardy (see also the chapter "Polar Roses," page 99).

Spring rose 'Maigold'®.

Examples of Varieties:
white: 'Bobbie James' (rambler, 118–197 in. [300–500 cm])
'Suaveolens' (shrub rose, 79–197 in. [200–300 cm])
cream: 'Albéric Barbier' (rambler, 118–197 in. [300–500 cm])
pink: 'Ferdy'® (shrub rose, 32 –39 in. [80–100 cm])
'Immensee'® (area rose, 12 –16 in. [30 –40 cm])
'Maria Lisa' (climber, 79 –187 in. [200 –300 cm])
'Max Graf' (area rose, 24–32 in. [60–80 cm])
'Raubritter' (shrub rose, 79–197 in. [200–300 cm])
Rosa sweginzowii 'Macrocarpa' (shrub rose, 79–197 in. [200–300 cm])
'Venusta Pendula' (rambler, 118–197 in. [300–500 cm])
red: 'Flammentanz'® (rambler, 118–197 in. [300–500 cm])
'Scharlachglut' (shrub rose, 59–79 in. [150–200 cm]) (also called 'Scarlet Glow')

The color yellow dominates in the especially early, once-flowering shrub roses, the so-called spring roses.

Spring Roses Beginning in May

The spring roses are once-flowering shrub roses that start blooming in May. They signal the beginning of the rose spring (see also "Spring Roses," page 68).

Examples of Varieties:
'Frühlingsgold' (yellow, shrub rose, 59–79 in. [150–200 cm])
Rosa hugonis (yellow, shrub rose, 79–118 in. [200–300 cm])
Rosa moyesii (red, shrub rose, 79–118 in. [200–300 cm])
Rosa sericea f. *pteracantha* (white, shrub rose, 79–118 in. [200–300 cm])

In addition, the native wild species *Rosa pimpinellifolia* can be regarded as a spring rose. Its imposing main flush of flowers appears at the beginning of the rose year, in May. It blooms less profusely again in the fall.

A transition to the next group is represented by the remontant spring rose 'Maigold' (yellow, shrub rose, 59–79 in. [150–200 cm]). Its imposing main flowering appears at the beginning of the rose year in May, and it blooms sparsely once more in the fall.

The Remontant Rose

On the threshhold of multiple flowerings stand the remontant rose varieties. They burst forth in June with an outstanding main flush of flowers and then bloom again less profusely from August until September.

Examples of Varieties:
'Dagmar Hastrup' (*Rosa rugosa* hybrid, 24–32 in. [60–80 cm]), 'Marguerite Hilling' (shrub rose, 59–79 in. [150–200 cm]), and 'Paul Noël' (rambler, 118–197 in. [300–500 cm])

The Repeat- and Early-Flowering Rose

Repeat-flowering varieties are today considered the embodiment of the modern, enormously prolific roses. In contrast with other shrubs, these varieties are able, in favorable light and temperature conditions, to produce ever new flower output after their first, main flowering.

Repeat- and Early-Flowering Varieties:
white: 'Schneeflocke'® (bedding rose, 16–24 in. [40–60 cm])
'Schneewittchen'® (shrub rose, 39–59 in. [100–150 cm]) (also called 'Iceberg')
cream: 'Karl Heinz Hanisch'® (hybrid tea, 24–32 in. [60–80 cm])
yellow: 'Bernstein Rose'® (bedding rose, 24–32 in. [60–80 cm])
'Lichtkönigin Lucia'® (shrub rose, 39–59 in. [100–150 cm])
'Polygold'® (bedding rose, 16–24 in. [40–60 cm])
'Sunsprite'® (bedding rose, 24–32 in. [60–80 cm])
pink: 'Centenaire de Lourdes' (shrub rose, 59–79 in. [150–200 cm])
'Frau Astrid Späth' (bedding rose, 16–24 in. [40–60 cm])
'Heritage'® (shrub rose, 39–59 in. [100–150 cm])
'Lavender Dream'® (area rose, 24–32 in. [60–80 cm])
'Matilda'® (bedding rose, 16–24 in. [40–60 cm])
'Silver Jubilee'® (hybrid tea, 24 –32 in. [60–80 cm])
red: 'Sarabande'® (bedding rose, 16–24 in. [40–60 cm])

As a rule, in the continental United States, two blooming periods occur, in June and August. The first flowering is the more vigorous. Depending on the variety, the transitions between the two are variable. Some roses flower continually until the August blossoming, others take a clearly discernible breather.

The Classic Repeat-Flowering Rose

Most repeat-flowering varieties bloom in two bursts—in June/July and in August/September. To coordinate the flowering of these varieties somewhat so they appear together in the garden, it is advisable to pinch back the young shoots from the beginning to the end of May, that is, about four weeks before the expected principal flowering. The exact procedure for pinching back is described on page 112 under the heading "Roses for Cutting."

The Repeat- and Late-Flowering Rose

In particular, the magnificent inflorescences of double-flowered roses from the area and shrub rose groups often appear very late. They need more time to develop their mighty clusters of flowers than do, say, the hybrid tea roses with their single blossoms. Thus, they may trail as much as three weeks behind the classic repeat bloomers in their production of flowers. This feature makes them ideally suited for filling in when the flowering of the earlier varieties comes to a halt. In addition, because they flower later in the fall, these roses prolong the rose garden season until the first frosts.

Repeat- and Late-Flowering Varieties:
white: 'Alba Meidiland'® (area rose, 32–39 in. [80–100 cm])
yellow: 'Ghislaine de Féligonde' (shrub rose, 59–79 in. [150–200 cm])
pink: 'Bella Rosa'® (bedding rose, 24–32 in. [60–80 cm])
'Diadem'® (bedding rose, 32–39 in. [80–100 cm])
'Eden Rose '85'® (shrub rose, 59–79 in. [150–200 cm])
'Flower Carpet'® (area rose, 24–32 in. [60–80 cm])
'Heidepark'® (bedding rose, 24–32 in. [60–80 cm])
'Lovely Fairy'® (area rose, 24–32 in. [60–80 cm])
'Magic Meidiland'® (area rose, 16–24 in. [40–60 cm])
'Pheasant'® (area rose, 24–32 in. [60–80 cm])
'Royal Bonica'® (bedding rose, 24–32 in. [60–80 cm])
'Super Dorothy'® (rambler, 118–197 in. [300–500 cm])
'The Fairy' (area rose, 24–32 in. [60–80 cm])
red: 'Fairy Dance'® (area rose, 16–24 in. [40–60 cm])
'Happy Wanderer'® (bedding rose, 16–24 in. [40–60 cm])
'Scarlet Meidiland'® (area rose, 24–32 in. [60–80 cm])
'Super Excelsa'® (rambler, 118–197 in. [300–500 cm])

Single Flower: *Rosa canina.*

Semidouble flower: 'Escapade'®.

Early-Flowering rose: 'Sunsprite'®.

Flower Forms

Each rose variety presents its visiting card with the form of its flowers. From the single wild rose to the very double rosettes of the romantic roses—roses offer a matchless wealth of forms.

The Single Rose Flower

The single flower is the original flower form of the rose. Four to seven petals surround the numerous stamens, which with their pollen are a rich source of food for insects. Single-flowered roses go well in natural gardens or as transition plantings from maintained cultivated gardens to fields. Thanks to intensive hybridization, besides the once-flowering wild roses are also repeat-flowering, very robust area, shrub, and climbing roses with single flowers.

Repeat-Flowering Area Roses:
'Apfelblüte'® (32–39 in. [80–100 cm])
'Ballerina' (24–32 in. [60–80 cm])
'Bingo Meidiland'® (16–24 in. [40–60 cm])
'Mozart' (32–39 in. [80–100 cm])
'Red Meidiland'® (24–32 in. [60–80 cm])
'Repandia'® (16–24 in. [40–60 cm])

Once-Flowering Area Rose:
'Immensee'® (12–16 in. [30–40 cm])

Native Wild Roses:
Rosa arvensis (32–39 in. [80–100 cm])
Rosa gallica (32–39 in. [80–100 cm])
Rosa majalis (59–79 in. [150–200 cm])
Rosa pimpinellifolia (32–39 in. [80–100 cm])
Rosa eglanteria (79–118 in. [200–300 cm])

Climbing Roses:
'Dortmund'® (repeat-flowering, 79–118 in. [200–300 cm])
'Maria Lisa' (once-flowering, 79–118 in. [200–300 cm])

Once-Flowering Shrub Roses:
'Frühlingsgold' (59–79 in. [150–200 cm])
Rosa hugonis (79–118 in. [200–300 cm])
'Borgogne'® (59–79 in. [150–200 cm])
Rosa moyesii (79–118 in. [200–300 cm])
Rosa sericea f. *pteracantha* (79–118 in. [200–300 cm])
'Scharlachglut' (59–79 in. [150–200 cm])

Repeat-Flowering Shrub Rose:
'Bischofsstadt Paderborn'® (39–59 in. [100–150 cm])

The Semidouble Rose Flower

The semidouble flower represents the transition from the original flower florm to the modern, double flower form. The varieties in this group are, so to speak, civilized wild roses. Not inferior to the wild roses as distributers of pollen, thanks to their many stamens, they also provide repeated spots of color in the garden with their multiple-petaled flowers (exceptions appear in parentheses).

Examples of Varieties:

Bedding Roses:
'Blühwunder'® (24–32 in. [60–80 cm])
'Dolly'® (24–32 in. [60–80 cm])
'Escapade'® (32–39 in. [80–100 cm])
'Heidepark'® (24–32 in. [60–80 cm])
'La Sevillana'® (24–32 in. [60–80 cm])
'Ricarda'® (24–32 in. [60–80 cm])
'Sarabande'® (16–24 in. [40–60 cm])
'Schleswig '87'® (24–32 in. [60–80 cm])
'Schneeflocke'® (16–24 in. [40–60 cm])

Area Roses:
'Flower Carpet'® (24–32 in. [60–80 cm])
'Lavender Dream'® (24–32 in. [60–80 cm])
'Marondo'® (24–32 in. [60–80 cm])
'Royal Bassino'® (16–24 in. [40–60 cm])
'Sommermärchen'® (16–24 in. [40–60 cm])
'Surrey'® (16–24 in. [40–60 cm])

Climbing Roses:
'Bobbie James' (once-flowering rambler, 118–197 in. [300–500 cm])
'Morning Jewel'® (79–118 in. [200–300 cm])
'Venusta Pendula' (once-flowering rambler, 118–197 in. [300–500 cm])

Shrub Roses:
'Centenaire de Lourdes' (59–79 in. [150–200 cm])
'Dirigent'® (59–79 in. [150–200 cm])
'Marguerite Hilling' (remontant, 59–79 in. [150–200 cm])
'Westerland'® (59–79 in. [150–200 cm])

The Double Rose Flower

Double roses are well-proportioned gems that exude true enchantment in any garden. In particular, the hybrid tea roses are an aesthetic treat for those who appreciate the delights of the garden. Noblesse oblige.

Since almost all hybrid tea, dwarf, and English Roses, very many bedding, climbing, and shrub roses bear double flowers, the mention of individual varieties in this chapter has been omitted. Instead, readers are referred to "The Rose Atlas" (pages 185–223).

Rose Inflorescences

Many bedding, climbing, and shrub roses are adorned with attractive inflorescences that do not loosely branch but consist of dense clusters of numerous small flowers, which add up to a true flower ball. Primarily, the modern, repeat-flowering area roses belong to this group. During their blooming season, as their name implies, they cover entire areas with a uniform carpet of flowers.

Examples of Varieties:

Single Inflorescences:
'Apfelblüte'® (32–39 in. [80–100 cm])
'Ballerina' (24–32 in. [60–80 cm])
'Bingo Meidiland'® (16–24 in. [40–60 cm])
'Marjorie Fair'® (24–32 in. [60–80 cm])
'Mozart' (32–39 in. [80–100 cm])
'Red Meidiland'® (24–32 in. [60–80 cm])
'Rush'® (shrub rose, 39–59 in. [100–150 cm])

Semidouble Inflorescences:
'Flower Carpet'® (24–32 in. [60–80 cm])
'Lavender Dream'® (24–32 in. [60–80 cm])
'Marondo'® (24–32 in. [60–80 cm])
'Sommermärchen'® (16–24 in. [40–60 cm])
'Surrey'® (16–24 in. [40–60 cm])

Double Inflorescences:
'Alba Meidiland'® (32–39 in. [80–100 cm])
'Fairy Dance'® (16–24 in. [40–60 cm])
'Ghislaine de Féligonde' (shrub rose, 59–70 in. [150–200 cm])
'Lovely Fairy'® (24–32 in. [60–80 cm])
'Mirato'® (24–32 in. [60–80 cm])
'Pheasant'® (24–32 in. [60–80 cm])
'Satina'® (16–24 in. [40–60 cm])
'The Fairy' (24–32 in. [60–80 cm])
'Super Dorothy'® (rambler, 118–197 in. [300–500 cm])
'Super Excelsa'® (rambler, 118–197 in. [300–500 cm])
'White Meidiland'® (16–24 in. [40–60 cm])

Rose inflorescences conjure up an abundance of flowers in the garden.

Growth Habits

Besides flower form, flower color, blooming rhythm, and height of growth, the value of a rose variety in a garden design crucially depends on its growth habit. Primarily because of the care and space required, the use of large group plantings of earlier days has given way to roses being used singly or in small groups, often together with perennials or other shrubs. In addition, the planting of a specimen shrub rose to function as a focal point in the home garden has come to be much more valued.

Only in recent times have the rose breeders dedicated increased attention to the growth form of the rose. Of all the characteristics of the rose, this has until now been the least regarded and shaped by hybridizing. We may look forward to trendsetting advances in this area in the next few years. Above all, the often gawky-looking hybrid teas may be freed of their form, which is reminiscent of a clothes rack. New shrub rose varieties may be given a round, densely foliaged growth form.

Professor Josef Sieber has offered guidelines for defining the different growth habits of the rose. The definitions of growth types that follow are based on his ideas and should be useful to the rose lover for guidance.

Low-lying growth, less vigorous

Low-lying growth, vigorous

Miniature form

Stiff, upright habit

Bushy shrub habit

Arching shrub habit

Climbing habit

■ Low-lying Growth Habit

In recent years, the low-growing roses, the **ground cover roses,** have experienced a veritable procession of triumphs. Some, but not all, of these varieties grow low and hug the ground. Each area planted with them, because of the dense foliage of these roses, is covered with a thick carpet of flowers by the third year after planting, at the latest.

Basically, these roses with low-lying, or procumbant, growth habits can be classified into two groups. The **less vigorous** varieties include 'Nozomi' (area rose, 16 to 24 in. [40 to 60 cm]), *Rosa repens* x *gallica* (area rose, 12 to 16 in. [30 to 40 cm]), and 'Snow Ballet' (area rose, 16 to 24 in. [40 to 60 cm]). The **vigorous** varieties with yard-long canes include 'Pheasant' (area rose, 24 to 32 in. [60 to 80 cm]), 'Immensee' (area rose, 12 to 16 in. [30 to 40 cm]), 'Magic Meidiland' (area rose, 16 to 24 in. [40 to 60 cm]), 'Marondo' (area rose, 24 to 32 in. [60 to 80 cm]), 'Repandia' (area rose, 16 to 24 in. [40 to 60 cm]), and 'Repens Alba' (*rugosa* hybrid, 12 to 16 in. [30 to 40 cm]). In addition, climbing roses like 'Super Dorothy' (rambler, 118 to 197 in. [300 to 500 cm]) or 'Super Excelsa' (rambler, 118 to 197 in. [300 to 500 cm]) grow flat without a support and then serve as ground cover or area roses.

■ Miniature Forms

Miniature roses form a small, compact bush. With a maximum growth height of 16 in. (40 cm), they clearly remain below knee height (24 in. [60 cm]). Because of their growth habit, they are suitable for group arrangments and especially for tub and box plantings.

Examples of Varieties
'Dwarfking', 'Guletta'®, 'Orange Meillandina'®, 'Peach Brandy', 'Pink Symphonie'®, 'Sonnenkind'®, (each 16–24 in. [40–60 cm])

■ Stiff, Upright Habit

This growth group includes the **hybrid tea roses.** With their sometimes sparse foliage and tall shrub structure, they sometimes look like flamingos that are wading through shallow water on thin legs. The upright growth of these varieties induces many gardeners to underplant them with perennials or other green plants. The results are often negative, for hybrid tea roses, in particular, know how to appreciate open soil.

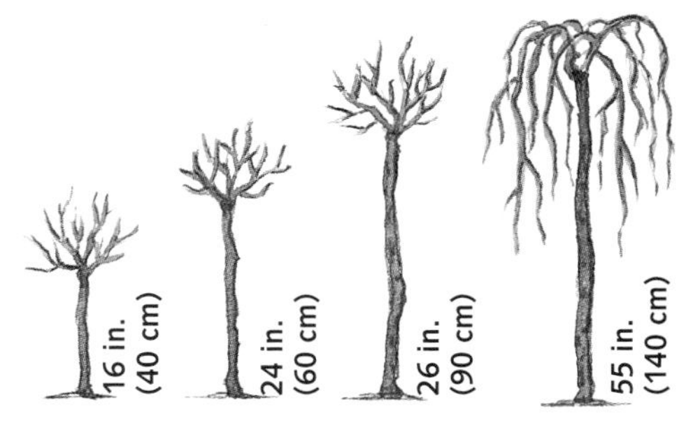

Special form: the tree rose

Examples of Varieties (Hybrid Teas):
'Aachener Dom'® (24–32 in. [60–80 cm])
'Banzai '83'® (32–39 in. [80–100 cm])
'Barkarole'® (32–39 in. [80–100 cm])
'Duftrausch'® (32–39 in. [80–100 cm])
'Elina'® (32–39 in. [80–100 cm])
'Fragrant Gold'® (24–32 in. [60–80 cm])
'Ingrid Bergman'® (24–32 in. [60–80 cm])
'Loving Memory'® (24–32 in. [60–80 cm])
'Paul Ricard'® (24–32 in. [60–80 cm])
'Peace' (32–39 in. [80–100 cm])
'Sunblest'® (24–32 in. [60–80 cm])
'The McCartney Rose'® (24–32 in. [60–80 cm])

A **bedding rose** that also has outstandingly stiff, tall growth is 'The Queen Elizabeth Rose' (32–39 in. [80–100 cm]).

Bushy Shrub Habit

Many **bedding** and **shrub** roses display a dense, beautifully shaped growth habit and develop into bushes of varying sizes. In the summertime, they scarcely permit a glimpse of bare wood, so thick is the mass of foliage and flowers. Thus, not without reason, they are among the most popular roses for the home garden, especially when a harmonious total garden picture is an important feature of the desired design. Some especially beautifully growing examples are listed in the box below.

Examples of Varieties:

Bedding Roses:
'Bella Rosa'® (24–32 in. [60–80 cm])
'Bonica '82'® (24–32 in. [60–80 cm])
'Escapade'® (32–39 in. [80–100 cm])
'Gartenzauber'® (16–24 in. [40–60 cm])
'Goldener Sommer '83'® (16–24 in. [40–60 cm])
'Goldmarie '82'® (16–24 in. [40–60 cm])
'Gruss an Aachen' (16–24 in. [40–60 cm])
'Happy Wanderer'® (16–24 in. [40–60 cm])
'La Paloma '85'® (24–32 in. [60–80 cm])
'La Sevillana'® (24–32 in. [60–80 cm])
'Leonardo da Vinci'® (24–32 in. [60–80 cm])
'Mariandel'® (16–24 in. [40–60 cm])
'Mountbatten'® (32–39 in. [80–100 cm])
'Nina Weibull' (16–24 in. [40–60 cm])

Shrub Roses:
'Dirigent'® (59–79 in. [150–200 cm])
'Schneewittchen'® ('Iceberg') (39–59 in. [100–150 cm])
'Vogelpark Walsrode'® (39–59 in. [100–150 cm])
'Westerland'® (59–79 in. [150–200 cm])

Area Roses:
'Lovely Fairy'® (24–32 in. [60–80 cm])
'Surrey'® (16–24 in. [40–60 cm])
'The Fairy' (24–32 in. [60–80 cm])

Wild Rose:
Rosa nitida (24–32 in. [60–80 cm])

Rampantly vigorous rambler: 'Bobbie James'.

Arching Shrub Habit

With some shrub roses, the dense habitus is further enchanced by the decorative arching or trailing canes. When these arching canes are covered with hoarfrost in the darkest time of the year, they provide garden interest of a uniquely beautiful nature.

Examples of Varieties:
'Abraham Darby'® (59–79 in. [150–200 cm])
'Borgogne'® (59–79 in. [150–200 cm])
'Centenaire de Lourdes' (59–79 in. [150–200 cm])
'Constance Spry' (59–79 in. [150–200 cm])
'Ferdy'® (32–39 in. [80–100 cm])
'Frühlingsgold' (59–79 in. [150–200 cm])
'Graham Thomas'® (39–59 in. [100–150 cm])
'Heritage'® (39–59 in. [100–150 cm])
'Raubritter' (79–118 in. [200–300 cm])

Climbing Habit

Among the group of roses with long, whiplike canes are—not surprisingly—the climbers. With a sturdy trellis to support them, they can cover walls and house walls, pergolas and gateposts in an unequaled sea of flowers.

Like lianas, the ramblers climb their way up on thin stems.

Heavy Flowering Depresses Drive to Grow: Basically, this means that the repeat-flowering climbing roses—like, for example, 'Ilse Krohn Superior', 'Lawinia', 'Morning Jewel', 'New Dawn', 'Rosarium Uetersen', 'Salita', 'Santana', and 'Sympathie'—with a height of 6 to 10 feet (2 to 3 m) are clearly less vigorous growers than the enormously vigorous once-flowering **ramblers** such as 'Albéric Barbier', 'Bobbie James', 'Flammentanz', 'Paul Noël', *Rosa* x *ruga*, and 'Venusta Pendula'. With their growth temperament, these roses can reach 20 feet (6 m) high and more. Their climbing habit remains remarkably loose and pleasing in these high flights and never looks stiff.

Special Form: The Standard

An artificially produced growth form is the tree or standard rose. The height of the crown is determined at the nursery by the height of the graft—16 inches (40 cm) (quarter standard), 24 inches (60 cm) (half standard), 36 inches (90 cm) (full standard), 55 inches (140 cm) (cascade or weeping). The rose variety grafted onto the wild trunk grows just exactly as it would on a stock grafted close to the ground—only at an elevated height. At these altitudes, many varieties are far less susceptible to disease than at ground level.

Growth Heights

The growth height of the desired variety is an important piece of information for every gardener. It affects the choice of the future site, the placement of the rose plant in front of, within, or behind existing beds, and the planning of any possible long-distance effects.

Although more than 30,000 different rose varieties exist, only a small selection are appropriate for culture as a tree rose. Outstanding example: 'Lawinia'®.

12–16 in. (30–40 cm):

procumbent area roses ('Immensee'®), **miniature roses** ('Guletta'®, 'Orange Meillandina'®, 'Pink Symphonie'®, 'Sonnenkind'®)

16–24 in. (40–60 cm):

bedding roses ('Amber Queen'®, 'Edelweiss'®, 'Goldener Sommer '83'®, 'Goldmarie '82'®, 'Mariandel'®, 'Sarabande'®), **area roses** ('Bingo Meidiland'®, 'Fairy Dance'®, 'Mirato'®, 'Nozomi'®, 'Royal Bassino'®, 'Satina'®, 'Surrey'®, 'White Meidiland'®)

24–32 in. (60–80 cm):

bedding roses ('Ballade'®, 'Bella Rosa'®, 'Bonica '82'®, 'Europas Rosengarten'®, 'Fragrant Cloud'®, 'La Paloma '85'®, 'Manou Meilland'®, 'Pussta'®, 'Ricarda'®, 'Rosali '83'®, 'Sommermorgen'®, 'Sunsprite'®, 'Träumerei'®), **hybrid tea roses** ('Aachener Dom', 'Fragrant Gold'®, 'Loving Memory'®, 'Silver Jubilee'®, 'Sunblest'®, 'The McCartney Rose'®), **ground cover roses** ('Ballerina', 'Flower Carpet'®, 'Lavender Dream'®, 'Palmengarten Frankfurt'®, 'Pheasant'®, 'Pink Meidiland'®, 'The Fairy', 'Wildfang'®)

32–39 in. (80–100 cm):

bedding roses ('Diadem'®, 'Make Up'®, 'Montana'®, 'The Queen Elizabeth Rose'®), **hybrid tea roses** ('Banzai '83'®, 'Barkarole'®, 'Deep Secret'®, 'Duftrausch'®, 'Elina'®, 'Peace', 'Polarstern'®), **area roses** ('Alba Meidiland'®, 'Apfelblüte'®, 'Dortmunder Kaiserhain'®, 'Mozart'), **native wild roses** (*Rosa arvensis, Rosa gallica, Rosa pimpinellifolia*), **shrub roses** ('Ferdy'®, 'IGA '83 München'®, *Rosa centifolia* 'Muscosa')

39–59 in. (100–150 cm):

shrub roses ('Astrid Lindgren'®, 'Bischofsstadt Paderborn'®, 'Graham Thomas'®, 'Kordes' Brilliant'®, 'Lichtkönigin Lucia'®, 'Polka '91'®, 'Romance'®, 'Schneewittchen'® ('Iceberg'), 'Vogelpark Walsrode'®)

59–79 in. (150–200 cm):

native wild roses (*Rosa majalis, Rosa scabriuscula*), **shrub roses** ('Abraham Darby'®, 'Centenaire de Lourdes', 'Dirigent'®, 'Eden Rose '85'®, 'Grand Hotel'®, 'Rugelda'®, 'Westerland'®)

79–118 in. (200–300 cm):

native wild rose (*Rosa eglanteria*), **repeat-flowering climbers** ('Compassion'®, 'Golden Showers'®, 'Ilse Krohn Superior'®, 'Lawinia'®, 'New Dawn'®, 'Rosarium Uetersen'®, 'Salita'®, 'Sympathie'®)

118–197 in. (300–500 cm):

ramblers (Albéric Barbier'®, 'Bobbie James'®, 'Flammentanz'®, 'Paul Noël'®, *Rosa* x *ruga*, 'Super Dorothy'®, 'Super Excelsa'®, 'Venusta Pendula')

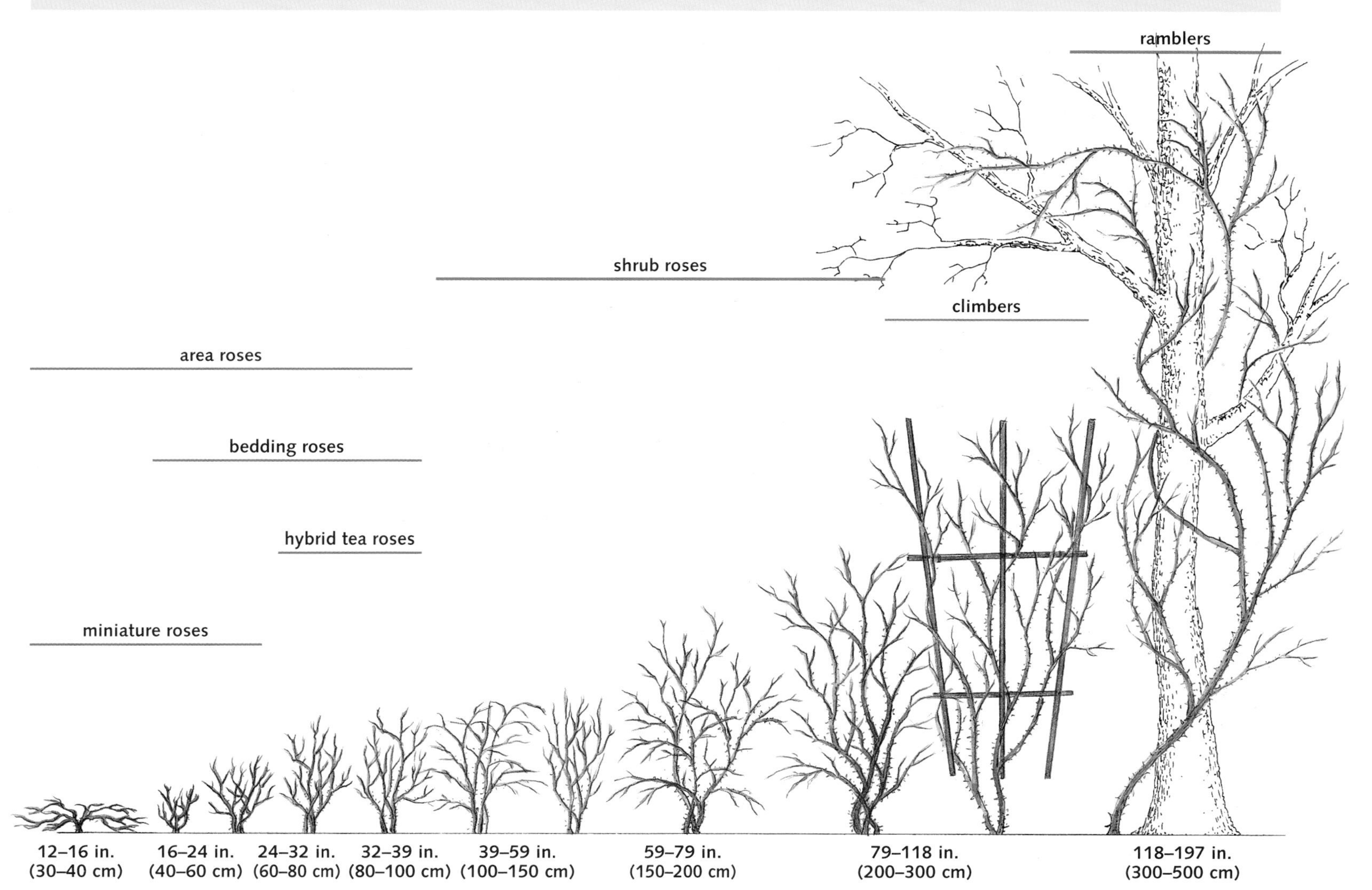

DESIGNING WITH ROSES

How can roses be used on the patio and in the garden? This depends on the desires and needs of the owner, for no rose can be used in just one way. On the contrary, the broad palette of possible uses—also in public green spaces—underlines the position of the rose as an extraspecial shrub.

The Design Use of the Rose

The rose holds a special position among the shrubs. In appropriate sites, roses cast a spell with their long summer flowering and their unique fragrance, which brings a special flair to the garden. Unarguably, the rose plays the leading role on many garden stages. As with any convincing production, however, the success of the performance of the queen of the flowers rises or falls with the quality of the supportiing cast. Wrong placement can take much of the charm from garden scenes that might otherwise earn applause.

The gardener is the director in the great theater of planting. His or her individual preferences decide the choice of plants, the use of their characteristics, and their advantageous combination. The breadth of variation of roses in terms of growth, color, flowering pattern, and hip production, with the numerous possible combinations that result, offer a creative challenge of a special sort.

One's own creativity should stamp a design with roses. Do not slavishly follow the self-styled plans from colorful garden magazines. The results often look quite different from the pictures when tried in one's own garden. Also, the number of plants per square foot said to be required in such plans is only a suggestion that no one need feel compelled to follow. It is also not absolutely necessary to make a strictly geometric placement of roses at precisely the same distance apart, like tin soldiers. The same number of roses can also be planted loosely in small groups, as if they were just making themselves comfortable there.

The larger the area of a planting of bedding roses, the more one must take into consideration the robustness of the variety.

The art of designing with plants consists of arranging them without the arrangement being noticeable. Luckily, ground rules make creating harmonious designs easier and help to arrange the ideas, often unclear at first, of how a bed or a section of the garden should look when finished. Probably the most important principle of harmony in working with roses is the color principle. You will benefit by taking some time to study it, perhaps in the wintertime. You should make use of information learned about the color principles when making your own garden plans.

'Goldmarie '82'®—One of the healthiest yellow bedding roses.

The Color Effect:

Yellow—warm, light color, harmonizes well with red

Blue—fresh, cool color, goes well with white, also harmonizes with orange red but not with violet red

Red—aggressive, very lively color, goes with many other colors, but be careful about plant neighbors in other shades of red; for example, avoid scarlet red next to violet red

Green—neutral color with yellow and red, also in several different shadings, brings freshness to the garden

White—intensifies color of all other colors, goes particularly well with crimson red, light pink, light blue, and violet

The Color Principle: Roses offer a broad scale of colors; colors lacking, such as blue, can be obtained with other plants. The basic colors of the color wheel—yellow, blue, and red—harmonize with each other without any problems. Colors that are next to each other on the wheel are more critical. Although they come from a basic color and apparently go well together, they may still swear at each other. Tension-producing contrasts are exciting, such as red with green, yellow with violet, blue with orange.

In designing with color, one must take into consideration not only the effect of the flowers and foliage, of course, but also the colors of the surroundings. A white rose in front of a white wall certainly creates little excitement. The same is true for a red rose in front of a brick wall.

Bedding Roses for the Garden

What are bedding roses? To be precise, the term *bedding roses* is a blanket term for **polyanthas** and **floribundas** (the latter are also called **polyantha hybrids**). By polyanthas, rose fanciers mean low-growing bedding roses with inflorescences bearing large blossoms consisting of clusters of many small flowers. They developed from crosses of *Rosa multiflora* and *Rosa chinensis.* The birth year of the polyanthas bears the date 1865, when the Englishman Robert Fortune sent cuttings of a less vigorous form of *Rosa multiflora* to Europe from Japan.

Through genetic crosses, the **polyantha hybrids** were developed from the polyanthas. In this century, by further crossing with hybrid tea roses, Svend Poulsen from Denmark created the first examples of this new group. The floribundas are large-flowered, winter-hardy, low, and repeat-flowering bedding roses whose elegant double flower form is reminiscent of that of the hybrid tea rose. Flowers of the floribundas sometimes look like miniature editions of the hybrid tea rose.

Understand it all? If not, do not worry. Because of the countless transitions within the original subgroups, even most rose professionals have abandoned the once-strict hierarchy and classified the corresponding types in the catalogs simply as bedding roses.

Nevertheless, some rose dealers also group hybrid teas and **miniature roses** with the bedding roses. This is of little practical value, however, since most miniature rose varieties demand enormous care in their bed arrangements because of their inadequate robustness. **Hybrid tea roses,** on the other hand, because of their tall, awkward growth habit, are not at all compatible with the bushy, compact growth of bedding roses and are only very limitedly suitable for colorful surface-covering

Double Bedding Roses

Variety	Color	Flowering Pattern	Height [in. (cm)]
'Amber Queen'®	yellow	repeat-flowering	16–24 (40–60)
'Ballade'®	pink	repeat-flowering	24–32 (60–80)
'Bella Rosa'®	pink	repeat-flowering, late	24–32 (60–80)
'Bernstein Rose'®	yellow	repeat-flowering, early	24–32 (60–80)
'Bonica '82'®	pink	repeat-flowering	24–32 (60–80)
'Chorus'®	red	repeat-flowering	24–32 (60–80)
'Diadem'®	pink	repeat-flowering, late	32–39 (80–100)
'Edelweiss'®	white	repeat-flowering	16–24 (40–60)
'Europas Rosengarten'®	pink	repeat-flowering	24–32 (60–80)
'Fragrant Cloud'®	red orange	repeat-flowering	24–32 (60–80)
'Frau Astrid Späth'	pink	repeat-flowering, early	16–24 (40–60)
'Gartenzauber '84'®	red	repeat-flowering	16–24 (40–60)
'Goldener Sommer '83'®	yellow	repeat-flowering	16–24 (40–60)
'Goldmarie '82'®	yellow	repeat-flowering	16–24 (40–60)
'Gruss an Aachen'	cream	repeat-flowering	16–24 (40–60)
'Happy Wanderer'®	red	repeat-flowering, late	16–24 (40–60)
'La Paloma '85'®	white	repeat-flowering	24–32 (60–80)
'Leonardo da Vinci'®	pink	repeat-flowering	24–32 (60–80)
'Make Up'®	pink	repeat flowering	32–39 (80–100)
'Manou Meilland'®	pink	repeat-flowering	24–32 (60–80)
'Mariandel'®	red	repeat-flowering	16–24 (40–60)
'Matilda'®	red	repeat-flowering, early	16–24 (40–60)
'Matthias Meilland'®	red	repeat-flowering	24–32 (60–80)
'Montana'®	red	repeat-flowering	32–39 (80–100)
'Mountbatten'®	yellow	repeat-flowering	32–39 (80–100)
'Nina Weibull'	red	repeat-flowering	16–24 (40–60)
'Play Rose'®	pink	repeat-flowering, early	24–32 (60–80)
'Pussta'®	red	repeat-flowering	24–32 (60–80)
'Rosali '83'®	pink	repeat-flowering	24–32 (60–80)
'Rose de Rescht'	pink	repeat-flowering	32–39 (80–100)
'Royal Bonica'®	pink	repeat-flowering, late	24–32 (60–80)
'Rumba'®	apricot/yellow	repeat-flowering	24–32 (60–80)
'Schöne Dortmunderin'®	pink	repeat-flowering	24–32 (60–80)
'Sommermorgen'®	pink	repeat-flowering	24–32 (60–80)
'Stadt Eltville'®	red	repeat-flowering	24–32 (60–80)
'Sunsprite'®	yellow	repeat-flowering, early	24–32 (60–80)
'The Queen Elizabeth Rose'®	pink	repeat-flowering	32–39 (80–100)
'Träumerei'®	pink	repeat-flowering	24–32 (60–80)
'Warwick Castle'	pink	repeat-flowering	24–32 (60–80)

Semidouble Bedding Roses

Variety	Color	Flowering Pattern	Height [in. (cm)]
'Blühwunder'®	pink	repeat-flowering	24–32 (60–80)
'Dolly'®	pink	repeat-flowering	24–32 (60–80)
'Escapade'®	lavender/white	repeat-flowering	32–39 (80–100)
'Heidepark'®	pink	repeat-flowering, late	24–32 (60–80)
'La Sevillana'®	red	repeat-flowering	24–32 (60–80)
'Märchenland'	pink	repeat-flowering	24–32 (60–80)
'Ricarda'®	pink	repeat-flowering	24–32 (60–80)
'Sarabande'®	red	repeat-flowering, early	16–24 (40–60)
'Schleswig '87'®	pink	repeat-flowering	24–32 (60–80)
'Schneeflocke'®	white	repeat-flowering, early	16–24 (40–60)

plantings. The attempt to use the scantly foliaged hybrid tea roses—the true queens of the roses with the utmost individuality—as area coverings, massive and monotonous without a royal entourage, seems almost less majestic.

In utter contrast, the bedding roses classified as floribundas and polyanthas can be seen as outstanding team players. They provide a grand area effect, particularly when they are planted in large groupings and at small distances apart in beds and borders. With their dense foliage, they shade the ground below and thereby hinder weed development.

Naturally, the **rule of thumb of robustness** holds especially true with the use of bedding roses to cover large areas. The larger the area of a planting of bedding roses, the more carefully sturdiness—besides growth habit and color—must be considered when choosing a variety.

All bedding roses develop well-branched canes. These are attention-getters. They have flower clusters with numbers of more or less double individual flowers in red, pink, white, and yellow. The first flush of flowers in the bedding roses almost always has a fantastic effect from a distance. Afterward most varieties take a breather for a bit to become a dominating, colorful accent in the garden once more in August.

When working with bedding roses, a playful design mood is needed. Straight lines, which are not present in nature, must not be used. Paths can pass the bedding roses in wide arcs. In principle, bedding rose plantings can be integerated into already-existing garden schemes without any problems.

The selection of varieties shown on page 49 gives preference to particularly weather-fast, vital varieties. Unfortunately, only a few bedding roses are also endowed with a fragrance. However, some varieties do offer the nose a lovely perfume, for example, 'Fragrant Cloud', 'Sunsprite', 'Manou Meilland', 'Rose de Rescht', 'Träumerei', and 'Warwick Castle'.

The Ornamental Rose Bed

Anyone who wants to install a rose bed can easily carry out his or her desire with the help of the bedding rose material in the tables. For example, as a central point in the middle of the bed, a pyramid with climbing plants would be suitable. In front of it could be bedding roses, staggered by height, of course, so that as a unit they form a pillar of flowers. The choice of color is left to the individual's taste. As a rule of thumb, groups of one variety should be planted. They should be grouped in no less than

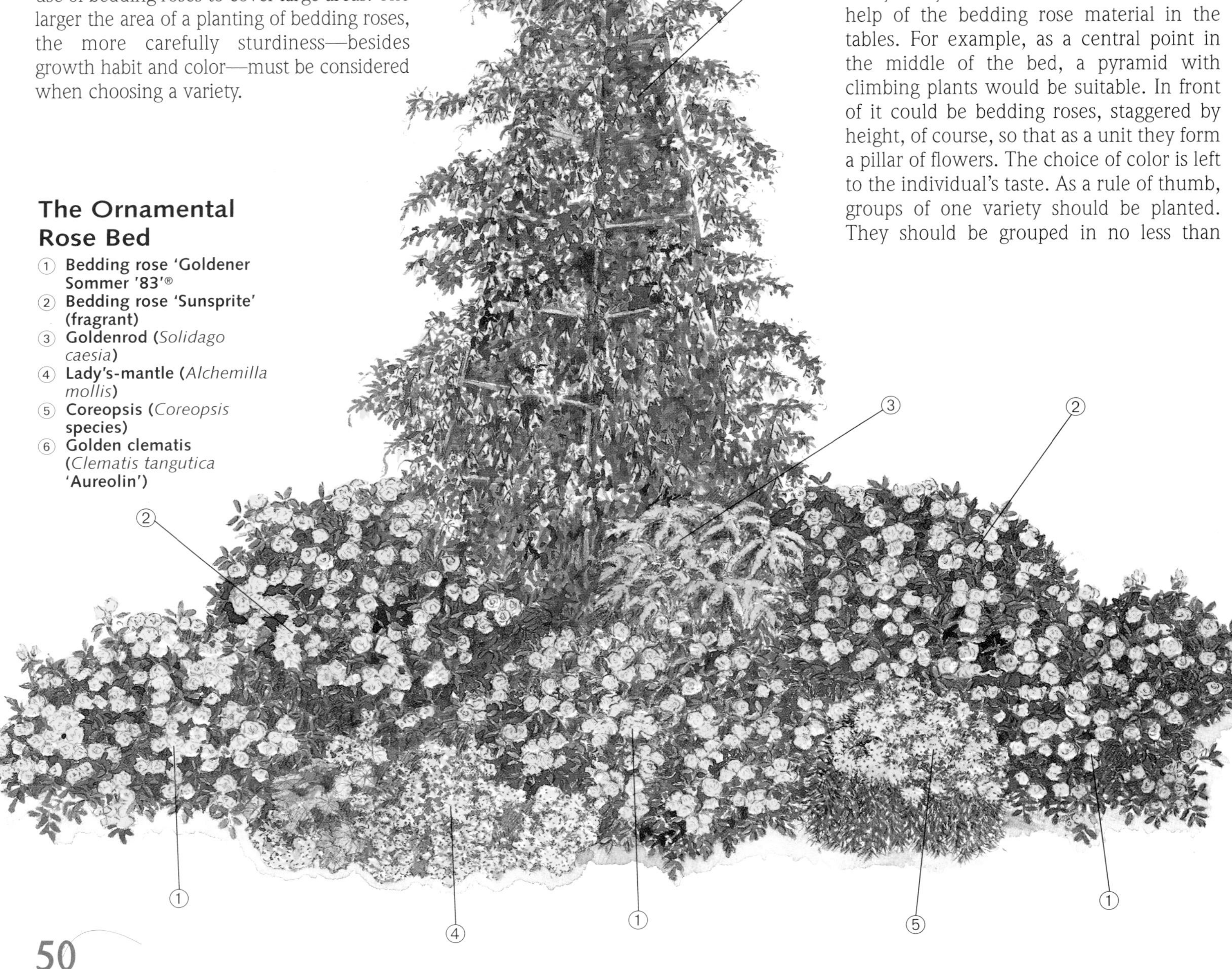

The Ornamental Rose Bed

1. **Bedding rose 'Goldener Sommer '83'®**
2. **Bedding rose 'Sunsprite' (fragrant)**
3. **Goldenrod (*Solidago caesia*)**
4. **Lady's-mantle (*Alchemilla mollis*)**
5. **Coreopsis (*Coreopsis* species)**
6. **Golden clematis (*Clematis tangutica* 'Aureolin')**

three, better in five, and ideally in seven levels. Constrasting neighbors are possible. However, tone-on-tone plantings, perhaps in the form of a yellow rose bed, as in our example, can be used.

For a yellow climber, a special treasure exists. The golden clematis (*Clematis tangutica* 'Aureolin') is a fascinating superclimber with—unusual for the genus *Clematis*—yellow flowers. These have their heads turned down like bells, are a much-sought-after source of nectar for bees, and make a show with their abundance of flowers from June to fall. In addition, their fruits are ornamented with silvery swirls of hairlike tails.

Use three to five plants of each per species and variety in groups. Use two *Clematis tangutica* for the wood or wrought-iron pyramid.

Climbing Roses

Variety	Color	Flower	Flowering Pattern	Height [in. (cm)]
'Compassion'®	orange pink	double	repeat-flowering	79–118 (200–300)
'Dortmund'®	red	single	repeat-flowering	79–118 (200–300)
'Golden Showers'®	yellow	double	repeat-flowering	79–118 (200–300)
'Ilse Krohn Superior'®	white	double	repeat-flowering	79–118 (200–300)
'Lawinia'®	pink	double	repeat-flowering	79–118 (200–300)
'Maria Lisa'	pink	single	once-flowering	79–118 (200–300)
'Morning Jewel'®	pink	semidouble	repeat-flowering	79–118 (200–300)
'New Dawn'	pearly pink	double	repeat-flowering	79–118 (200–300)
'Rosarium Uetersen'®	pink	double	repeat-flowering	79–118 (200–300)
'Salita'®	orange	double	repeat-flowering	79–118 (200–300)
'Santana'®	red	double	repeat-flowering	79–118 (200–300)
'Sympathie'	red	double	repeat-flowering	79–118 (200–300)

Ramblers

Variety	Color	Flower	Flowering Pattern	Height [in. (cm)]
'Albéric Barbier'	cream	double	once-flowering	118–197 (300–500)
'Bobbie James'	white	semidouble	once-flowering	118–197 (300–500)
'Flammentanz'®	red	double	once-flowering	118–197 (300–500)
'Paul Noël'	pink	double	remontant	118–197 (300–500)
'Rosa x rugosa'	pink	semidouble	once-flowering	118–197 (300–500)
'Super Dorothy'®	pink	double	repeat-flowering	118–197 (300–500)
'Super Excelsa'®	crimson rose	double	repeat-flowering	118–197 (300–500)
'Venusta Pendula'	pink/white	semidouble	once-flowering	118–197 (300–500)

Climbing Roses and Ramblers for Wall and Pergola

Climbing roses in the home garden make fairy tales come true. Anyone who carefully reads the Brothers Grimm's *Briar Rose* receives a revelation of the design possibilities of this gloriously colored adventurer among the roses: "But around the castle a hedge of briars began to grow, which every year grew higher and finally enclosed the castle entirely and grew over it, so that nothing more of it was to be seen, not even the banners flying from the roof."

Of course, many a year would have to pass for such a castle to disappear behind climbing roses and rampant ramblers. They can climb faster up a pergola, a house wall, a carport, a garden lamp, a gatepost, or an intricate pyramid. By the third year after planting, at the latest, climbing roses create a fairy tale ambiance in even the darkest garden corners.

The term **climbing roses** is used as a general term for roses that resemble shrub roses but, unlike them, develop long canes. Climbing roses cannot climb on their own; they need help to ascend walls and pergolas. Climbing roses that put forth particularly long, soft, thin canes are called **ramblers.** With their yard-long canes, they can grow in light, old trees to heights of 20 to 23 feet (6 to 7 m), greening large pergolas and imposing gateposts. From the botanical point of view, these ramblers are members of the so-called straddlers, that is, they are always on the lookout for a hold with their prickly canes. If the search is successful, they hook their canes in firmly.

In addition to these differences that depend on growth type, climbing roses are classified according to their flowering pattern. There are **once-** and **repeat-flowering varieties.** Repeat-flowering varieties often display weaker growth vigor—in contrast with the wildly rampant, usually once-flowering, liana-like ramblers—rarely growing more than 10 feet (3 m) high. Knowledge of their flowering patterns provides the rose lover with a basis for undertaking appropriate pruning measures, which are presented in detail in "A Course in Rose Pruning," beginning on page 152.

Control cane growth: The more horizontally climbing rose canes are trained, the more flowering side wood they develop.

Climbing roses always grow toward the light. As they go, they cover everthing in their path with a colorful ribbon of flowers. The garden lover can determine the direction of cane growth. He or she either guides them up on a trellis, loosely tied, or lets them hang down from the top of a wall. If the master climbers find no support for their gloriously colored embrace, they spread their long canes like a ground cover over stones and slopes, as sometimes the species 'Super Dorothy' and 'Super Excelsa' will do.

The really fascinating thing about climbers is that in spite of their long canes, they always work in even the smallest gardens and in the smallest plant areas—provided the way to the heights is open. More and more urban dwellers are taking advantage of this to make their house walls

'Venusta Pendula': A once-flowering rambler with liana-like growing power.

look attractive even on the street side. This is even possible with climbing roses if the house wall directly abuts the sidewalk—without a front garden. About 3 square feet (0.25 m^2) of free soil surface and deep soil are sufficient for them to transform a spot where it had not seemed possible to introduce anything green at all.

Climbing roses love walls in southeast or southwest situations. Hot south walls are unsuitable, for there the roses quickly suffer from spider mites and mildew. However, a very suitable site could also be problematic if a broad overhang reduces the air exchange and promotes the trapping of humidity. To promote the flowering of climbing roses at the viewer's eye level, it is a good idea not to plant between but under the windows. Thus, the canes can be compelled to grow horizontally early on and encouraged to develop flower-bearing side shoots. Horizontal tying is also better for high-yield flower production than pruning when training a climbing rose along a pergola or a canopy. Pruning climbing roses too often ultimately leads only to production of too much wood. However, the rose in the home garden is supposed to be enjoyed as a flowering plant and not as a forest shrub. Therefore, as a rule, climbing roses are only thinned out (see page 154).

Rose Arches: Special design effects can be achieved in the garden with rose arches, behind which there opens another, surprise-filled world when someone passes through the arch. For small arches, rather weakly vigorous, repeat-flowering varieties are suitable; even ramblers can be included among the choices for larger arches. Regardless of the variety chosen, one should not forget roses' spines when designing rose arches. The distance between the posts should thus be wide enough for someone to pass through without any danger of injury. The same goes, of course, for pergolas and other trellis arrangements. If not enough space is available, almost-thornless varieties such as 'Maria Lisa' can be used. Rose arches are an artistic design frame that require the gardener to do the work of tying and training during the first years. The desired effect does not develop in the first year, either. However, the necessary patience will be rewarded in the third year with lush rose glory.

The rambler 'Bobbie James' helps thinly leaved trees achieve a second flowering.

Trellis: A climbing support for roses must be sturdy and solidly anchored. Trellises of wood or plastic-coated wire are suitable. The minimum distance from the wall should be 3 in. (8 cm). This serves to ensure adequate ventilation and makes the work of tying easier. Galvanized metal or wrought iron has proven to be especially useful. It is certainly somewhat more expensive. In addition, using galvanized metal requires more installation time than if one used other trellis materials. However, a trellis created from galvanized metal lasts for a small eternity. If the metal or iron structure has a grillwork-like structure, it will allow comfortable bending of the canes in all directions. With expert installation, such structures, even after many years, will also bear the not-inconsiderable burden of the weight of flower-laden, rain-soaked canes of a large climbing rosebush.

■ Roses for Hedges, Fences, and Specimen Plantings

Roses for dense hedges, roses as fence crowners, roses for solo roles—in these design areas, the rose comes into its own.

All roses are shrubs. However, the term **shrub roses** designates those varieties and species that attract attention with their vigorous growth in height and breadth. The flowers may be single or semidouble, but they may also be double and resemble hybrid teas. In catalogs, shrub roses are occasionally subdivided into the once-flowering **park roses** and the repeat-flowering **ornamental shrub roses.**

Once-Flowering Park Roses: Among the classic park roses are 'Frühlingsgold', 'Maigold', *Rosa hugonis, Rosa sericea f. pteracantha,* 'Sharlachglut', and others. Once-flowering park roses also include species and varieties with one-time flowers in spring or early summer, vigorous growth, abundant hip production, and uncommon frost hardiness. The name *park roses* indicates the place where they can best be planted—in parks and extensive garden

Patience rewarded: After three years at the most, climbing roses give pergolas a lush frame of flowers.

installations. There they unfold their lush glory undisturbed and grow freely. They are ideally suited for underplanting large trees, as single shrubs, or in loose hedges, which can easily become 10 to 13 feet (3 to 4 m) high.

Repeat-Flowering Ornamental Shrub Roses: The constant flowering of the ornamental shrub roses, among which number many English Roses and old roses, inhibits their growth. Therefore, planting these roses as a somewhat larger bedding rose, even in small garden corners, is possible without any problem. Because of their repeated flowering, their great sturdiness, and their manageable size—most are 6 1/2 feet (2 m) tall at maximum—repeat-flowering ornamental shrub roses shine as appropriate specimen plantings in the garden. Therefore, they are also emphasized in the variety table (see page 56).

Shrub Roses for Specimen Planting: Anyone who ever visits Insel Mainau gets an idea of what mountains of flowers ornamental shrub roses can develop into when the gardener takes into consideration their traits. The world-renowned shrub rose varieties like 'Westerland' and 'Schneewittchen' ('Iceberg'), for instance, show themselves to be especially growth happy. The special thing about these splendid ornamental shrubs is that they have usually developed from a single plant. They have been allowed to grow for many years without pruning except for a little thinning out from time to time. Besides their robust flower production and primitively vigorous frost hardiness, shrub roses have another impressive characteristic, namely developing a free-growing, compact, and self-contained inflorescence. Many modern shrub roses develop into round ornamental shrubs. With their dimensions and the protraction of their flowering season, these roses easily surpass the classic ornamental shrubs of our gardens. When planted together, ornamental shrub roses and ornamental shrubs are ideal partners. To learn more about foliage shrubs, turn to the section entitled "Rose Combinations" and begin reading on page 89.

Human-sized shrub roses are suitable for free-growing rose hedges in large gardens or as fence roses for rosy barriers.

If the subject is of the suitability of shrub roses for **specimen plantings,** this should not rule out their use in **group plantings.** The term *specimen planting* actually refers to the free, exposed placement of one plant or several rosebushes. If such a rose grove or a single shrub is planted in the middle of a lawn, the soil around the roses must remain clear at intervals of about 20 in. (50 cm). Anyone who wishes to mulch this so-called tree circle (see the table entitled Shrub roses, which appears below) should never pile the mulch too high. Moles find an inviting, cuddly warm nest in there. With the delicious rose roots being present, moles even have a richly laden table.

Shrub Roses for Fences: In the gardens of past generations, fences fulfilled the important function of being a barrier to protect the garden vegetables from the browsing of wild animals or from the unwanted scratching for food by chickens and other domestic animals. The fences of the kitchen garden were relatively low so that, while protection was guaranteed, the neighbors were able to cast curious looks at the lush glory of flowers beside the vegetables.

Fences also played an important role in the legal terminology of the Middle Ages. During that time, only enclosed land was under protection. Fences permitted a legal handhold against vegetable thieves and other unwanted garden visitors.

Whether wood, woven, wire netting, or elegant wrought-iron boundaries—diverse forms of fences enclose the home garden. In the present day, their first function is to offer protection and mark boundaries. With shrub roses functioning as fences, protection is offered by imposing thorns and beauty results from constant blooms all summer long, a design feature that is functional and at the same time aesthetically appealing. Blooming shrub roses that climb the pickets of a fence or over a wall recall the romance of old kitchen gardens. At the same time, these blooming, climbing shrub roses make crossing them—thanks to their unique armament—a risky adventure.

Shrub Roses for Hedges: For one thing, a hedge created entirely of roses assumes the same protective function as a fence—without shutting out one's neighbors. Such a fence also provides a living retreat reservation for many creatures. Even low rose hedges are ecologically sound fences that establish clear boundaries but, at the same time, signal colorful open friendship to the external world.

Rose hedges—like other hedges—perform countless life-preserving tasks. They equalize temperature extremes. Numerous creatures use the increased humidity and the damp coolness produced by increased dew development within the tangles of branches. Rose hedges decrease the wind velocity. They are living dust filters and sound dampers. They clean the air and produce oxygen daily through their leaf activity. Rose hedges created from shrub roses offer birds and other animals nest protection and living space. When hips are present, such hedges also provide a favorite source of nutrition to birds and other animals (see page 134).

In the following tables, repeat-flowering rose varieties are introduced that are suitable for loose, free-growing hedges. Not only are ornamental shrub roses considered but—for lower hedges—also varieties from the group including area and bedding roses; here again 'Escapade', 'La Sevillana', and 'Pink Meidiland' stand out particularly as those with abundant production of hips. Among the varieties suitable for lower hedges is also the one called 'Rose de Rescht'. Lovers of old roses can create a rosy small hedge with it. However, the close plant spacing required for a hedge also produces a greater susceptibility to fungal disease. As a result, a hedge created with shrub roses takes a somewhat greater amount of care. The ideal location of a hedge is sunny; also, robust, repeat-flowering rose hedges have no place under dripping trees. Naturally, varieties with similar growth heights and forms can be mixed together to create colorful hedges.

Shrub Roses

Variety	Color	Flower	Flowering Pattern	Height [in. (cm)]
'Armada'®	pink	double	repeat-flowering	39–59 (100–150)
'Astrid Lindgren'®	pink	double	repeat-flowering	39–59 (100–150)
'Bischofsstadt Paderborn'®	red	single	repeat-flowering	39–59 (100–150)
'Centenaire de Lourdes'	pink	semidouble	repeat-flowering	59–79 (150–200)
'Dirigent'®	red	semidouble	repeat-flowering	59–79 (150–200)
'Dornröschenschloss Sababurg'®	pink	double	repeat-flowering	39–59 (100–150)
'Eden Rose '85'®	pink	double	repeat-flowering, late	59–79 (150–200)
'Elmshorn'	red	double	repeat-flowering	59–79 (150–200)
'Ferdy'®	pink	double	once-flowering	32–39 (80–100)
'Freisinger Morgenröte'®	orange	double	repeat-flowering	39–59 (100–150)
'Ghislaine de Féligonde'	yellow	double	repeat-flowering	59–79 (150–200)
'Grand Hotel'®	red	double	repeat-flowering	59–79 (150–200)
'Gütersloh'®	red	double	repeat-flowering	39–59 (100–150)
'IGA '83 München'®	pink	double	repeat-flowering	32–39 (80–100)
'Ilse Haberland'®	pink	double	repeat-flowering	39–59 (100–150)
'Kordes' Brillant'®	orange	double	repeat-flowering	39–59 (100–150)
'Lichtkönigin Lucia'®	yellow	double	repeat-flowering, early	39–59 (100–150)
'Marguerite Hilling'	pink	semidouble	remontant	59–79 (150–200)
'Polka '91'®	amber	double	repeat-flowering	39–59 (100–150)
'Raubritter'	pink	double	once-flowering, late	79–118 (200–300)
'Rödinghausen'®	red	double	repeat-flowering	39–59 (100–150)
'Rokoko'®	cream yellow	double	repeat-flowering	39–59 (100–150)
'Romanze'®	pink	double	repeat-flowering	39–59 (100–150)
'Rosendorf Sparrieshoop'®	pink	double	repeat-flowering	39–59 (100–150)
'Rosenresli'®	pink	double	repeat-flowering	39–59 (100–150)
'Scharlachglut'	red	single	repeat-flowering	59–79 (150–200)
'Schneewittchen'® ('Iceberg')	white	double	repeat-flowering, early	39–59 (100–150)
'Vogelpark Walsrode'®	pink	double	repeat-flowering, early	39–59 (100–150)
'Westerland'®	apricot	semidouble	repeat-flowering	59–79 (150–200)

Before installing a hedge, it is helpful to take a look around the neighborhood. Minimum distances to the property line vary according to zoning regulations in each community. You should be aware of them. Finding out about them from the local zoning office or consulting the appropriate literature will spare later conflicts with the neighbors.

Ramblers as Pruned, Formal Hedges: Rambler rose varieties such as, say, 'Bobbie James' have also demonstrated their value as fence or hedge roses. They are a space-saving, very pliant, colorful hedge alternative for constricted situations where space is simply not available for free-growing shrub-rose hedges.

The rambler is planted directly against a stable wire mesh fence, which it quickly covers. After several years, the greened fence can be shaped into a hedge about 20 inches (50 cm) deep and 6 1/2 feet (2 m) high. To accomplish this, the rose—and other foliage shrub hedges as well—are pruned into the desired shape with pruning shears or electric hedge clippers.

Tall Hedge Roses (about 3 to 6 1/2 feet [1 to 2 m])

Variety	Color	Flower	Flowering Pattern
'Astrid Lindgren'®	pink	double	repeat-flowering
'Bischofsstadt Paderborn'®	red	single	repeat-flowering
'Centenaire de Lourdes'	pink	semidouble	repeat-flowering
'Dirigent'®	red	semidouble	repeat-flowering
'Dornröschenschloss Sababurg'®	pink	double	repeat-flowering
'IGA '83 München'®	pink	double	repeat-flowering
'Ilse Haberland'®	pink	double	repeat-flowering
'Lichtkönigin Lucia'®	yellow	double	repeat-flowering, early
'Polka '91'®	amber	double	repeat-flowering
'Schneewittchen'® ('Iceberg')	white	double	repeat-flowering, early
'Westerland'®	apricot	semidouble	repeat-flowering

Low Hedge Roses (about 2 to 3 feet [0.6 to 1 m])

Variety	Color	Flower	Flowering Pattern
'Ballerina'	pink/white	single	repeat-flowering
'Escapade'®	lavender/white	semidouble	repeat-flowering
'La Sevillana'®	red	semidouble	repeat-flowering
'Marjorie Fair'®	red	single	repeat-flowering
'Mountbatten'®	yellow	double	repeat-flowering
'Pink Meidiland'®	pink/white	single	repeat-flowering
'Rose de Rescht'	deep pink	double	repeat-flowering

'Westerland'®: A shrub rose variety that—left unpruned—develops into a round, leafy, beautifully formed ornamental shrub.

Area Roses for Low-Maintenance Plantings

The term **area rose** is new. It embraces the **ground cover** and **small shrub roses,** whose varieties were usually separated from one another only with difficulty. The name *area rose* descibes the uses of these varieties. They are able to suppress weeds over large soil surfaces and supply low-maintenance coverage.

However, area roses are not miracle workers. In their first years, they need care and must receive support from the gardener in their battle against weeds.

Area roses should not be confused with ground covers, which are suitable for underplanting of trees and large shrubs that cast considerable shadow. The confusion arises from the double meanings of the term *ground cover.* On the one hand, it means plants that serve as anchors for slopes and as substitutes for lawns. On the other hand, the term *ground cover* also means plants suitable for underplanting of large woody plants. For the latter purpose, shrubs that do not have matted roots are used, never roses for the long term. At most, under young trees with still small, barely shading crown areas, a gardener can consider covering the area with robust rose varieties for some years. Under large, mighty trees, however, roses receive too little sun and too much moisture, once from rain and then from the dripping from the crown. As a consequence of this high humidity, the incidence of many fungal diseases increases, especially spot anthracnose.

The area roses include rose varieties with diverse growth forms. The ground-covering varieties with flat, prostrate canes as well as varieties with an arching growth habit like small shrubs offer possibilities for splendidly colorful design of open spaces.

Area roses are vigorous growers and are acknowledged to be outstandingly **robust** and **winter hardy.** Varieties like 'Bingo Meidiland', 'Bonica '82', 'Flower Carpet', 'Lavender Dream', 'Magic Meidiland', 'Mirato', 'Palmengarten Frankfurt', 'Pink Meidiland', 'Marjorie Fair', 'Repandia', and 'Surrey' document breeding advances that have produced vital, disease-resistant roses that maintain themselves without use of pesticides and fungicides.

Based on their stable genetic structure, area roses are considered easily satisfied hard workers among the roses. Their high **stress tolerance**—varieties like 'Immensee' even tolerate automobile tires driving over their canes—was discovered by municipal park departments at the beginning of the 1980s. At that point in time, the expensive care of hybrid tea and bedding roses in public gardens was turning out to be increasingly unaffordable. In addition, the leaf vigor of many old rose classics left more to be desired from year to year. The area roses filled the gap outstandingly.

A selection of time-tested varieties is presented in the table above. It lists twenty-five different varieties of area roses. The table includes information about the color, flower, height range (in both inches and centimeters), and growth habit of each of these varieties.

Area roses are not only an easy-to-manage rose material for many plant situations like public gardens, industrial areas, noise-protecting walls, parking areas, slopes, and many more. These plants (primarily the single to semidouble flowering varieties) also provide a high-yielding source of pollen for honeybees, bumblebees, and insects for months at a time. The dense leaf cover of the area roses not only successfully keeps down undesirable weed growth but, at the same time, offers secure breeding and nest protection to birds and protected shelter to many small mammals and other creatures. In addition, the hips of the fruit-bearing varieties fill the pantries of many animals during the winter months.

Area Roses

Variety	Color	Flower	Height (in. [cm])	Growth Habit
'Alba Meidiland'®	white	double	32–39 (80–100)	bushy
'Ballerina'	pink/white	single	24–32 (60–80)	arching
'Bingo Meidiland'®	pink	single	16–24 (40–60)	bushy
'Bonica '82'®	pink	double	24–32 (60–80)	bushy
'Flower Carpet'®	pink	semidouble	24–32 (60–80)	bushy
'Immensee'®	pink	single	12–16 (30–40)	flat, very vigorous
'Lavender Dream'®	lavender	semidouble	24–32 (60–80)	low bushy
'Lovely Fairy'®	pink	double	24–32 (60–80)	bushy
'Magic Meidiland'®	pink	double	16–24 (40–60)	flat, very vigorous
'Marjorie Fair'®	red	single	24–32 (60–80)	arching
'Max Graf'	pink	single	24–32 (60–80)	flat, very vigorous
'Mirato'®	pink	double	16–24 (40–60)	bushy
'Nozomi'®	pearly pink	single	16–24 (40–60)	flat, less vigorous
'Palmengarten Frankfurt'®	pink	double	24–32 (60–80)	bushy
'Pheasant'®	pink	double	24–32 (60–80)	flat, very vigorous
'Pink Meidiland'®	pink/white	single	24–32 (60–80)	arching
'Red Meidiland'®	red	single	24–32 (60–80)	low bushy
'Repandia'®	pink	single	16–24 (40–60)	flat, very vigorous
'Royal Bassino'®	red	semidouble	16–24 (40–60)	bushy
'Satina'®	pink	double	16–24 (40–60)	bushy
'Scarlet Meidiland'®	red	double	24–32 (60–80)	arching
'Sommermärchen'®	pink	semidouble	16–24 (40–60)	bushy
'Surrey'®	pink	semidouble	16–24 (40–60)	bushy
'The Fairy'	pink	double	24–32 (60–80)	bushy
'White Meidiland'®	white	very double	16–24 (40–60)	widely spreading

Note Although single varieties of area roses can also be used in very large numbers for planting of very large areas, there should not be a repeat of the misguided planning of green monoculture that took place in the 1970s. The larger the area under design, the more diverse the varieties that should be planted. Use of a single variety is not only regarded as bleak and unsatisfying from a design point of view, it also encourages the spread of disease and pests. Varied-species combinations of area roses with, for example, perennials represent the better alternative in any case.

In Short: Area roses combine the highest functionality while fulfilling ecological demands of contemporary landscape design. Increasingly, their use is no longer limited to public areas. Instead, many home and community gardeners take advantage of this advance in rose breeding. They ornament smaller areas, tubs, and planters with varieties from this robust rose group.

Own-Root Area Roses: Anyone who wants to plant area roses over a large area should make sure to get own-root plants when buying. Own-root roses are propagated by cuttings, as a rule, and they lack the grafted understock (see the section about propagation by soft cuttings, page 174). In this way, the formation of troublesome suckers is avoided. Own-root area roses are allowed to grow undisturbed

'Palmengarten Frankfurt'® covers areas with an abundance of pink flowers.

for three to four years before they are rejuvenated in early spring by machine, perhaps with a mowing bar or with hedge clippers, to a height of about 12 in. (30 cm). By the following summer, they are exuberant again with a lush production of flowers. Yearly pruning back would not be only time consuming but would notably decrease the functionality of the area rose. Only when they are allowed to develop undisturbed can area roses—as they should—cover the ground with their foliage, leaving no gap.

Covering the areas between the roses with bark mulch no higher than 2 in. (5 cm) after planting has proven to be beneficial. The increased oxygen requirement through the microbial activity must be taken into account. More details on this subject can be found in the section on "Practice" under the heading "Ground Cover Through Mulching" (pages 157–158).

Own-root area roses like 'Lovely Fairy'® used as low-maintenance street shrubs add color to the city.

A planting of many different species of area roses and perennials considerably reduces the amount of care needed.

Removing spent rose blossoms in summer is unnecessary with area roses because they are good at **self-cleaning.** This is a further advantage for rose lovers who would prefer to doze in the hammock during the warm days rather than to putter around the rose bed.

The retail trade offers self-rooted area roses with a small root ball, which is either in a small plastic pot or a recycled-paper pot that can be planted with the rose. Unfortunately, the selection is still limited in spite of the described advantages of self-rooted roses. The reason for this is that in comparison with grafted, superstrong rosebushes, area roses propagated by cutting and on their own roots show less woody growth. The plant dealers know from experience that many gardeners prefer to purchase sturdy rose plants first. Therefore, plant dealers exclusively offer grafted plants. However, when specifically asked to, it is not difficult for the dealer to also provide own-root roses like the thousands used in public gardens.

■ Portable, Frost-Hardy Patio Roses

Portable, frost-hardy roses for the patio and balcony—planted in pots, tubs, baskets, or larger balcony planters—allow the long-term decorating of sites where otherwise a profusion of roses would not be possible.

However, mobility does not satisfy only the design creativity of the terrace and balcony fans. It also diminishes the amount of care required. If a situation proves to be unsuitable because, for instance, the heat collects in front of a hot south wall, the patio roses can be placed in another spot. Thus spider mite, mildew, and company will be deprived of the paradise-like environment in which they develop. (See also the section "The Rose Site—A Place in the Sun," pages 142–144, in the chapter entitled "Practice").

Naturally, a patio rose will also thrive better in a situation that is appropriate for a rose than in a shady spot. However, the risk of parasitic attack by the worst scourge of rose lovers—pot anthracnose—is markedly diminished by the distance between garden level soil and a terrace or balcony. Scarcely any soilborne resting spores are splashed up from the terrace surface by rainwater, as is constantly the case in the garden. Thus, even rose groups such as the miniature roses can be used well in designs on the patio.

In contrast, when planted in the garden, these same rose groups do not have a chance of surviving unless they are copiously treated with fungicides and pesticides.

However, roses, being outstandingly deep-rooted plants, are not tub shrubs par excellence. The shape and size of containers and balcony planters should thus receive special attention.

Container Shapes, Container Materials: An ideal for the rose with its depth-seeking, long roots is the tall, extended cylinder shape, preferably with a stabilizing bulge at the bottom to decrease top-heaviness. It represents the best possible compromise between pots that are wide at the top—which are easy to repot and water—and the stable, bulging shapes, which do, however, create problems for repotting.

Container Placement

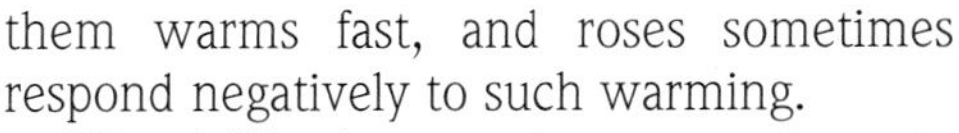

In front of south walls, roses can suffer from too strong a reflected light and react with leaf burn or spider mite infestation.

In garden shops, you can find containers of all kinds made of all possible materials:

- ***Terra-cotta:*** The advantage of terra-cotta pots is adequate ventilation of the soil in them, but a disadvantage is the high rate of water evaporation through the container wall and the associated danger of the soil drying out. High-fired, quality, hand-thrown pots in particular tend to have this water-absorption problem less and are frost hardy. Over the years, high-quality pots become ever more valuable because of the patina that develops.
- ***Plastic:*** Pots made of plastic are easy to transport and they save water, but they provide very little insulation. The soil in them warms fast, and roses sometimes respond negatively to such warming.
- ***Wood:*** Wooden containers are resistant to blows and breakage, insulate well (especially planters of treated oak), and have a high durability. The manifestations of weathering after some years can be problematic.
- ***Willow baskets:*** These baskets have a warm, rustic appearance, but they also are hardly protective against the cold and are only limitedly durable. After two to three years, they begin to dissolve. They are lined with perforated, water-permeable black plastic.
- ***Stoneware and other pottery:*** These materials provide a clean, visually appropriate container solution for the terrace with little water loss and high durability. Among them are the glazed East Asian containers.
- ***Natural stone and artificial stone planters:*** Old feeding troughs of sandstone or granite look nice, but they are only limitedly portable.
- ***Galvanized washtubs:*** These are wonderful plant containers for romantic arrangements. However, care must be taken to provide sufficient drainage.

Potting Soil: Patio roses often remain in their pots for many years. The planting soil must therefore have enough moderately moist air and nutrients available for the roots of the heavy-feeding rose. Numerous soil mixtures are available in garden centers, including special potting soil, and any one of them may be used.

If you work up to 10 percent of perlite or broken slate into premixed soil or into your own mixture, it improves the soil structure for the long term.

You should avoid soils with a high loam or clay portion since these elements move in the substrate and stop up the drainage holes in the container. Loamy garden soil, often recommended in the literature, is ruled out as potting soil for terrace roses for the same reason. Soil with a very high peat content tends to sink in the pot. If it ever dries out, this type of soil is very difficult to moisten again. Besides, avoiding peat contributes to

Window box with 'Pink Symphonie'®.

Pot roses reproduced from cuttings (top, bottom) are suitable for the smallest containers, but they are not frost-hardy roses for the terrace. A crowning glory for a container-plant arrangement on a terrace is created by the weeping standard of 'Super Excelsa'® (pictured at left).

the conservation of this valuable natural raw material.

Also unsuitable are straight compost and already fertilized balcony planter soil. On the other hand, soils that have a high wood fiber content, up to 30 percent, have proven to be very good for patio roses.

Potting and Repotting: Roses of all available forms can be planted in containers. Most important of all is to choose a container that is sufficiently large. Bare-root roses must be able to spread their roots in the pot freely and without bending. The balls of container roses should be about 4 in. (10 cm) from the pot walls when they are sufficiently watered. To prevent later problems with overwintering, make sure the height and breadth of the pot are at least 16 in. (40 cm) each.

A drainage layer of broken pieces of clay pots, perlite, and/or Styrofoam pieces (some gardeners like to recycle those plastic peanuts that are used for shipping) are scattered on the bottom of the pot up to the level of about 5 percent of the inside container height.

It is a good idea to use, especially in pots with larger drainage holes, a fine mesh wire (like window or door screen material) directly over the drainage hole to prevent perlite (if it is used as a drainage layer) and soil from draining out of the container when a plant is watered. The mesh is available at most hardware stores. Just cut out a piece to fit over the bottom of the container so it fully covers the drainage hole area.

After the open spaces around the rose roots are filled in with the soil, water the pot thoroughly. When the soil has settled after this, an edge with a height of about 1 in. (3 cm) should be left for watering.

As a rule, pot roses must be **repotted** after two or, at the latest, three years. After this time, the roots have burrowed their way through the entire area. The hollow spaces necessary for soil aeration are also lost to root growth. In addition, the salt damage resulting from intense culture in the container becomes increasingly visible because the amount of soil left can now breathe out the toxic metabolic products of the roots in only very limited volume.

Repotting terrace roses in ever larger containers eventually reaches its limits. As an alternative to this, it is possible to cut down the present root ball all around with a knife or by using a strong water stream to remove all the soil completely from the roots. The prerequisite for both is that the rose be in a condition of absolute dormancy. After the root ball is hosed off, the thick roots and the above-ground canes are cut back. The now-rejuvenated rose is repotted. Even when it is being transplanted into a larger pot and the ball is not being cut down or hosed off, the matted roots should be loosened by hand or with a knife.

If terrace roses remain in one pot for years, getting them out of there can be difficult. Depending on the vigor of their growth, they will usually have compacted themselves in the container.

In containers of clay or wood, it helps to dampen the soil before pulling the plant out. However, just the opposite is true with containers of stoneware or plastic—the soil should be allowed to dry out. If the roots have worked themselves into the wickerwork of willow baskets, usually the only thing to do is to cut the basket.

Watering: Depending on the container material and size,the roots of the rose are often exposed to changing soil moisture. Steady, even soil humidity would be the ideal—but in any case, the alert rose lover should not let his or her roses dry out. In most cases, however, too much of a good thing occurs instead—the roses are drowned. However, sogginess damages the important root hairs in a very short time. Inevitably, air-loving soil organisms are lost. The result is root rot, which virtually suffocates the rose.

To allow excess water to run off easily, the soil must have sufficiently large drainage holes. Besides, the pot can be set on narrow strips with a height of about 0.5 in. (1 cm).

Caution is needed when using saucers for pots that are always outdoors, because rainwater collects in them. With saucers that are quite flat, the risk of the soil's becoming too wet from standing in water is diminished.

Anyone who irrigates with water containing a lot of lime must reckon with chlorosis from iron deficiency. This can be countered with iron fertilizers like Sequestrin or Fetrilon (see page 158).

Watering systems can be precisely regulated to provide for optimal watering of terrace roses. They are available from garden stores and mail-order companies. By means of humidity sensors, so-called tensiometers, the water content of the potting soil is determined and correspondingly watered as necessary. However, the water-filled tensiometer must be protected from freezing in winter.

Fertilizing: Slow-release fertilizers have proven their value. They are placed near the roots so that root growth takes place more in the center of the container. The bigger the distance of the roots from the container wall, the less the danger of their injury by frost.

Frost Protection: Terrace roses are among the winter-hardy container plants. As such, they can—with the help of certain protective measures—remain outdoors year round. This avoids the burdensome and difficult job of removing them from the terrace in the fall. Since they are winter hardy, this helps many garden lovers who lack suitable frost-free storage places for the containers.

In mild climate zones, all roses may be overwintered out-of-doors without any

Water Drainage

Important: Faultless water drainage through the drainage layer of clay shards . . .

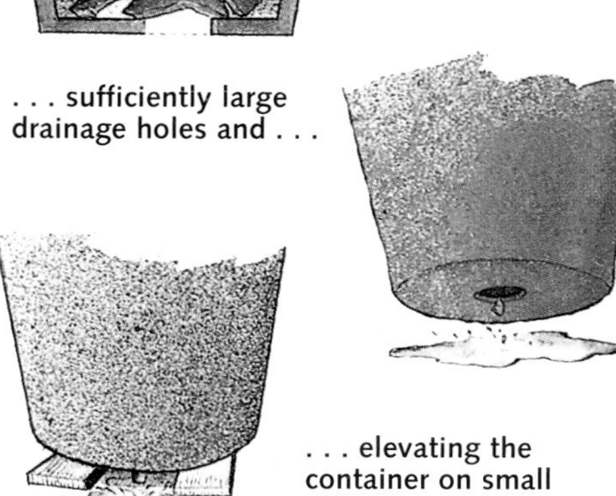

. . . sufficiently large drainage holes and . . .

. . . elevating the container on small strips.

problems. However, rose lovers who live in decidedly frosty areas need not rule out terrace roses. They should—as with those roses planted in the garden—choose especially hardy, once-flowering polar roses. These are described in the section, "The Rose for Special Situations," which begins on page 99.

What does *winter hardiness* mean? The winter hardiness of a rose describes the ability of the above- and below-ground plant parts to survive long periods of freezing temperatures without damage. The more active the roots and the less mature the canes are, the more frost-sensitive the plant. For example, anyone who uses a high-nitrogen fertilizer on his or her container roses after July 15 ought not to be surprised when they do not develop the necessary hardness of wood to be winter hardy. Also, too small a container allows roots to congeal shocklike—without gradual transition—and thus freeze to death.

A danger for a dormant root that is frost hardy occurs when it is activated by a period of mild weather in winter. Then it reacts very sensitively to the renewed onset of harsh cold.

This danger can be avoided by not overwintering the container in a place where the winter sun will warm it particularly fast.

Basically, **frost protection measures** are recommended for terrace roses so that they will survive even outstandingly frigid winters when even the most winter-hardy shrubs are affected. This frost protection consists of a strong, water-resistant layer of insulation, about 4 in. (10 cm) thick, around the container that makes it difficult for frost to penetrate the container wall quickly. Frost protection allows a slow, root- and container-protecting freezing and thawing of the bush.

You can make the insulation layer yourself or use ready-made solutions available at a garden store. For example, a jacket of coir reinforced with wire mesh is easy to apply:

- The jacket is placed at a distance of 4 in. (10 cm) around the pot, which has been set onto boards. It should also stick up over the upper edge of the pot for about 4 in. (10 cm).

Frost Protection for Container Plants

Insert the pot into a cage filled with leaves.

- The empty space between jacket and pot is filled with fine birch, beech, or apple leaves or leaves of ornamental shrubs.
- To keep the leaves from reacting with the humic acid set free on the upper surface of the pot, the addition of a layer of plastic film between leaves and container is recommended.
- A net over the upper surface of the leaves keeps them from being blown away by the wind.

Also important is **protection** of the above-ground canes from **winter sun and drying out.** For this, covering the plant with sacking, ball-wrapping cloth, or spruce branches is best.

Of course, potted roses can also be overwintered with the help of other methods. For instance, the pot can be put into a larger container like a bricklayer's tub. Here, too, the space between containers is then filled with leaves. The following is very important: The outer container must have water drainage holes so that no moisture can collect.

Another practice method consists of wrapping the rose pot with bubble wrap to make a sturdy protective blanket about 4 in. (10 cm) thick. The blanket should be allowed to stick up over the top. This space must be filled with fine leaves.

Another possible but quite time-consuming winter-protection measure for the pot is to sink it into the garden. This is recommended only for bodybuilding fans with lots of room in the garden.

Devices for Moving Containers: Garden stores offer a number of devices that make transporting tub plants easier. Moving larger containers may become necessary, for instance, if a garden party requires lots of space and the plants must be crowded together or if a particularly beautiful terrace rose needs to find a spot in an exposed location in the garden for a particular occasion.

The easiest to handle are **pot trolleys,** which basically resemble the furniture rollers for which they are named. The pot is simply put onto the trolley and pushed to the desired place. When choosing one, it is important to be sure that the wheels can be locked and that no water can collect on the trolley.

With a **pot porter,** a large pot can be moved easily by two people, and one can keep a safe distance from prickly specimens like climbing or shrub roses.

The **carrying strap,** made of plastic webbing straps, works similarly and can be stored without taking up much space when it is not in use.

Designing with Patio Roses: The great design advantage of terrace roses, like other potted plants, lies in their mobility. They can be placed here today, there tomorrow—just

Arrangements of entirely different flowering terrace roses create particularly beautiful garden ornaments.

Standard roses are a friendly and rosy welcome greeting at gate and front door.

as one pleases. In addition, the terrace roses can be assigned completely different design functions.

Climbing roses trained on trellises and poles, for example, can serve as flowering screens for privacy and rosy protection from wind. They can be used wonderfully to divide the terrace into different areas or to protect those sitting there from public view.

Particularly beautiful container roses also can be used as temporary garden ornaments. They reveal their full decorative capabilities as a gaily colored flowering rose sculpture freestanding on the lawn or in front of a green shrubbery backdrop. When the flowers have faded, they can be moved into a less clearly visible corner of the garden and replaced with another terrace rose or another container plant.

Climbing roses, shrub roses grown on pyramids, and also half and full standards can be positioned as friendly welcoming gatekeepers at the left and right of the front door.

Tree roses—quarter, half, and full standards and also weeping standards—in pots unfurl a special charm when they are underplanted with appropriate perennials that are decidedly inferior to the standards in their growing power. Doing so thus produces different flowering levels.

The following perennials have shown themselves to be very good for this purpose: winter savory (*Satureja montana*), low campanula species (*Campanula portenschlagiana*), and Japanese pachysandra (*Pachysandra terminalis*).

Both acid-loving heath plants from the Ericaceae family and ground covers with aggresive root growth are unsuitable.

For full standards in very large tubs of at least 50 quarts (50 l) in volume, area roses may also be considered as underplantings. These include 'Magic Meidiland', 'Alba Meidiland', and 'The Fairy'. Furthermore, weeping tree roses can be used. They unfold their gigantic crowns at a stately 71 inches (180 cm) high (including the container). 'Paul Noël', for example, is a variety that will develop into such an imposing rose tree.

Of course, garden roses with the appropriate preparation and care are not the only plants proven to be frost hardy in containers. Numerous other garden shrubs have been grown over the years with good success in tubs and can be wintered out-of-

Climbing roses on trellises—a flowering privacy screen.

Rose Varieties for Pots

Variety	Class	Color	Flower	Growth Habit
'Abraham Darby'®	shrub	apricot	double	arching
'Ballerina'	area	pink/white	single	arching
'Blühwunder'®	bedding	pink	semidouble	upright
'Bonica '82'®	bedding	pink	double	bushy
'Eden Rose '85'®	shrub	pink	double	upright
'Flower Carpet'®	area	pink	semidouble	bushy
'Ghislaine de Féligonde'	shrub	yellow	double	arching
'Graham Thomas'®	shrub	yellow	double	arching
'Heritage'®	shrub	pink	double	arching
'Ilse Krohn Superior'®	climbing	double	white	arching
'La Sevillana'®	bedding	red	semidouble	bushy
'Lawinia'®	climbing	pink	double	arching
'Leonardo da Vinci'®	bedding	pink	double	bushy
'Lichtkönigin Lucia'®	shrub	yellow	double	upright
'Louise Odier'	shrub	pink	double	arching
'Marjorie Fair'®	area	red	single	arching
'Mirato'®	area	pink	double	bushy
'Mountbatten'®	bedding	yellow	double	bushy
'New Dawn'	climber	pearly pink	double	arching
'Othello'®	shrub	red	double	upright
'Palmengarten Frankfurt'®	area	pink	double	bushy
'Paul Ricard'®	hybrid tea	amber	double	upright
'Pink Symphonie'®	dwarf	pink	double	upright
'Play Rose'®	bedding	pink	double	bushy
'Raubritter'	shrub	pink	double	arching
'Romanze'®	shrub	pink	double	upright
'Rosarium Uetersen'®	climber	rose	double	arching
'Rose de Rescht'	bedding	red	double	upright
'Schneewittchen'® ('Iceberg')	shrub	white	double	bushy
'Sunsprite'®	bedding	yellow	double	upright
'Surrey'®	area	pink	semidouble	bushy
'The Fairy'	area	pink	double	bushy
'Westerland'®	shrub	apricot	semidouble	bushy

Roses for Troughs

Variety	Class	Color	Flower	Growth Habit
'Bella Rosa'®	bedding	pink	double	bushy
'Bonica '82'®	bedding	pink	double	bushy
'Flower Carpet'®	area	pink	semidouble	bushy
'La Sevillana'®	bedding	red	semidouble	bushy
'Lovely Fairy'®	area	pink	double	bushy
'Mirato'®	area	pink	double	bushy
'Palmengarten Frankfurt'®	area	pink	double	bushy
'Royal Bonica'®	bedding	pink	double	bushy
'Sommermärchen'®	area	pink	semidouble	bushy
'Sommerwind'®	area	pink	semidouble	bushy
'Sunsprite'®	bedding	yellow	double	upright
'Super Dorothy'®	rambler	pink	double	arching
'Super Excelsa'®	rambler	crimson rose	double	arching
'Swany'®	area	white	double	arching
'The Fairy'	area	rose	double	bushy
'White Meidiland'®	area	white	very double	widely spreading

The red-leaved Japanese maple is a noble container shrub of the first order.

Decorative standard roses in pots with low-growing campanulas as a colorful underplanting.

Dwarf forms of valuable evergreen shrubs are rewarding pot plants, even for smaller containers.

Yew can be shaped for mobile formal shrubbery.

doors without any problems. Here is a selection:

▶ **Foliage Shrubs:** paperbark maple (*Acer griseum*), box elder (*Acer negundo* 'Aureomarginatum'), Siberian pea tree (*Caragana arborescens* 'Lorbergii'), winged spindle tree (*Euonymus alata*, *Euonymus planipes),* witch hazel 'Westerstede' (*Hamamelis* variety), privet 'Atrovirens Compact' (*Ligustrum vulgare* variety), common lilac (*Syringa vulgaris* varieties).

▶ **Shade Trees for the Terrace:** Norway maple (*Acer platanoides* 'Globosum'), Young's weeping birch (*Betula pendula* 'Youngii'), hawthorn (*Crataegus* x *lavallei* 'Carrierei').

▶ **Evergreen Shrubs:** dwarf balsam fir 'Picolo' (*Abies balsamea* variety), Norway spruce (*Picea abies* 'Acrocona'), Serbian spruce 'Pendula' (*Picea omorika* variety), pine (*Pinus leucodermis*), blue dwarf stone pine (*Pinus pumila* 'Glauca'), American arborvitae 'Smaragd' (*Thuja occidentalis* variety), pillar American arborvitae (*Thuja occidentalis* 'Emerald').

The tables on pages 65 and 67 offer selected rose varieties that, on the basis of their growth habit, vitality, and winter-hardy root systems, are suitable for pots, tubs, boxes, and hanging baskets.

In addition, the quarter, half, full, and weeping standard roses listed in the section "Standard Roses" (pages 123 through 125) will thrive in containers of the appropriate size.

Conservatory Roses: For more than a hundred years, wealthy villa owners have cultivated a certain fragrant rose in their elegant conservatories and greenhouses that is still available commercially today: the yellow climbing tea rose 'Maréchal Niel'. This ensured the provision of roses for the buttonhole year-round. Anyone who has a greenhouse can have fun keeping this variety in large enough pots (it is not winter hardy out-of-doors).

■ Roses with Noteworthy Prickles—And Those Without Prickles

What would a rose be without prickles, without this defensive outgrowth that imprints the character of the rose like nothing else and yet can be strikingly different from rose to rose?

Every species, every variety has its own form of prickliness, and some of them are armed to the teeth.

Some shrub roses adorn themselve with a cloak of thorns so aesthetically appealing, so decorative that this ornament in itself makes planting this variety worthwhile.

One of the most striking of these is *Rosa sericea* f. *pteracntha* (white, single, 79–118 in. [200–300 cm]), which develops only four petals and attracts attention with brilliant red, winglike spines up to 1 in. (3 cm) long (see the small picture on page 29).

Also fascinating is *Rosa sweginzowii* 'Macrocarpa' (pink, single, 79–118 in. [200–300 cm]), which besides striking prickles has marvelous hips to offer.

A hot tip for all lovers of the romance of winter hoar frost is the once-flowering shrub rose 'Ferdy' (pink, double, 32–39 in. [80–100 cm]). Row after row of fine, pointed prickles line up in a bristly coating

The Rose Garden on the Terrace

Container roses can create floral magic on the terrace. The valuable containers can be arranged according to desire and mood.

1. **Shrub rose 'Schneewittchen'® ('Iceberg'), a dream all in white.**
2. **Planting of an old feed trough with the overhanging varieties (e.g., 'Super Dorothy'® or 'Super Excelsa'®).**
3. **Winged spindle tree** *(Euonymus alata),* **decorative foliage shrub with triple assets: winged corky bark, brilliant red fall foliage, and greenish yellow flowers in May that attract the bees.**
4. **Hanging basket with 'Pheasant'®.**
5. **Privacy screen and space dividers with climbing rose 'Rosarium Uetersen'®.**
6. **Cheery yellow box elder** *(Acer negundo* **'Aureomarginatum'), unique yellow-white variegated foliage, yellowish white, bee-attracting flowers from March to April, winged seeds (nose nippers) in hanging clusters in September.**
7. **Fragrant rose 'Westerland'® tied up on a pyramid.**
8. **Half standard 'Bonica '82'®, underplanted with winter savory** *(Satureja monatana).*

Roses for Boxes

Variety	Class	Color	Flowers	Growth Habit
'Guletta'®	miniature	yellow	double	double
'Orange Meillandina'®	miniature	orange-red	double	double
'Peach Brandy'®	miniature	apricot	double	double
'Pink Symphonie'®	miniature	pink	double	double
'Sonnenkind'®	miniature	yellow	double	double
'Zwergkönig '78'®	miniature	red	double	double

Roses for Hanging Baskets

Variety	Class	Color	Flowers	Growth Habit
'Alba Meidiland'®	area	white	double	bushy
'Flammentanz'®	rambler	red	double	arching
'Flower Carpet'®	area	pink	semidouble	bushy
'Marondo'®	area	pink	semidouble	arching
'Mirato'®	area	pink	double	arching
'Pheasant'®	area	pink	double	arching
'Scarlet Meidiland'®	area	red	double	arching
'Snow Ballet'®	area	white	double	arching
'Super Dorothy'®	rambler	pink	double	arching
'Super Excelsa'®	rambler	crimson pink	double	arching
'Swany'®	area	white	double	arching
The Fairy	area	pink	double	bushy

'Ghislaine de Féligonde' has thornless canes, as if it were created for children's hands.

whose tangles trap the frost crystals in winter and refract the winter light into glittering facets.

However, for lovers of tamer roses that have almost no prickles at all, there are also recommended varieties. The climbing rose 'Maria Lisa' (pink, single, 79–118 in. [200–300 cm]) has only a few prickles, as if it were made for children's hands.

Roses with Fall Coloring

Admittedly, it is not common to plant roses in the garden because of their fall coloring. The leaves of the varieties introduced during this season do not attain the brilliance of the classic fall-colored woody plants like witch hazel, shadblow, and Japanese maple. Nevertheless, the small but good *Rosa rugosa* hybrids do provide striking yellow fall coloring, while the fiery red of *Rosa nitida* is close to spectacular.

Reddish Fall Coloring

Class	Variety/Species
Rambler	'Bobbie James'
Climber	'Ilse Krohn Superior'®
Wild rose	*Rosa nitida*
Shrub rose	*Rosa sweginzowii* 'Macrocarpa'
Rambler	'Super Dorothy'®
Rambler	'Super Excelsa'®

Yellowish Fall Coloring

Class	Variety/Species
Area	'Alba Meidiland'
Rugosa hybrids	'Dagmar Hastrup,' 'Foxi,'® 'Yellow Dagmar' 'Hastrup,'® Pierette,'® 'Polareis,'® 'Polarsonne,'® 'Schnee-Eule'®
Area	'The Fairy'

Spring Roses

Wilhelm Kordes is considered the inventor of spring roses since in the 1920s, he began the breeding of the entire group of spring roses. To do so, he bred various hybrids with forms of the Scotch rose (*Rosa pimpinellifolia*), which gave the hybrids their resistance to fungus disease and their tolerance of arctic winters. Wilhelm Kordes gave the varieties that start blooming at the end of May names incorporating the German word for spring (*Frühling*), e.g., 'Frühlingsgold'.

Today, the term *spring roses* is established as a synonym for all rose species and varieties that bloom especially early. These roses have shrublike forms that open in the spring.

Heath Gardens: Heath gardens display plants from the family of the heaths and heathers (Ericaceae).

Classic heath garden companions are broad-leaved evergreens and conifers. However, spring roses can add a pleasant surprise. In the first place, they get along in relatively poor, acid heath soil. Secondly, their growth habit makes them fit in as lead plants in large heath collections.

Besides the spring roses, varieties such as 'The Fairy' or 'Alba Meidiland' are good for the areas around the entry to heath gardens.

Fall magic: A flower of 'Graham Thomas'® amid colorful fall foliage.

The fall foliage of certain rose varieties bids a joyously colorful farewell to summer.

Shrub rose 'Robusta' with bright fall leaf color.

Spring Roses

Variety/Species	Color	Flowers	Height (in. [cm])	Growth Habit
'Frühlingsgold'®	yellow	single	59–79 (150–200)	arching
'Maigold'®	yellow	double	59–79 (150–200)	upright
Rosa hugonis	yellow	single	79–118 (200–300)	arching
Rosa moyesii (grafted)	red	single	79–118 (200–300)	upright
Rosa pimpinellifolia	cream	single	32–39 (80–100)	bushy
Rosa sericea f. *pteracantha*	white	single	79–118 (200–300)	upright

'Frühlingsgold'®—a successful crossing of *pimpinellifolia* hybrids with hybrid tea roses.

The Ecological Value of the Rose

The term *ecology* is often erroneously equated with conservation. Ecology is, however, a branch of science that concerns itself—objectively—with the study of interactions between creatures and the environment and that describes them precisely.

Ecology occurs day in, day out, constantly and continually all over the earth and, of course, independent of any agency. There is no good or bad ecology, just as there is not any good or bad gravity. We can, however, with the help of the conservation laws, exert an influence on the **complexity** and **equilibrium** of an ecosystem.

In general usage, *ecology* has currently become a buzzword, increasingly estranged from its real meaning. No one really knows very clearly anymore what the original sense of the word was. We therefore make do with vague definitions where whatever is called ecological is somehow considered natural. To try to go against that today is something like fighting windmills. Therefore, understand that when this text states that a plant has an especially high ecological value, it is discussing the contribution that plant makes to other creatures within the framework of the natural cycle. In this sense, creatures can be, for example, deer that graze on the shoots of shrubs, bees that feast on flower pollen and fruit nectar, or birds that seek protection for their nestlings in rose shrubbery. The ecological value of a plant is thus widely inclusive, relative, and determined by the goals of the plant users.

Pollen-dispersing rose blossoms are an important food source for insects.

While rose gardeners may wish for the unsullied rose aesthetic, which rules out the natural relationships in the home garden, environmentally conscious rose gardeners do not immediately let go a blast of fungus and aphid spray for a slight fungus infection. Instead, they recognize this attractiveness of the rose foliage to other creatures as the ecological value of their roses in the causal natural cycle. The rose cannot only take but must also give. Only thus can nature function independently in the garden.

The line between aesthetics and ecology is thus fluid. Each gardener must establish it for himself or herself. Those who can bring both poles into proper equilibrium will be able to unite both pleasure and easy maintenance in harmony in their private refuge.

Hips, essential winter food for numerous bird species.

The following pages show that many ecological values for humans and animals are to be found in roses. By choosing the appropriate species and varieties, one can very definitely influence the ecological value of his or her garden and create a recipe for the harmonious relationship between aesthetics and ecology.

The dog rose *(Rosa canina)*, originally indigenous to the edges of forests and in hedges.

Rosa scabriuscula

Rosa majalis **(Cinnamon Rose)**

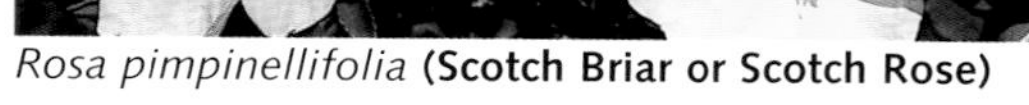

Rosa pimpinellifolia **(Scotch Briar or Scotch Rose)**

Rosa eglanteria **(Eglanteria, Sweet Briar)**

Rosa arvensis **(Field Rose)**

Rosa gallica **(French Rose)**

Roses in Combination

In many rose books, little is found on the subject of combining roses with other shrubs, perennials, annuals, bulbs, and grasses. Many gardens display the rose all by itself in an often wearisome monotony. These gardens always have the same symmetrical bed arrangements. Loose, lively arrangements of roses, perennials, and other green things are lacking almost entirely. Is it simply the want of knowledge of plants or is it a fear that a royal household might somehow damage the queen of the flowers?

Numerous cultivated varieties among the roses are available to create richly varied combinations. The resulting diversity of species is not only easy on the eye, but is also an important requirement for lower maintenance and for wild garden design. Moreover, rose combinations correspond with the natural behavior of many rose species, which in the wild appear as borderline plants at the edges of light woods.

■ . . . With Perennials

Roses can be combined with perennials at any time of the garden year for a feast of color.

When working with both plant groups, the gardener has at his or her disposal a palette of color from pastel to multicolored that allows any garden to be designed from spring to fall—nonstop. In particular, perennials can be used to introduce into the garden the blue shades missing in roses.

However, so that the good-neighbor relationship of roses and perennials is not endangered by a competitive battle, where only the most aggressive species survive, it is necessary to bear in mind the very **different feeding requirements** of both groups. Roses and perennials are never mixed but are always planted next to one another—be it in groups, swaths, rows, or individually. Being next to one another permits the need-appropriate feeding and care so that partners do not become rivals.

First of all, if one visualizes the life rhythms of perennials, it will quickly become clear why roses and perennials can live in one "house" but are better accommodated in separate "apartments." Perennials are plants that live for a number of years. In spring they put out new growth explosively and forcefully. They accomplish flowering and seed formation within a few months, sometimes even within weeks. For this reason, many perennials have a very high need for feeding in the spring, exactly at the same time roses are just getting ready to put out new shoots—"pawing the ground," as it were. If perennials planted right next to roses were to withdraw nutrients in their momentary ravaging hunger, this would give the roses a bad start.

In addition to competing for nutrients, some perennials and roses do not make good companions because of their competing growth habits. Shrub, area, and landscape roses develop dense foliage that makes an underplanting impossible. Also, because roses need an open soil so that one can hill them in regions with cold winters, perennials cannot grow too close or they would just get in the way of this process. Also, for pruning, one needs to be able to get close to the rosebushes, which would be difficult if perennials were growing around the plants.

In short, the queen of the flowers does not accept just any companion, but she gladly lets herself be surrounded by cavaliers who keep their distance. If they do not, clippers and spades help.

An **exception** occurs with only awkwardly shaped hybrid tea roses or old roses with thin foliage and loose shrub structure. These prickly "clothes stands" can be superbly underplanted with ground-covering perennials like winter savory (*Satureja montana*) or low campanula species (*Campanula portenschlagiana* is recommended; *C. pocharskyana*, on the

In a poem, Friedrich Hebbel described the juxtaposition of a climbing rose and a white lily:

Rose and Lily

The rose loves the lily,
It stands at her feet,
Soon the glow pales in her fairest leaf,
It falls, in order to greet her.
The lily marks it well,
She likes the little leaf.
The wind blows it away; leaf after leaf
He scatters afar.
But the rose never ceases,
Lets fall ever new leaves.
She greets and greets herself almost to death;
But not one of them hits its mark.
The last one the lily catches
And closes herself tightly together.
Now the leaf glows in her throat
As if it were a heart full of flames.

other hand, is not since it grows too vigorously). Their usefulness lies not only in the aesthetic appearance but also in notable weed suppression. Some plants that are not suitable are creeping plumbago (*Ceratostigma plumbagoides*) and Acaena (*Acaena buchananii*), whose matted roots interfere with the rose and can actually strangle it.

Perennials are best able to crowd up against the shrub roses. Because of their vigor and height, shrub roses create a point of interest in the middle of a perennial border as a solitary point of color. For example, white phlox (*Phlox*) species, which bloom in August, get on well with once-flowering shrub roses. White can also mediate between rival colors and bring somewhat different colors of red roses to a fiery radiance.

Delphiniums are classic companion plants for bedding and shrub roses. Spikes of delphinium in many shades of blue increase the beauty of roses. The world-famous perennial breeder Karl Foerster has poetically described his varieties with names like heavenly, gentian, cornflower, night, and sea-green blue. As a brilliant blue background for low-growing bedding roses, delphiniums even take over a lead-plant function; there can be no more talk of a framing companion. In fact, the **growth form** of roses and perennials is an important design feature. While upright perennials create interesting, tension-rich contrasts with low-growing roses, shrublike rose varieties can enter into an ideal design symbiosis with arching perennials.

Other, often-seen blue rose companions are catnips (*Nepeta*) and sages (*Salvias*), which—in contrast to delphinium—favor rather lean soil conditions. Accompanying blue-flowering herbs, like lavender, are discussed in the section about roses with herbs (pages 85 through 94).

The color of the heavens summoned to the garden: blue salvia surrounds the rose blossoms.

Fleabane, a rosy cavalier with lush annuals.

Clumps of lady's-mantle are a tried-and-true yellow-flowered foreground planting for rose beds.

TIP Delphinium fans cut back the spent flower stalks to within about a handbreadth of the ground. The plant wastes no strength on seed formation but sends up new stalks and provides delight in late summer with a repeat flowering.

Besides **white** and **blue, pale yellow** is the third ideal perennial color for use with roses. Yellows are good for tone-on-tone designs and go wonderfully with dark-red bedding and shrub roses. Desert candles (*Eremurus*) are good as yellow candles in the background. The lady's-mantle (*Alchemilla mollis*) deserves a special recommendation. Its yellow veil of flowers in June and July is surrounded by an extremely attractive foliage that possesses an unusual characteristic—it gives up water droplets at the edges of the leaves. Botanists call this phenomenon *guttation*. When the sun catches these water droplets, they glitter like stars in the evening sky.

Silver-leaved perennials are worth gold as accompaniments for roses. Their gray leaf color expresses their hunger for light and recommends them as neighbors for the sun-loving rose. A favorite position is held by the lavender cotton (*Santolina chamaecyparissus*), which is a fragrant indoor decoration when dried. Besides this are the everlasting flower (*Anaphalis triplinervis*) with white clusters of flowers, the catnip (*Nepeta*), lavender, and artemisias (*Artemisia stelleriana* and *A. schmidtiana* 'Nana').

Plantain lillies (*Hosta*) introduce decorative foliage colors to their "apartment sharing" with roses. The spring gold plantain lily (*Hosta fortunei* 'Aurea') forms dense clumps and new growth is golden yellow. *Hosta sieboldiana,* the blue-leaved plantain lily, is attractive with violet-blue foliage. *Hosta undulata* 'Univitata', the midsummer plantain lily, surprises with its small heart-shaped leaves that are white-and-green striped. In July, plantain lilies bloom in a subdued violet with pink and prefer a rich, humus-containing soil. They will tolerate dryness for a short time; however, in full sun, they react sensitively.

Shasta daisies, yarrow, and delphinium are perennials whose needs are compatible, allowing a multiple-species rose garden.

Walls in the garden are very beautiful when designed to set off roses and perennials. In any case, roses and stone go well together. Cushion perennials are good perennial material if they are planted before and behind roses but not in the groups of roses.

Aubrieta (*Aubrieta*—blue or violet flowers from April to May, cushions up to 4 in. [10 cm] high, cut back after flowering), madwort (*Alyssum*—yellow flowers from April to May, up to 8 in. [20 cm] high, loose cushion), and perennial candytuft (*Iberis*—white flowers from April to May, about 4 in. [10 cm] high, loose cushiony carpet) are all possibilities for walls. They are allowed to creep over the edges of walls, together with the rose varieties that, with their overarching growth, paper walls with flowers. The conclusion of the spring flowering of the cushion perennials occurs in June. Appropriate rose species are suggested in the section "The Wallflowers—Trailing Roses for Walls" (page 108).

The perennial color **red** is the only one that is dangerous in combination with roses because then too many shades of red are mixed. Therefore, if you use red perennials, do not use ones that flower at the same time as the rose.

Some more **general notes** follow. Perennials and roses both love a pruning in the spring. This means that not only designing but plant maintenance jobs can be combined.

Perennials are increasingly being offered for sale in nurseries in **recycled-paper pots** or similar disposable containers. Although these plants appear at first glance not to be particularly dusted off because of moss growth or because the pot has already started to deteriorate, you do not have to discard the packing. The eco-pot can be planted with the plant with only the outside torn away from the pot a bit. This facilitates the pathway of the new root hairs into the surrounding soil and promotes the disintegration of the eco-pot in the soil.

Taller perennials like delphinium, monkshood, or lavatera (*Lavatera*) are grateful for a support. They tend to flop in the wind and rain; simple constructions of bamboo stakes and string prevent this. Timely application of these support measures is important, for they should grow with the

plants so that the plants retain the form that is typical for their species. If the perennials have already set flowers and are already bent over from the burden of them, they will no longer stand upright even after being tied up (too) late.

Runner-Forming Roses: In a curious way, roses may become troublesome in perennial beds. Some rose species like *Rosa nitida* or *Rosa gallica* turn up unexpectedly, with their runners in perennial and flower beds. This urge to spread can be countered by planting grafted rosebushes.

The Rose-and-Perennial Bed

The suggestion for a plant combination in pink, white, and blue attempts to combine a low-maintenance variety of species with space-saving garden design. The vertical is used with climbing roses on a stable pergola while at its feet grow perennials and bushy rose varieties of various heights and with differing flowering times. Candytuft and catnip open the promenade of flowers in May, the campanulas and baby's breath follow in June. After that, the roses dominate until fall. Blue asters bring the flowering year to a close in October along with the last rose blooms.

> **Planting Tip**
> Usually perennials are sold in pots and containers made of plastic. The delicate earth ball can be removed without injuring it by turning the pot upside down plant and all, lightly striking the edge on a hard surface, and then gently drawing out the root ball.

Rose-and-Perennial Bed

① **1 Climbing rose 'New Dawn'**
② **1 Climbing rose 'Super Dorothy'®**
③ **1 Climbing rose 'Rosarium Uetersen'®**
④ **1 Shrub rose 'Schneewittchen'® ('Iceberg')**
⑤ **2 Bedding roses 'Bonica '82'®**
⑥ **1 Climbing rose 'Ilse Krohn Superior'®**
⑦ **4 Campanulas**
(Campanula portenschlagiana)
⑧ **5 Baby's breath**
(Gypsophila **species or varieties)**
⑨ **4 Asters, blue**
(Aster novi-belgii **varieties)**
⑩ **5 Perennial candytuft**
(Iberis **species)**
⑪ **5 Everlastings**
(Anaphalis triplinervis)
⑫ **4 Bedding roses 'Sommermärchen'®**
⑬ **5 Catnips**
(Nepeta x faassenii)

Blue- to Violet-Flowered Perennials

Plant Name [USDA Zone]	Height (in. [cm])	Months of Flowering	Uses	Situation
Balloon flower (*Platycodon grandiflorus*) [4–9]	20 (50)	VII–VIII	bedding	sunny
Italian aster (*Aster amellus species*) [4–9]	20–24 (50–60)	VII–VIII (–IX)	bedding, cutting	sunny
Marjoram (*Origanum* species) [3–9]	8–16 (20–40)	VII–IX	bedding	sunny
Monkshood (*Aconitum napellus*) [4–8]	43 (110)	VI–VII	bedding, edging shrubbery, lead plants	partly sunny
Fleabane (*Erigeron* hybrids, e.g., 'Adria') [4–9]	32 (80)	VI–VII, IX	terraces, bedding, reblooms	warm, sunny
Salvia (*Salvia nemerosa* varieties like 'Blue Hill' [4–9]	16 (40)	VI–VIII	bedding, reblooms after cutting back	sunny, long-term bloomer
Campanula (*Campanula* species, e.g., *C. persicifolia, C. glomerata*) [4–9]	4–32 (10–80)	VI–VII (–VIII)	cushion perennial, cutting	sunny, with taller bedding roses
Catnip (*Nepeta* x *faassenii*) [3–9]	10 (25)	V–IX	terraces, rock gardens, cut back after flowering	sunny
Globe thistle (*Echinops ritro* 'Veitch's Blue') [3–9]	32 (80)	VII–IX	terraces, beds	sunny
Lavender (*Lavandula* species and varieties) [5–9]	16 (40)	VI–VII	terraces, beds, pots	warm
Delphinium (*Delphinium* hybrids, e.g., 'Völkerfrieden', 'Finsteraarhorn', 'Sommernachtstraum', 'Moerheimii') [3–9]	32–59 (80–150)	VI–IX	bedding, cutting, lead plants, reblooms after cuting back	sunny

Yellow-Flowered Perennials

Plant Name [USDA Zone]	Height (in. [cm])	Months of Flowering	Uses	Situation
Lady's-mantle (*Alchemilla mollis*) [3–8]	12–16 (30–40)	VI–VII	bedding, cutting	undemanding
Goldenrod (*Solidago caesia*) [3–9]	24 (60)	VII–VIII	borders	sunny
Coreopsis (*Coreopsis* species) [3–10]	10 (25)	VI–VIII	borders	sunny
Desert candle (*Eremurus* species and hybrids) [5–9]	59–79 (150–200)	V–VI	borders, cutting	sunny

White-Flowered Perennials

Plant Name [USDA Zone]	Height (in. [cm])	Months of Flowering	Uses	Situation
Fleabane (*Erigeron* hybrids like 'Sommerneuschnee') [4–9]	24 (60)	VI–VII, IX	terraces, bedding, reblooms	warm, sunny
Campanula (*Campanula latifolia* 'Alba', *C. persicifolia* 'Alba') [4–9]	32–39 (80–100)	VI–VII (–VIII)	cutting	sunny
Asters (*Aster novae-angliae* and *A. novi-belgii* varieties) [4–9]	39–55 (100–140)	IX–X	borders	sunny
Aster dumosus 'Schneekissen' [4–9]	10 (25)	VIII–IX	cushions, slopes, terraces	sunny
Madonna lilies (*Lilium candidum*) [4–9]	32–47 (80–120)	VI–VII	borders	warm
Everlasting (*Anaphalis triplinervis*) [3–9]	10 (25)	VII–VIII	terraces, bedding, topping walls	warm, sunny
Baby's breath (*Gypsophila* species and varieties) [3–8]	32 (80)	VI–VIII	bedding, slopes, terraces	sunny

Pink-Flowered Perennials

Plant Name [USDA Zone]	Height (in. [cm])	Months of Flowering	Uses	Situation
Musk mallow (*Malva moschata*) [3–9]	28 (70)	VI–IX	bedding, cutting	out of direct sun
Regal lily (*Lilium regale*) [4–9]	32–59 (80–150)	VII	borders	sunny
Tree mallow (*Lavatera thuringiaca*) [4–10]	50 (150)	VII–IX	borders	sunny

Red-Flowered Perennials

Plant name [USDA Zone]	Height (in. [cm])	Months of Flowering	Uses	Situation
Red cranesbill (*Geranium sanguineum*) [4–11]	12 (30)	V–VIII	ground cover	sunny
Asters (*Aster novae-angliae* and *A. novi-belgii* varieties) [4–9]	39–55 (100–140)	IX–X	borders	sunny

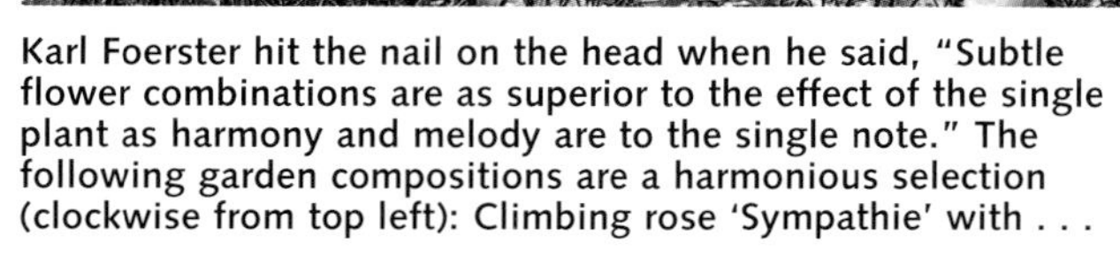

Karl Foerster hit the nail on the head when he said, "Subtle flower combinations are as superior to the effect of the single plant as harmony and melody are to the single note." The following garden compositions are a harmonious selection (clockwise from top left): Climbing rose 'Sympathie' with . . .

. . . lady's-mantle *(Alchemilla mollis),* betony *(Stachys grandiflora),* rosemary, and scented geranium (*Pelargonium* species); colorful variety of roses dance around the sculpture; roses and lamb's-ears *(Stachys byzantina);* bedding rose 'Sunsprite'® and yellow chamomile; shrub rose 'Schneewittchen'® ('Iceberg') and campion; roses with marjoram, phlox, yellow chamomile, foxglove, and clover; 'Bonica '82'® and 'Pascali'®, surrounded by verbenas.

Red oriental poppies and California poppies (*Eschscholzia*) constrast prettily with the white shrub rose 'Schneewittchen' ('Iceberg')

A standard 'Rosarium Uetersen' raises its magnificent flower-laden head.

■ . . . With Annuals and Bulbs

Annuals are joyous rose companions that can colorfully augment groups of roses. To some rose lovers, the liveliness of the annuals next to the glowing elegance of the rose flowers appears all too garish—they categorically avoid combining annuals and roses. This is certainly a very personal, subjective reaction. In this regard, we must agree with Frederick the Great that everyone may be happy in his own way.

In any case, because of the completely different rhythms of long-lived roses and short-lived annuals, it is advisable first to get some experience with both patterns of behavior. Filling empty spaces with annuals in a bed that formerly contained exclusively rose beds, for instance, is a way to start. Every now and again it happens that a rosebush fails. Because of

the phenomenon of disease when roses are replanted in the same place, so-called replanting disease (see page 144), a replacement rose grows well at first but then visibly fails. However, if the soil in this spot is improved with some garden compost and some appropriate annuals are put in—carefully avoiding the rose roots with the several-tined spading fork—the advantages of the annuals are revealed even to the staunchest rose lover.

Furthermore, these advantages go beyond their use in design. The inclusion of annuals increases the wealth of species and decreases the impact of attacks of aphids and rose leafhoppers, to name only two. (Unfortunately the rose leaf roller wasp appears unaffected by annuals, so obviously annuals do not attract the natural enemies of this wasp variety.)

What annuals go with what roses? The beginner should first limit himself or herself to white- or blue-flowered annuals and work creatively with these. For instance, white bedding roses harmonize outstandingly with tall, delicate, elegant cosmos (*Cosmos bipinnatus*). However, if this mixture is not pleasing, the next year another companion can be chosen for the roses. Low annuals like sweet alyssum (*Lobularia*) decorate the ground very charmingly under standard roses. In small groups, the attractive silvery leaves of the common groundsel (*Senecio*) go with all colors of roses. *Nigella* looks very pretty with yellow bedding roses.

Nigella **seed themselves easily in the rose bed.**

Growing One's Own: By starting their growth indoors in seed-starting kits, simple plastic houses, or tunnels, the blooming time of annuals can be advanced considerably. Whether you use homegrown annuals or purchased plants, the same thing holds true. However, only plant after May 15, when night frosts can no longer be expected.

Planting **bulbs** next to roses has triggered many a bitter argument between garden lovers and rose enthusiasts. Some rose lovers warn as a matter of course against planting tulips, narcissi, and crocuses in the vicinity of roses and would not even tolerate *Scilla* and snowdrops. Since repeat-flowering roses do not start blooming until June, however, bulbs can provide a wealth of flowers in spring until the rose flowers make their entrance onto the garden stage again.

Annuals as Rose Companions

Plant Name	Flower Color	Height (in. [cm])	Notes
Pheasant's-eye (*Adonis aestivalis*)	many colors	12–20 (30–50)	time from seeding till flowering varies
Flossflower (*Ageratum* varieties)	blue, pink, white	6–8 (15–20)	numerous varieties; water well
Alkanet (*Anchusa capensis*)	blue	10–12 (25–30)	full sun
Snapdragon (*Antirrhinum majus*)	purple and yellow	8–39 (20–100)	half-shrub where winters are mild
Swan River daisy (*Brachycome iberidifolia*)	blue	8 (20)	no blazing sun
Bush violet (*Browallia speciosa*)	blue, white	6–10 (15–25)	full sun, warm
Cosmos (*Cosmos bipinnatus*)	rose red	39–47 (100–120)	flowers VII–X
Blue lace flower (*Trachymene coerulea*)	sky blue	24 (60)	sun, warm
Dahlia (*Dahlia* hybrids) (tubers can be overwintered indoors)	yellow, white, violet	up to 16 (40)	sun
California poppies (*Escholzia californica*)	many colors	10–20 (25–50)	flowers VI–X
Snow-on-the-mountain (*Euphorbia marginata*)	white bracts	28–39 (70–100)	flowers VII–X
Gazania (*Gazania* hybrids)	white, yellow, orange	6–10 (15–25)	full sun
Globe candytuft (*Iberis umbellata*)	lavender or purple	8–13 (20–30)	flowers VI–VIII
Edging lobelia (*Lobelia erinus*)	blue, reddish violet, white	4–12 (10–30)	flowers V–X
Sweet alyssum (*Lobularia maritima*)	white, pink, lavender	4–6 (10–15)	flowers VI–VIII, use as edging
Nemesia (*Nemesia* varieties)	blue, red	8–20 (20–50)	many varieties
Love-in-a-mist (*Nigella damascena*)	blue, pink, white	20 (50)	flowers VI–IX
Annual phlox (*Phlox drummondii*)	many colors	4–20 (10–50)	flowers VII–IX
Common mignonette (*Reseda odorata*)	greenish yellow, red stamens	6–24 (15–60)	flowers VII-IX
Dusty miller (*Senecio maritima*)	yellow	up to 32 (80)	decorative, gray tomentose leaves
Vervain (*Verbena bonariensis*)	light blue	up to 79 (200)	does not like any wetness
Zinnia (*Zinnia angustifolia* 'White Star')	white	16 (40)	warm

The autumn crocus fills holes, a color salute at a flower-poor time.

Colorful tulip variety in front of (but not in) a rose bed.

The same thing is true for bulbs as for the perennials and the annuals. "Anything goes, as long as you like it," if the characteristics and requirements of the particular plant group are taken into account. Tulips and narcissi, for instance, do not belong in the middle of an existing rose bed. In the first place, during early summer, they have just developed an abundant mass of leaves after having flowered. This might create a highly humid microclimate within the foliage of the roses, which would not dry out until evening, and thus might promote fungus disease. In the second place, the bulbs would definitely be injured by cultivation and hilling. The solution is simple. Place the bulbs in front of or beside the rose groupings. There they can be planted at the ideal depth. Their foliage can be left in place over the summer and allowed to die down naturally—an important prerequisite for lush flowering the next spring.

Roses combined with perennials and bulbs.

Fragrant plants mixed together: roses and lavender.

■ . . . With Herbs

Combining roses and herbs links the attractive decorativeness with the usefulness of plants. Both roses and herbs were important elements of the live home pharmacy in the cloister gardens of the Middle Ages. To this day, fresh herbs from the garden have remained popular ingredients for refining and cooking food. With some rosebushes planted next to them, it is only a small step to one's own little cloister garden.

The gentle colors of old roses are especially pretty with herbs. Shrub roses are reinforced in their naturalness by herb arrangements. To allow the roses to be fed without the herbs receiving too much feeding at the same time and thus decreasing in aroma, it is again recommended that both plant groups be used next to one another.

Herbs not only please the eye, but some, like **lavender,** have the reputation

of keeping aphids away from roses by using their aromatic ways. However, this works only when an appropriate quantity is planted both around and very close to the roses. With its gray foliage, which indicates good heat tolerance, the lavender goes well with the sun-loving rose. When planted beside the scentless bedding roses, lavender also provides a pleasant note of fragrance in rose-perennial beds.

Principally climbing, shrub, area, and bedding roses can be combined with herbs. Rosebushes that are already deeply rooted are particularly suitable. Why? They can survive noticeable dry spells surprisingly well. They need not be watered constantly but, at most, watered thoroughly now and then during very hot spells. The dry soil in the upper layers comes in handy for many herbs with southern origins and provides for a very tasty harvest from the rose herb bed.

Rosy Herb Garden

Our design suggestion ensures the home supply of herbs in the smallest possible area, creates a fragrant pot ambience, and uses a rose arch as a means of giving height to the garden dimensions. The named herb species belong to the basic assortment used in the kitchen and that thrive problem free in the garden. Anyone who has visits from night snails can protect his or her herb bed with snail fences against unwanted guests. Of course, the design of the herb corner may be of a formal, geometric nature, bordered with a framing planting of compact-growing hedging box 'Blauer Heinz'. This is introduced in the section about roses and broad-leaved evergreens (page 92).

> **TIP** Experienced gardeners cut lavender back about halfway in the middle of August after the flowers are finished. This promotes new growth, and the plants come through the following winter better.

Most herbs come from southern regions and—like the rose—prefer sunny situations. Stepping stones laid down in the herb bed facilitate access to the individual plants for maintenance and harvesting work. These stones also store the warmth of the day, which they then give up in the evening, fostering the herbs. The herb bed should be established close to the house so that one has access to the aromatic ingredients quickly and uses them as fresh as possible.

■ . . . With Grasses and Bamboos

Grasses: The grasses provide structure and discipline in designing with roses. They are light seekers like the roses, but

Rosy Herb Garden

1. **Lavender** *(Lavandula angustifolia)*
2. **Common thyme** *(Thymus vulgaris)*
3. **Peppermint** *(Mentha x piperita)*
4. **Lemon verbena, in pot** *(Aloysia triphylla)*
5. **Rosemary, in pot** *(Rosmarinus officinalis)*
6. **Parsley** *(Petroselinum* **species)**
7. **Garden sage** *(Salvia officinalis)*
8. **Lemon balm** *(Melissa officinalis)*
9. **Tarragon** *(Artemisia dracunculus)*
10. **Bay, as full standard in pot** *(Laurus nobilis)*
11. **Chives** *(Allium schoenoprasum)*
12. **Summer savory** *(Satureja hortensis)*
13. **Oregano** *(Origanum vulgare)*
14. **Climbing rose 'New Dawn'**

Herbs for Use with Roses

Plant Name [USDA Zone]	Height (in. [cm])	Flowering Time, Color	Uses	Notes
Chives (*Allium schoenoprasum*) [2–10]	8–12 (20–30)	from June, reddish pink	salads, cottage cheese, soups	do not cook in; perennial, for edging
Lemon verbena (*Aloysia triphylla*) [Grow as annual]	up to 79 (200)	from June, whitish pink	sweet dishes, iced tea, baked goods, fragrant pillows, bath additions	non-winter-hardy shrub, leaves have lemon scent
Dill (*Anethum graveolens*) [Annual (A)]	up to 47 (120)	June to August, yellow	salads, soups, raw vegetables	annual, warm, sunny, for flower tufts—as Bavarians wear on hats!
Tarragon (*Artemisia dracunculus*) [4–10]	up to 51 (130)	from June, pale green	sauces, soups, poultry	perennial, also semishade
Sweet woodruff (*Galium odoratum*) [4–9]	up to 12 (30)	May, white	punch, juices, fragrant potpourris, flavoring in May wine	perennial, good ground cover in shade
Bay (*Laurus nobilis*) [8–11]	in South, over 33 ft. (10 m)	from May, yellowish white	soups, sauces, marinades, vinegar	pot plant, formal shrubbery
Lavender (*Lavandula angustifolia*) [5–9]	up to 20 (50)	from July, blue	sauces, fish, stews, fragrant potpourris	perennial subshrub, full sun, dry
Lemon balm (*Melissa officinalis*) [4–9]	up to 36 (90)	from July, pale blue	salads, soups, cottage cheese, poultry, fish, tea	perennial, sunny
Peppermint (*Mentha* x *piperita*) [3–9]	up to 24 (60)	from July, pale pink	salads, sauces, soups, raw vegetables, tea	perennial, even semishade
Marjoram (*Origanum majorana*) [A]	up to 20 (50)	from July, white, pinkish red	meat dishes, cottage cheese, casseroles, potato dumplings	annual, sunny
Oregano (*Origanum vulgare*) [3–9]	up to 20 (50)	from July, pink, red, white	fish, meat, pizza, sauces	perennial, sunny
Parsley (*Petroselinum* species) [A]	up to 12 (30)	from June, greenish yellow	soups, sauces, cottage cheese, meat, vegetables, salad	curly or particularly flavorsome smooth-leaved varieties, grow as annual
Rosemary (*Rosmarinus officinalis*) [7–9]	up to 28 (70)	from April, blue violet	meat, broiled dishes, vegetables, vinegar	perennial subshrub, sunny, needlelike leaves
Common sage (*Salvia officinalis*) [4–9]	up to 28 (70)	from May, blue violet	meat, broiled dishes, vegetables, tea	perennial subshrub, sunny
Summer savory (*Satureja hortensis*) [A]	up to 12 (30)	from July, white, rose violet	meat, game, stews, beans	annual, also semishade
Common thyme (*Thymus vulgaris*) [4–9]	up to 16 (400)	from May, violet	meat, sausage, broiled dishes, pâtés, vegetables, cheese, vinegar	perennial subshrub, sunny

they do not cross paths in the root region. During the summer, they offer a long-lasting, constant picture and radiate composure and calm. During the cold season, many species fill the flowerless rose beds with their evergreen-to-yellow or white-variegated blades and bring life to the dull gray of winter.

Low-growing species like blue fescues (*Festuca cinera* or *F. gautieri*) are ideally suited for enclosing bedding rose groups. Taller grasses planted beside shrub and wild roses give these a frame and structure.

Tried-and-true rose grasses such as blue oat grass or fountain grass show off best in restricted locations. They also provide linearity even in small home gardens.

From time to time, after five years at the latest, the clumps should be dug up and divided.

Bamboo: The bamboo also belongs to the large family of grasses. What makes it supergrass is the fact that it does not die back in the winter like other grasses. This gives the gardener the chance of working it into the garden design like broad-leaved evergreens and conifers—as a firm constant throughout the entire year. The gardener and rose lover must know about the division of bamboo into clump-forming or runner-forming species and varieties, the latter often with very invasive tendencies. Nurserymen's rhizome gates made of strong sheeting or steel rings are obtainable commercially. When buried around the plant, these keep the formation of runners under control and prevent wandering of underground shoots throughout the garden. If they are allowed to have their own way, rhizomes will not even stop at the terrace foundations themselves.

Bamboos bloom at long intervals, which can be as long as 120 years. After flowering

Fall Salute in October: Gracefully upright grasses (here Miscanthus) and fall asters at the end of the rose year.

Rose Grasses

Grass [USDA Zone]	Leaf Color	Height (in. [cm]) of Foliage/Flowers	Notes
Silver ear/spike grass (*Achnatherum calamagrostis*) [6–11]	light green	up to 24/32 (60/80)	tail-like flower panicles, arching, sunny location
Feather reed grass (*Calamagrostis* x *acutiflora* 'Karl Foerster') [5–9]	yellowish green blades	39/63 (100/160)	clumplike, densely sprouting, sunny location
Japanese sedge grass (*Carex morrowii* 'Variegata') [5–9]	red-brown blades	16/20 (40/50)	semishady
Tufted hair grass (*Deschampsia caespitosa*) [4–9]	dark-green leaf clumps	up to 24/39 (60/100)	sunny to semishady, loves moist soil
Blue fescue (*Festuca cinerea*) [4–8]	steel-blue leaf clumps	6/8 (15/20)	sunny to semishady, low cushions
Blue fescue (*Festuca gautieri*) [4–9]	green leaf clumps	each 6 to 8 (15 to 20)	lawnlike, low, for sunny locations
Hakonechloa (*Hakonechloa macra*) [4–9]	bluish	up to 12/24 (30/60)	mild winter situation
Blue oat grass (*Helichotrichon sempervirens*) [4–8]	bluish green	up to 24/39 (60/100)	evergreen, dense clumps
Miscanthus (*Miscanthus sinensis*) [5–9]	greenish	59/79 (150/200)	monumental giant grasses
Miscanthus sinensis 'Gracillimus' [6–9]	light-green	47/59 (120/150)	sunny location
Zebra eulalia (*Miscanthus sinensis* 'Zebrinus') [6–9]	yellow-striped	47/59 (120/150)	densely bushy, sunny location
Moor grass (*Molinia arundinacea*) [5–8]	greenish	up to 20/79 (50/200)	golden October coloring
Switch grass (*Panicum virgatum* varieties) [5–9]	green, also reddish gray	up to 47/67 (120/170)	moderately strong grower
Fountain grass (*Pennisetum compressum*) [6-9]	greenish	20/28 (50/70)	decorative flower spikes in fall
Spodiopogon (*Spodiopogon sibiricus*) [5–9]	gray-green	20/39 (50/100)	cutting grass
Feather grass (*Stipa barbata*) [5–11]	yellowish green	12/32 (30/80)	sunny

and fruit devlopment, the bamboo dies out. Some species grow herblike, others—in their East Asian habitat—grow as tall as 115 feet (35 m).

Young plants, in particular, should be protected in winter with a mulch covering. Older specimens are characteristically frost hardy. Therefore, as a rule, spring planting in April is preferred. During their growing period, the bamboos respond positively to organic fertilizer and good, moist soil. However, sogginess injures them. For use in conjunction with roses, the clump-forming bamboos with their "controlled" growth and those of the runner-producing species with weak-to-moderate vigor are suitable.

The following selection of bamboos go primarily with shrub and wild roses:

▶ *Fargesia murielae* 'Phoenix' (umbrella bamboo variety): Evergreen, medium-strong, clumplike growing bamboo of the new generation, culms olive green, broad-spreading growth form, height 8 to 13 feet (2.5 to 4 m).

▶ *Fargesia nitida:* Medium-strong, clump-like bamboo, culms reddish brown, forming a densely bushy shape, height 8 to 13 feet (2.5 to 4 m).

▶ *Indocalamus tesselatus:* Grows about 3 feet (1 m) tall, runner-forming bamboo species; pretty leaves.

▶ *Phyllostachys aurea* (golden bamboo): Runner-forming, medium-tall bamboo species, dense foliage, green-yellow culms with thick joints, height 8 to 10 feet (2.5 to 3 m).

▶ *Phyllostachys flexuosa* (zigzag bamboo): Runner-forming, medium-tall bamboo species, culms curving wavy, growing zigzag, green, later yellow-black flecked, height 7 to 16 feet (2 to 5 m).

▶ *Phyllostachys nigra* (black bamboo): Weak runner-forming, moderately tall bamboo species, culms thick, green, later shining brown-black, height 10 to 20 feet (3 to 6 m).

▶ *Phyllostachys viridis* 'Mitis': Runner-forming, medium-tall bamboo species with green culms, later colored dull yellow, height 16 to 23 feet (5 to 7 m).

▶ *Pleioblastus humilis* var. *pumilus:* Runner-forming, low bamboo species, culms green, dense, growing stiffly upright, self-contained growth habit, height 20 to 32 inches (50 to 80 cm).

Shrub rose 'Eden Rose '85'®, surrounded by a filigree of grass blades.

Spiky best selection: fountain grass 'Hameln'.

Crimson structuring element: switch grass 'Rotbraun'.

▶ *Sasa kurilensis:* Runner-forming bamboo species, height 5 feet (1.5 m); decorative leaves.

▶ *Sasa palmata* var. *nebulosa:* Runner-forming bamboo species, height 5 to 7 feet (1.5 to 2 m); strikingly large, attractive foliage.

■ . . . With Foliage Shrubs

Foliage shrubs are the plants that imprint our gardens the longest. No other plant group displays so many different leaf, growth, fruiting, or flowering characteristics. This wealth of forms, which we often see as ornament, is not just a luxury that nature provides for itself but is always based on function.

When the gardner knows this, he or she can derive the appropriate light and location requirements from particular characteristics—for example, the leaf or the growth habit. Hard, leathery, and/or small leaves can indicate heat tolerance, slit or pinnated foliage is often very well adapted for wind. Older shrubs that have foliage all the way to the inside may be considered shade tolerant, shrubs or trees with leaves only on their periphery signal hunger for the sun. Red foliage also tolerates much sun. Bluish-silvery foliage indicates high drought tolerance. On the other hand, bright-green shrubs want much soil moisture, while yellow-leaved foliage shrubs tend to scald with uneven amounts of moisture in the soil.

Leaf humus is a particularly valuable kind of humus that strengthens a shrub's inner powers of resistance provided that it is used in the garden and not stuffed into the trash can. Large leaves should be composted before being spread underneath smaller shrubs. Under larger shrubs and trees, large leaves can be left to rot right there in piles up to about 2 in. (5 cm) thick.

When planting shrubs in front of or next to roses, be careful to leave sufficient distance—depending on the vigor of the plant's growth—between them. The light conditions for the roses should not be impaired by the development of the shrub. Planting a shrub backdrop on the east and/or the north side of the garden works out to be ideal. It halts the flow of cold air and, above all, protects the roses from cold east winds, which can severely affect them.

The torture of deciding which shrubs to use with roses in one's own garden is relieved by a few **basic considerations.** The choice of bright-green spring bloomers presents no difficulties since they are not in visual competition with the roses. Shrubs with colored leaves are eye-catching and therefore should not be placed close to a rose. The more the shrubs' leaves and/or flowers can distract from the roses in bloom in summer, the farther away they should be planted from the queen of the flowers—from the point of view of color play and contrast as well. The more aggressively the shrub grows, the bigger its distance from the rose should be. Trees ultimately become so large that they must be guaranteed enough space from the beginning.

Anyone who has large rose beds can use small shrubs very well to break up the monotony and visually separate the individual rose groups from one another. Also suitable as treats to the eye are small tree forms of valuable foliage shrubs, for instance the small almond tree or the grafted ornamental cherry 'Brilliant'.

▶ *Acer shirasawanum* 'Aureum': Wonderful specimen shrub with golden-yellow long-distance effect and East Asian ambience for nonsunny locations and moist soils, up to 79 in. (200 cm) tall, growth funnel shaped, leaves golden yellow and incised; also for pots and troughs.

▶ Serviceberry (*Amelanchier lamarckii*): Robust shrub giant offering best privacy function for sunny to nonsunny places, growth upright, leaves bronze when new, impressive fiery vermilion coloring in fall, flowers from April to May in white clusters, attracts bees, dark-red edible berries beginning in August, bird food shrub; also for containers.

▶ Butterfly bush in varieties (*Buddleia davidii* hybrids): Its intense flower fragrance attracts butterflies to the garden from great distances, upright growing, flower panicles up to 12 in. (30 cm) long, different varieties in white, pink, red, or violet starting in July; nectar- and pollen-producing bee food for sunny to semishady locations; also for pots, severe cutting back in spring ensures full flowering annually.

▶ Beautyberry (*Callicarpa bodinieri* var. *giraldii*): Up to 79 in. (200 cm) tall ornamental shrub with lilac-colored fruits in fall, yellow-orange fall coloring, delicate pink flowers with fragrance from July to August, bee food, bird food shrub; also for pots.

▶ Mountain dogwood (*Cornus nuttallii*): The most beautiful flowering dogwood, collector's item, large shrub with oval leaves that in fall glow red, flowers (botanically, bracts) in May in striking cream white, for sunny to nonsunny situations.

▶ Broom (*Cytisus nigricans* 'Cyni'): Summer flush of bloom in yellow, up to 39 in. (100 cm) tall, heat tolerant, flowers smelling of honey, abundant bee food source for sunny locations; also for pots.

▶ Deutzia (*Deutzia gracilis*): Very pretty and lacy deutzia, up to 32 in. (80 cm) tall, with delicate white flowers from May to June; source of nectar and pollen for bees for sunny to semishady locations; hard cutting back in spring increases urge to flower.

▶ Pearlbush (*Exochorda macrantha* 'The Bride'): Pearlbush with the largest individual flowers for sunny to nonsunny locations, white curtain of panicles in May, up to 39 in. (100 cm) tall, leaves oval lanceolate; also for pots.

▶ Forsythia (*Forsythia* garden varieties): Spring bloomer with yellow long-distance effect, up to 99 in. (250 cm) tall, also for pots, as specimen, or in hedges in sunny to nonsunny locations; severe cutting back of the oldest flowering shoots and thinning out after flowering increases flower output next season; variety tip: 'Weekend' already flowers as young plant, radiant yellow; a wonderful new dwarf forsythia is 'Mélée d'Or', which only becomes 39 in. (100 cm) tall and blooms magnificently.

▶ Fothergilla (*Fothergilla major*, syn. *F. monticola*): Collector's item with heart-shaped, blue-green leaves that color

Unique bright-pruple fruit ornament: Beautyberry with berries.

A duet of lacy climber and shrub rarity for those in the know: plainly and gently. An alpine clematis *(Clematis alpina)* winds itself through a fothergilla.

Combine the climbing rose 'Flammentanz' together with a rhododendron in front of a conifer backdrop.

glowing orange to crimson in the fall, greenish white flower spikes that attract attention in May before the foliage shows; also for pots.

▶ Wild hydrangea 'Annabelle' (*Hydrangea arborescens* variety): Wonder hydrangea with gigantic white balls of flowers up to 10 in. (25 cm) in diameter, up to 39 in. (100 cm) tall; flowers from July to September; loves damp soil; for sunny to shady locations.

▶ Japanese rose (*Kerria japonica* 'Pleniflora'): Shrub with carnation-like, yellow flower rosettes from April to June for sunny to nonsunny locations, up to 59 in. (150 cm) tall; ideal in mixed flower hedges, on slopes, on fences.

▶ Beautybush (*Kolkwitzia amabilis*): Pale-pink flowering shrub, which marks the transition between spring and summer; growth arching, up to 79 in. (200 cm) tall; bee food for sunny to nonsunny locations; for hedge, specimen plantings; also for pots.

▶ Star magnolia 'Royal Star' (*Magnolia stellata* varieties): Spring magnolia with snow-white flowers for sunny garden locations, shrub up to 79 in. (200 cm) high, flowering lavishly while still young, specimen, also for pots, perhaps one of the most elegant spring bloomers in March to April.

▶ Flowering crab apples (*Malus* varieties): Attractive flowers, foliage, and fruit; depending on variety, flowers double or single, pink, red, or white, fruits yellow, red, or multicolored, bee feeder and nesting shrub for birds for sunny locations; variety selection: 'Charlottae' has delicate pink flowers, large, greenish yellow fruits; 'John Downie' has white flowers, very large fruits.

▶ Purple-leaf sand cherry (*Prunus* x *cistena*): Flowering plum for the smallest sunny to nonsunny sites, growing barely over 59 in. (150 cm) tall; dark-red foliage; light-pink flowers beginning in April; also in pots, very drought tolerant.

▶ Ornamental cherry 'Brilliant' (*Prunis kurilensis* varieties): Magnificent bloomer, either as a shrub or as a standard, with even young plants lushly covered with white flowers; specimens or in groups, for cut flowers and also for pots.

▶ Japanese flowering cherry (*Prunus serrulata* 'Amanogawa'): Flowering beacon

Recommended variety of *ceanothus:* The dark-blue 'Gloire de Versailles'.

The evergreen *Viburnum daviddi* is easily recognizable by its deeply indented leaf nerves.

in April/May, delicate whitish pink double flowers adorn a slender columnar growth form, also for small positions in sunny areas; specimens or in groups.

Pillar-forming shrubs are generally suitable for use as special focal points and serve to partition the garden.
▶ Flowering almond (*Prunus triloba*): Rose red, densely doubled flowers announce the spring in April, shrub up to 59 in. (150 cm) tall or as standard; severe pruning after each flowering promotes development of new flowers; also for pots.
▶ Japanese spiraea 'Little Princess' (*Spiraea japonica* varieties): Little flowering princess with whitish pink flowers from June to July, 12 to 16 in. (30 to 40 cm) tall; provides nectar and pollen for bees; for areas, slopes, small hedges; also for pots.
▶ Dwarf spiraea 'Shirobana' (*Spiraea japonica* varieties): Individual white and pink flowers offer a changing play of color on one inflorescence, up to 24 in. (60 cm) tall; provides nectar and pollen for bees; also for pots.
▶ Fall lilac (*Syringa microphylla* 'Superba'): Miniature lilac with fragrance, only 39 in. (100 cm) tall; violet-pink flowers in May, weaker second flowering in fall; specimens or in groups for small areas in a sunny location; provides nectar and pollen for bees.
▶ Viburnum (*Viburnum bodnantense* 'Dawn'): Winter-flowering shrub with fragrance, grows 79 in. (200 cm) and more, red fall coloring; also for pots.

Blue Shrubs: Plants that are blue always play a special role in combinations with roses. This will be the case so long as there are no really blue roses. When considering the choice of blue-flowered shrubs available, blue roses are not really even needed. Blue shrubs, along with their colleagues from the areas of perennials and annuals, provide breadth and depth in the garden. The perennial legend—Karl Foerster—even wrote an entire book about the blue treasures of the garden. Caution is needed. The effect of blue is cool; in combination with gray it is even cold. However, when red and pink roses are added, perceptible warmth comes into the play of colors. Besides the recognized classics like hydrangeas, butterly bushes, and lilacs in the blue to lilac-purple varieties, the following blue boys deserve attention:
▶ Bluebeard (*Caryopteris* x *clandonensis* 'Kew Blue'): Fragrant, dark-blue flowers in late summer attract bees and other insects; small shrub with gray-green foliage, ideal rose companion; pruning in spring increases flower output; also for pots.
▶ New Jersey tea (*Ceanothus* varieties): Summer and fall bloomer, up to 39 in. (100 cm) tall; foliage dark green, blue panicles of flowers from July to frost, hard pruning in spring promotes heavy flowering; also for pots; variety selection: 'Gloire de Versailles' with blue flowers.
▶ Russian sage 'Blue Spire' (*Perovskia atriplicifolia* variety): Late-summer bloomer for sunny situations, flowers from July to October, inflorescence in blue, up to 39 in. (100 cm) tall; valuable source of nectar and pollen in a time when they are hard to come by; pruning in spring increases flower production; also for pots; the silver-gray foliage looks outstanding with roses.
▶ Rose of Sharon (*Hibiscus syriacus* garden varieties): Late-summer bloomer with mallow or Malva-like flowers, up to 99 in. (250 cm) tall, for sunny situations. in red, violet, pink, white; also for containers; blue variety: 'Blue Bird'.

Evergreens: Shrubs are termed *evergreen* if they retain their leaves for more than two growing seasons. However, evergreens also go through a period of leaf fall; often in the middle of summer their older leaves turn yellow and drop. This is a completely normal occurrence and—insofar as the gardener notices it at all—is no reason for concern.

Evergreens can be used in designing with shrubs as calming, green poles in the middle of cycles of changing color play. Rose beds can be delineated by them. They establish a (semishady) background. Above all, they bring green to wintery gray.
▶ Evergreen barberry (*Berberis candidula*): Robust small shrub; yellow flowers from May to June, thorned, up to 20 in. (50 cm) tall, dark-blue berries in fall; bee food; also for pots and troughs.
▶ Common boxwood (*Buxus sempervirens* var. *arborescens*): Green raw material for geometrically clipped figures (topiary), without pruning over an adult's height, leathery, shining leaves, very shade tolerant; also in containers, especially terra-cotta; the best pruning time is the middle of June.

▶ Box 'Blue Heinz' (*Buxus sempervirens* variety): Compact, slow-growing selection, which, with its dark-green, blue-overcast shoots, is ideal to use as boxwood-enclosing hedge.

▶ Blue hollies (*Ilex* x *meserveae* varieties): Enormously winter-hardy varieties. Bluish, glossy shoots and leaves, ornamental red berries in 'Blue Princess'; up to 118 in. (300 cm) tall, also for pots and troughs; Christmas decoration.

▶ Honeysuckle (*Lonicera nitida* 'Elegant': Up to 36 in. (60 cm) tall, bright-green shrub for surfaces and slopes.

▶ Oregon grape (*Mahonia bealei*): Most beautiful mahonia species, up to 59 in. (150 cm) tall, pale-yellow flower clusters with fragrance from February to April, dark-blue fruit from July; also for containers.

▶ Japanese Pieris (*Pieris japonica* varieties): Flowering shrub, even in shady situations, up to 39 in. (100 cm) tall, depending on variety; white to pink, rarely red panicles from April to May; also for containers; variety tips: 'Debutante', 'Flamingo', 'Forest Flame', 'Purity', 'Valley Rose', 'Variegata', 'Valentine's Day'.

▶ Firethorn (*Pyracantha* garden varieties): For a privacy screen, ground cover (also in full shade), for impenetrable hedges, thorny branches, 39 to 99 in. (100 to 250 cm) tall, white flower umbels in May, fruit color varying in orange, red, and yellow; offers nectar and pollen to bees, nesting for birds.

▶ Evergreen viburnum (*Viburnum davidii*): Flowering and fruiting shrub for connoisseurs, leaves with deeply indented nerves, in June large, whitish pink flower umbels, blue-frosted fruits.

Decorative Little Kitchen Garden with a Romantic Flair

Historically, the kitchen garden followed the monastery gardens of the Middle Ages. The geometric bed shapes were taken over,

Decorative Little Kitchen Garden with a Romantic Flair

1. **Full standard rose 'Raubritter'**
2. *Rosa centifolia* **'Muscosa'**
3. **Lavender** *(Lavandula angustifolia)*
4. **Bedding rose 'Leonardo da Vinci'®**
5. **Bedding rose 'Rose de Rescht'**
6. **Shrub rose 'Heritage'®**
7. **Shrub hydrangea 'Annabelle'**
8. **Delphinium 'Völkerfrieden'** *(Delphinium* **varieties)**
9. **Boxwood 'Blauer Heinz'** *(Buxus)*
10. **Rose balls**
11. **Russian sage 'Blue Spire'** *(Petrovskia atriplicifolia* **varieties)**
12. **Ceanothus 'Gloire de Versailles'** *(Ceanothus* **varieties)**
13. **Cushion aster** *(Aster dumosus* **'Schneekissen')**
14. **Catnip** *(Nepeta x faassenii)*
15. **Tree peony 'Reine Elisabeth'** *(Paeonia suffruticosa* **hybrid)**
16. **Monkshood (***Aconitum***)**

with boxwood framing providing additional emphasis of the formal bed character. Paths divided the area into rectangles or squares, thus making crosses, which are a reminder of Christian teachings.

Small pathways in cross form with a circle in the middle in the design suggestion offered here allow the gardener to reach every part of the beds easily. Like almost no other garden form, kitchen gardens link the useful and the decorative in a traditional and, at the same time, very colorful manner. Typical are wooden fences covered with peas and vetch as an external enclosure. This clearly defines one's own little treasure but never hides it so completely that neighbors are unable to still cast an envious glance inside.

Standard roses and glass rose balls serve as focal points. Someone who has enough space can also add a bench or arbor.

■ . . . With Conifers

Roses and conifers—why not? The bizarre growth habit of one conifer creates lively contrasts in the garden, while the dark green of another provides calmness in the design. With their ever-blue, ever-yellow, or evergreen needles, the evergreen conifers say good-bye to wintertime melancholy and bring color to the roseless cold season.

The choice of conifers appropriate for combining with roses is vast. Here we are primarily introducing compact dwarf forms, the best-colored needled selections, and exquisite specimens that can provide structure in the garden. The latter, however, must be sited far enough away from the roses. At first, as young plants coming from the nursery, they look tiny.

> **TIP** Ground-covering conifers can also be used between rosebushes. In larger gardens, wild roses go best with imposing yews and pines.

However, as soon as they have gained a foothold, they increase continually year by year. The distance between the rose beds and the drip line of conifers should always be 3 feet (1 m) at the minimum.

Vigorous shrub roses like *Rosa moyesii* go well with vital conifers in large gardens.

No space problems arise when working with **dwarf forms.** With their usually dark-green color, these miniature editions of their imposing species colleagues bring tranquility to even the smallest areas that, if the designer has been carried away by enthusiasm, may be planted all too colorfully. Their green really harmonizes with all rose colors. The bedding roses, because of their growth form, are especially suitable as partners for these needled Lilliputians.

In ignorance, conifers are frequently disqualified as exotics and thus considered ecologically valueless. In this regard, many varieties serve in the spring as invisible yet plentiful producers of pollen for bees, bumblebees, and other insects. With their year-round green, as in hedges, they offer many birds and other animals security, even in winter, since predators cannot see into them. Conifers are very low maintenance and undemanding. Caterpillars, for instance, avoid them, preferring native leaves as food.

▶ White fir (*Abies balsamea* 'Nana'): Flattened sphere, compact dwarf form; produces pollen for bees; sunny locations; also for pots, troughs, and balcony boxes.
▶ Lawson cypress (*Chamaecyparis lawsoniana* 'Dart's Blue Ribbon'): Bluish needle coloration for valuable hedges or specimens; also for pots.
▶ Lawson cypress (*Chamaecyparis lawsoniana* 'Minima Glauca'): Dense and globular dwarf form, soft, almost cuddly needles; also for pots and troughs.
▶ Hinoki false cypress (*Chamaecyparis obtusa* 'Nana Gracilis'): Valuable natural bonsai without need for any pruning, very shade tolerant but also thrives in sunny locations, dwarf shrub with fresh green needles; also for pots and troughs.
▶ Silver sawara false cypress (*Chamaecyparis pisifera* 'Boulevard'): Blue-green to blue-white needles, dense growth form; also for pots and troughs.
▶ Sawara false cypress 'Sungold' (*Chamaecyparis pisifera* variety): Golden-

yellow needles, weakly growing, sun tolerant; also for pots and troughs.
▶ Chinese juniper (*Juniperus chinensis* 'Blaauw'): Shrub with upright growth, gray-blue needles, fruits, for sunny location, robust; also for pots.
▶ Chinese juniper (*Juniperus chinensis* 'Old Gold'): Yellow, compact shrub with trailing growth tips; also for pots.
▶ Creeping juniper 'Repanda' (*Juniperus communis* variety): Flat cushion of needled branches with dark-green, silver-striped needles; also for pots.
▶ Blue rug juniper (*Juniperus horizontalis* 'Wiltonii'): Absolutely flat, slow-growing ground cover, ideal for overgrowing stones and rocks.
▶ Dwarf Japanese garden juniper (*Juniperus procumbens* 'Nana'): Dwarf, low-growing cushion shrub; also for pots.
▶ Savin (*Juniperus sabina* 'Tamariscifolia'): Robust, surface-green, star-shaped, broadly spreading, blue-green needles; also for pots.
▶ Singleseed juniper 'Blue Carpet' (*Juniperus squamata* variety): Steel-blue greener of surfaces, densely covering, for sunny sites, robust, undemanding; also for pots.
▶ Singleseed juniper (*Juniperus squamata* 'Blue Star'): Compact growth habit with bluish needles; also for pots and troughs; one of the most beautiful conifer shrubs for accompanying roses.
▶ Eastern red cedar (*Juniperus virginiana* 'Skyrocket'): Slender, upright growth, blue-green needles, truly cypress-like, winter-hardy variety; also for pots and troughs.
▶ Japanese larch (*Larix kaempferi* 'Blue Ball'): Broadly globular dwarf form, needles bluish, exquisite conifer for connoisseurs.
▶ Norway spruce (*Picea abies* 'Echiniformis'): Dwarf shrub, seldom growing more than 0.75 in. (2 cm) per year, hedgehog-like growth habit, green needles; also for pots and balcony boxes.
▶ Norway spruce (*Picea abies* 'Little Gem'): Nestlike growth, bright-green needles, for sunny locations; also for troughs, balcony boxes.
▶ Dwarf Norway spruce (*Picea abies* 'Pygmaea'): Compact, globular, extremely slow-growing conifer in fresh green; also for pots and troughs.
▶ White spruce (*Picea glauca* 'Echiniformis'): Hedgehog-like dwarf form with blue-green needles.

▶ White spruce (*Picea glauca* 'Laurin'): Dwarf spruce, ideal in troughs and balcony boxes, green treasure for those in the know.
▶ Colorado spruce (*Picea pungens* 'Glauca Globosa'): Nestlike, broadly rounded growth habit with steel-blue needles.
▶ Bristle-cone pine (*Pinus aristata*): Bushy conifer shrub that is striking because of the white sap secretions on its needles; a pollen-producing shrub that attracts bees.
▶ Bosnian pine (*Pinus leucodermis*): The most beautiful upright pine species for the home garden; broadly bushy growth with dark-green, elegant needles, bark attractively patterened like snakeskin; Provides food for bees.

TIP All pines remain compact if the young shoots are removed at the end of May. Many dwarf pine forms are included among the winter-hardy woody shrubs for pots.

▶ Scotch pine (*Pinus sylvestris* 'Fastigiata'): Conifer with blue-green needles and columnar growth habit, ideal between roses; also for large pots and troughs.
▶ Mugho pine (*Pinus mugo* 'Mops'): Low-growing greener for surfaces with broad, triangular growth habit, hardly over 32 in. (80 cm) high; needles green, very close together; goes with roses, also for troughs.
▶ Eastern white pine (*Pinus strobus* 'Radiata'): Slow-growing shrub with bluish green needles; bee-nurturing shrub; pretty with roses.
▶ English yew (*Taxus baccata* 'Fastigiata'): Dense columnar growth habit, dark-green needles; bee-nurturing shrub; also in pots; the yellow cultivar (*Taxus baccata* 'Fastigiata Aureomarginata') sports ornamental yellow-green needles.
▶ Prostrate English yew (*Taxus baccata* 'Repandens'): Ground-covering, flat-growing dwarf yew for sunny to fully shady locations, needles dark green.
▶ Eastern arborvitae 'Sunkist' (*Thuja occidentalis* variety): Golden-yellow needled gnome with triangular growth habit; also for pots.
▶ Canadian hemlock (*Tsuga canadensis* 'Nana'): Pretty, dark-green dwarf form, becoming scarcely knee-high; also for pots and troughs.

Evergreen garden dwarf: Globular pine 'Mops'.

Pretty with roses: Norway spruce 'Echiniformis'.

Blue-needled cushion: Dwarf juniper 'Blue Star'.

Rose Bed with Conifers

1. **Round box** *(Buxus)*
2. **Shrub rose 'Ghislaine de Féligonde'**
3. **Bosnian pine** *(Pinus leucodermis)*
4. **Hostas of various colored-leaved varieties** *(Hosta)*
5. **Climbing rose 'Maria Lisa'**
6. **Alaska cedar** *(Chamaecyparis nootkatensis* **'Pendula')**
7. **Bamboo** *(Fargesia nitida)*
8. **Mugho pine** *(Pinus mugo* **'Mops')**
9. **White spruce** *(Picea glauca* **'Echiniformis')**
10. **Singleseed juniper 'Blue Carpet'** *(Juniperus squamata* **variety)**
11. **Eastern red cedar** *(Juniperus virginiana* **'Skyrocket')**
12. **Rambler 'Venusta Pendula'**

Rose Bed with Conifers

Where children play, something is always going on, people meet, communication occurs. The design for such a place must be appropriate all year round. Our suggestion therefore constructs an evergreen scenario around a children's swing set, with conifers, other evergreens, and roses. To prevent child injury from the rose thorns, the only varieties used are ones that are almost thornless. Naturally, other varieties could also be used, but they should then be separated from close child contact by foreground plantings of other shrubs. In addition, the swing set should be set far enough from trees and stumps to avoid injury to children.

A striking, rosy exclamation point is a tree stump, which creates an aesthetically meaningful element. It is woven around, liana-like by the ramber 'Venusta Pendula'. Naturally, a thinly foliaged tree or large shrub would perform the same support function. A bamboo creates an exciting jungle atmosphere. Obviously, a child's garden bed could still be fitted in so that even the youngest can grow crops and harvest them.

■ . . . With Large-Flowered *Clematis* and Other Flowering Climbers

The *Clematis*—common names include virgin's bower, leather flower, and vase vine—displays many parallels to roses. On the basis of its vigor and doubled flowers, people have called it queen of the climbing plants. Also, like the queen of the flowers, a reputation for being fragile and somewhat difficult in culture clings to *Clematis.* It is true that the *Clematis,* like the rose, needs deeply cultivated soil and reacts with extreme sensitivity to pooled water. Therefore, it pays to improve pure loam or clay soils before planting.

A distinction is made in the groups of winter-hardy, woody *Clematis* between wild species and the large-flowered *Clematis* varieties. For the gardener, the latter have the highest ornamental value because of their large flowers, up to 8 in. (20 cm) across, and their habit of blooming more than once. They cover elegant trellises, latticework, fences, and also thinly foliaged trees with green.

Clematis varieties gladly hold their heads up to the sun, but they prefer to let their feet dangle in the cool shade. That defines the light requirements of the *Clematis* in full. It also suggests the chance of combining with roses, for the development of shade in the root region of *Clematis* can be achieved by, say, a foreground planting of flat-growing area roses like 'Nozomi'.

A further possibility is the common planting of climbing roses, especially ramblers, and large-flowered *Clematis,* perhaps letting them grow through a thinly leaved tree. The *Clematis* is planted on the cool, shaded side of the tree, the rambler rose on the side of the trunk toward the sun. Both climbers should be guaranteed sufficient distance from the trunk so that the tree roots will not compete with them too much.

The *Clematis* is a bee food supplier, and the birds use its fluffy seed heads to cushion their nests. The main flowering season varies according to variety, but it begins in May and continues into the fall.

When planting flowers of *Clematis* and roses that bloom at the same time, color

English Roses and *Clematis,* dreamily combined. A tip: The rose garden at St. Albans (near London) displays countless *Clematis*-rose combinations.

harmony should be taken into consideration. Concurrently flowering *Clematis* and climbing or shrub roses are among the most exciting flowering high points of the entire garden year. Both groups offer manifold contrasting color and flower forms, such as some large-flowered *Clematis* beside single wild rose blooms or light flower colors next to dark.

It leaves even 'Nelly Moser' behind: The *Clematis* variety 'Dr. Ruppel'.

Choice of Varieties (flower color, flowering time):
'Dr. Ruppel' (strong rose with crimson stripes, early to midsummer)
'Ernest Markham' (dark violet, midsummer to fall)
'Gypsy Queen' (purple blue, midsummer to fall)
'Hagley Hybrid' (pink with purple-red center band, midsummer to early fall)
'Jackmanii' (violet purple, midsummer to early fall)
'Koenigskind' (royal blue, early summer to fall)
'Lady Betty Balfour' (purple with yellow stamens, late summer to fall)
'Lasurstern' (violet blue with yellow stamens, early summer)
'Mme. le Coultre' (white, early to midsummer)
'Rouge Cardinal' (ruby red, midsummer to early fall)
'The President' (dark violet, early to midsummer)

Note
Clematis are very fragile and break easily; therefore they are invariably staked when offered commercially. Care is thus needed when unpacking, transporting, and especially during planting.

***Clematis* effortlessly cling to arbors and pergolas, provided that their tendrils find a support (wire or string, for instance) that prevents slipping of their lush but thin stems.**

Besides the *Clematis* there are **other climbers** for combining with roses—in the smallest spaces. Be it a bare wall, a light post, a light tree, a shed, or an old climbing rose—a small space for a superclimber can be found anywhere. The following is a selection:

▶ Tara vine (*Actinidia arguta*): Climbing shrub with gooseberry-like fruits, high vitamin content, edible fruit from October, yellow fall coloring; for pergolas, grows in thin-crowned trees, also for pots.

▶ Five-leaf akebia (*Akebia quinata*): Climbing plant with leaves divided fingerlike and brownish violet flowers from May; cucumber-like fruits from August; for pergolas, trellises, also for pots.

▶ Dutchman's pipe vine (*Aristolochia durior*): Climbing plant with pipelike flowers and heart-shaped leaves; for pergolas, also for pots.

▶ Red trumpet vine (*Campsis* x *tagliabuana* 'Mme Galen'): Vine with feathery leaves and orange-red trumpetlike flowers from June to September, needs a warm location in a sunny situation; also for pots.

▶ Climbing hydrangea (*Hydrangea anomala* ssp. *petiolaris*): Climbing shrub for conoisseurs, with huge flower panicles up to 10 inches (25 cm) across; for walls, pergolas, also for pots.

▶ Jasmine (*Jasmine nudiflorum*): Brings yellow light power into wintery gardens since it blooms from February to March, sometimes even starting in November; food source for bees; also suitable for use in pots.

▶ Honeysuckle (*Lonicera* x *brownii* 'Dropmore Scarlet'): Orange-red-flowered vine for pergolas and trellises.
▶ Honeysuckle (*Lonicera* x *heckrottii*): Climber with yellow-and-red bicolored flowers from June to September; fragrant; also for pots.
▶ Woodbine 'Serotina' (*Lonicera periclymenum* variety): Lushly flowering honeysuckle with pure white flowers that turn yellow to red as they fade.
▶ Wisteria (*Wisteria sinensis*): Very vigorously twining superclimber with blue racemes up to 12 in. (30 cm) long from May to June; for walls, pergolas, lattices, also for pots; **caution:** should not get too close to roses; Wisteria vines need a very strong and sturdy support since their mature weight will break insubstantial supports.

The Rose for Special Situations

Sometimes home gardens lie in regions that are characterized by extreme climates or siting conditions and do not present optimal situations for roses. However, rose varieties with special characteristics are available for these situations provided that inappropriate care, in addition, does not exacerbate the extremes of the location.

■ Polar Roses for Frosty Situations

Anyone who has a home garden or a vacation home with a terrace at elevations over 197 feet (500 m) above sea level or in markedly frosty situations (hollows) need not avoid roses. Naturally, the choice is limited. Nonetheless, varieties in white, pink, red, and even yellow can be recommended for large and small planting sites. In particular, the once-flowering shrub and climbing roses and the *rugosa* hybrids are known to be outstandingly winter hardy.

Once-Flowering Shrub and Climbing Roses: The "defect" of their flowering pattern is inherent in once-flowering roses. Strangely, a characteristic that is considered utterly normal in other flowering shrubs is found to be a disadvantage in them. Who ever thought of ruling out the rhododendron just because it blooms only once a year? Everything occurs in its season. Once-flowering roses do their magic before the repeat-flowering varieties, as a rule. The blooming season for once-flowering roses, which can be up to five weeks long, can offer countless design possibilities for plantings. Once-flowering shrub and climbing roses are unique in the best sense of the word (in the table on page 101, these are marked with an asterisk).

Besides the well-considered choice of variety, special emphasis must be placed on appropriately caring for polar roses.

Fertilizing: Using nitrogen-rich fertilizers sparingly so as to develop the necessary maturity of the wood is the top rule. Always use organic fertilizers in winter or spring so that conversion will have been achieved by early summer for best growth promotion. Fertilizers, be they organic or inorganic, must come into contact with sufficient soil moisture to be able to be utilized. If they remain on the surface of the soil because of summer drought and are incorporated only with late-summer downpours, a fall spurt of growth with fatal frost damage will result.

Winter Pruning: Delay pruning until the end of April to avoid frost damage to butter-soft early canes with the consequent weakening of the plant. With repeat-flowering varieties, pruning in summer should be avoided because it retards the chronological development of hips of the variety. These should be allowed a natural developmental rhythm appropriate to the shortened growing season of the location. Again, this also better serves the maturing of the wood. First-class care in a frost-exposed location means a secure outcome for a variety that is in itself frost hardy.

Winter Protection: It goes without saying that bedding roses, in particular, must be guaranteed optimal winter protection (through hilling and by covering with branches).

'Bonica '82'® possesses legendary winter hardiness.

Winter is here: Frost coats the delicate flowers of 'Schneewittchen' ('Iceberg')

Frosted hips of a hedge rose say farewell to autumn.

Roof Garden Roses for Breezy Heights

If one believes in the many plant recommendations for roof gardens, almost everything that is green is suited to this unusual location. These suggestions have completely disregarded the fact that roof gardens can be of very differing natures as places for plantings. Thus a garden over a garage can be a continuation of the real garden at the same height level, with similar care requirements and similar plantings. The other extreme is created by the roof garden on an exposed high floor that towers over the real garden area. This form of roof garden is subjected to all the inclemencies of the weather without protection and tends to have limited soil conditions. These factors result in an extremely limited choice of plants. Classified in between is the sunken, shaftlike roof garden surrounded by walls,

Polar Roses (*= once-flowering)

Variety/Species	Color	Flowers	Height (in. [cm])	Growth Habit
'Bonica '82'®	pink	double	24–32 (60–80)	bushy
'Dortmund'®	red	single	79–118 (200–300)	arching
'Ferdy'®*	pink	double	32–39 (80–100)	arching
'Flammentanz'®*	red	double	118–197 (300–500)	unsupported, flat-growing
'Flower Carpet'®	pink	semidouble	24–32 (60–80)	bushy
'Foxy'®	pink	double	24–32 (60–80)	bushy
'Frau Dagmar Hartopp'	pink	single	24–32 (60–80)	upright
'IGA '83 München'®	pink	double	32–39 (80–100)	upright
'Magic Meidiland'®	pink	double	16–24 (40–60)	flat, vigorous
'Marguerite Hilling'	pink	semidouble	59–79 (150–200)	bushy
'Pierette'®	pink	double	24–32 (60–80)	bushy
'Pink Grootendorst'	pink	double	39–59 (100–150)	upright
'Pink Meidiland'®	pink/white	single	24–32 (60–80)	arching
'Play Rose'®	pink	double	24–32 (60–80)	bushy
'Polareis'®	pink	double	24–32 (60–80)	bushy
'Polarsonne'®	red	double	24–32 (60–80)	bushy
'Repens Alba'* *(Rosa x paulii)*	white	single	12–16 (30–40)	flat, vigorous
*Rosa moyesii** (grafted)	red	single	79–118 (200–300)	upright
*Rosa sericea f. pteracantha**	white	single	79–118 (200–300)	upright
Rosa sweginzowii 'Macrocarpa' *	pink	single	79–118 (200–300)	upright
'Rosarium Uetersen'®	pink	double	79–118 (200–300)	arching
'Scharlachglut'*	red	single	59–79 (150–200)	upright
'Schnee-Eule'®	white	double	16–24 (40–60)	upright
'Schneewittchen'® ('Iceberg')	white	double	39–59 (100–150)	bushy
'Surrey'®	pink	semidouble	16–24 (40–60)	bushy
'Venusta Pendula'*	pink/white	semidouble	118–197 (300–500)	unsupported, flat-growing
'Yellow Dagmar Hastrup'®	yellow	semidouble	24–32 (60–80)	upright

situated a little bit higher than the paved inner courtyard.

As a rule, roses on the roof are exposed to stronger and longer sunshine. With the first rays of the sun, the workday begins for the rose and ends only with the last of the evening sun. Many heat-sensitive rose companions, such as pachysandra (*Pachysandra*) or colored-leaved shrubs, drop out here. The heat does not much bother robust rose varieties that have sufficiently moist soil in the summer. Much more danger lurks in the cold season. The sun-loving rose can be quickly induced to sprout by the strong winter sun. The severe cold returning at night then strikes the active rose—sometimes with devastating results. Therefore, roses in a roof garden must be well covered with branches or burlap during the frost period. However, the gardener must be careful not to cause additional warming because of too much protection.

On top of the world: Roses for the roof garden.

Area Roses for Roof Gardens

Variety	Color	Flowers	Height (in. [cm])	Growth Habit
'Alba Meidiland'®	white	double	32–39 (80–100)	bushy
'Ballerina'	pink/white	single	24–32 (60–80)	arching
'Bingo Meidiland'®	pink	single	16–24 (40–60)	bushy
'Bonica '82'®	pink	double	24–32 (60–80)	bushy
'Flower Carpet'®	pink	semidouble	24–32 (60–80)	bushy
'Lovely Fairy'®	pink	double	24–32 (60–80)	bushy
'Magic Meidiland'®	pink	double	16–24 (40–60)	flat, vigorous
'Marjorie Fair'®	red	single	24–32 (60–80)	arching
'Mirato'®	pink	double	16–24 (40–60)	bushy
'Palmengarten Frankfurt'®	pink	double	24–32 (60–80)	bushy
'Pheasant'®	pink	double	24–32 (60–80)	flat, vigorous
'Pink Meidiland'®	pink/white	single	24–32 (60–80)	arching
'Red Meidiland'®	red	single	24–32 (60–80)	low bushy
'Royal Bassino'®	red	semidouble	16–24 (40–60)	bushy
'Satina'®	pink	double	16–24 (40–60)	bushy
'Scarlet Meidiland'®	red	double	24–32 (60–80)	arching
'Sommermärchen'®	pink	semidouble	16–24 (40–60)	bushy
'Surrey'®	pink	semidouble	16–24 (40–60)	bushy
'The Fairy'	pink	double	24–32 (60–80)	bushy
'White Meidiland'®	white	very double	16–24 (40–60)	broad-spreading

Own-root area roses are very good for anchoring slopes.

A further extreme condition in the roof garden is the high evaporation rate, which is caused not only by the beaming down of the sun but also by the constant blowing of the wind. Thirsty roses are susceptible to pests and disease. For this reason, the soil must always be kept sufficiently moist. Automatic watering systems, which are available from plant supply houses and garden centers, are very helpful for this.

Because of the limited soil depth in roof garden areas, mainly own-root rose plants are considered for this purpose. They will even thrive at a shallower planting depth—in contrast with grafted roses—and make effective use of even the upper soil layers with their fine root systems to maintain themselves.

Bank Roses for Steep Locations

Greening and anchoring slopes was the original job of the area roses. When planted as summer-blooming substitutes for the monotonous slope plantings of cotoneaster *(Cotoneaster)*, which offered only green leaves, the true career of the hard worker among the roses began. The first positive experiences were gathered with 'Max Graf' and later with 'The Fairy'. Happily, rose breeders recognized this development early. Within very few years, a wide assortment of suitable bank roses was available in the trade. Thus today, with prudently selecting varieties and combining them with other ground covers, a monotony of repeated plants can be avoided.

What was described for roof gardens also holds true here—all slopes are not the same. A slope can just as well be a gently graded rock garden as a terracing within several beds. The deciding factor for planting is the incline of the slope and the size of the area. Area roses' characteristic of being able to form new roots on canes that lie directly on the ground equips these plants to help anchor slopes. However, they cannot perform miracles. Slopes that are too steep must be reinforced by supporting walls in addition. This is not only practical but, with the use of decorative masonry, very

Area Roses for Slopes

Variety	Color	Flower	Height (in. [cm])
'Alba Meidiland'®	white	double	32–39 (80–100)
'Apfelblüte'®	white	single	32–39 (80–100)
'Heidetraum'®	pink	semidouble	24–32 (60–80)
'Immensee'®	pink	single	12–16 (30–40)
'Magic Meidiland'®	pink	double	16–24 (40–60)
'Marondo'®	pink	semidouble	24–32 (60–80)
'Max Graf'	pink	single	24–32 (60–80)
'Mirato'®	pink	double	16–24 (40–60)
'Palmengarten Frankfurt'®	pink	double	24–32 (60–80)
'Pheasant'®	pink	double	24–32 (60–80)
'Pink Meidiland'®	pink/white	single	24–32 (60–80)
'Repandia'®	pink	single	16–24 (40–60)
Rosa repens x *gallica*	pink	single	12–16 (30–40)
'Royal Bassino'®	red	semidouble	16–24 (40–60)
'Satina'®	pink	double	16–24 (40–60)
'Surrey'®	pink	semidouble	16–24 (40–60)
'The Fairy'	pink	double	24–32 (60–80)

Rugged bank roses, whose canes can . . .

attractive aesthetically at the same time. On slopes that are hard to reach for pruning and maintenance, own-root plants should always be used. Thus tiresome suckers will not need to be removed, and pruning need be done only every four or five years for rejuvenation.

Bikini Roses—Heat-Tolerant Roses for Hot Locations

Some roses like it hot. Although all roses are sun lovers, a special classification is used for roses that thrive in south-facing sites with enormously high levels of sun exposure. A determinant of how much heat a variety may be able to endure is the smallness of the leaves. The smaller the leaf, the less the transpiration—which translates into water evaporation. Small foliage is found particularly in varieties from the group of area roses. In addition, these varieties cover the ground with their dense foliage and thus preserve the soil moisture. Besides water balance, the heat fastness of the flowers plays a role in a variety's qualifications as a sunbather. Red-flowered varieties, in particular, are bleached by persistent sun radiation. When compared with their former brilliance, excessive sun exposure causes them to be only a vapid love letter instead of a fiery declaration of love. White roses are especially heat fast.

form new roots on contact with the ground: 'Immensee'® (above, front) and 'Flower Carpet'®.

It has cut a fine figure on steep inclines for many years: 'The Fairy'.

Bikini Roses

Variety	Color	Class	Flower	Height (in. [cm])
'Alba Meidiland'®	white	area	double	32–39 (80–100)
'Apfelblüte'®	white	area	single	32–39 (80–100)
'Ballade'®	pink	bedding	double	24–32 (60–80)
'Bella Rosa'®	pink	bedding	double	24–32 (60–80)
'Bingo Meidiland'®	pink	area	single	16–24 (40–60)
'Chorus'®	red	bedding	double	24–32 (60–80)
'Dortmund'®	red	climber	single	79–118 (200–300)
'Flower Carpet'®	pink	area	semidouble	24–32 (60–80)
'Lavender Dream'®	lavender	area	semidouble	24–32 (60–80)
'Montana'®	red	bedding	double	32–39 (80–100)
'Nozomi'®	pearly pink	area	single	16–24 (40–60)
'Palmengarten Frankfurt'®	pink	area	double	24–32 (60–80)
'Pink Meidiland'®	pink/white	area	single	24–32 (60–80)
'Romanze'®	pink	shrub	double	39–59 (100–150)
'Rosarium Uetersen'®	pink	climber	double	79–118 (200–300)
'Royal Bonica'®	pink	bedding	double	24–32 (60–80)
'Scarlet Meidiland'®	red	area	double	24–32 (60–80)
'Schneewittchen'® ('Iceberg')	white	shrub	double	39–59 (100–150)
'Snow Ballet'®	white	area	double	16–24 (40–60)
'Sunsprite'®	yellow	bedding	double	24–32 (60–80)
'Super Dorothy'®	pink	rambler	double	118–197 (300–500)
'Super Excelsa'®	crimson pink	rambler	double	118–197 (300–500)
'Surrey'®	pink	area	semidouble	16–24 (40–60)
'The Fairy'	pink	area	double	24–32 (60–80)
'White Meidiland'®	white	area	very double	16–24 (40–60)

Roses for Part Shade

Variety	Color	Class	Flower	Height (in. [cm])
'Aachener Dom'®	pink	hybrid tea	double	24–32 (60–80)
'Bobbie James'	white	rambler	semidouble	118–197 (300–500)
'Bonica '82'®	pink	bedding	double	24–32 (60–80)
'Christopher Columbus'®	orange	hybrid tea	double	24–32 (60–80)
'Dortmund'®	red	climber	single	79–118 (200–300)
'Flammentanz'®	red	rambler	double	118–197 (300–500)
'Flower Carpet'®	pink	area	semidouble	24–32 (60–80)
'IGA '83 München'®	pink	shrub	double	32–39 (80–100)
'Immensee'®	pink	area	single	12–16 (30–40)
'Kazanlik'	pink	shrub	semidouble	59–79 (150–200)
'La Sevillana'®	red	bedding	semidouble	24–32 (60–80)
'Mirato'®	pink	area	double	16–24 (40–60)
'New Dawn'	pearly pink	climber	double	79–118 (200–300)
'Nozomi'	pearly pink	area	single	16–24 (40–60)
'Paul Noël'	pink	rambler	double	118–197 (300–500)
'Peace'®	yellow/red	hybrid tea	double	32–39 (80–100)
'Pheasant'®	pink	area	double	24–32 (60–80)
Rosa rugosa	pink	rambler	semidouble	118–197 (300–500)
'Repandia'®	pink	area	single	16–24 (40–60)
'Romanze'®	pink	shrub	double	39–59 (100–150)
'Schneewittchen'® ('Iceberg')	white	shrub	double	39–59 (100–150)
'Surrey'®	pink	area	semidouble	16–24 (40–60)
'The Fairy'	pink	area	double	24–32 (60–80)
'The Queen Elizabeth Rose'®	pink	bedding	double	32–39 (80–100)
'Venusta Pendula'	pink/white	rambler	semidouble	118–197 (300–500)

■ Rain-Fast Roses for Regions with High Rainfall

In regions with above-average quantities of rain, perhaps in mountainous country, the very first consideration in selecting rose varieties is good resistance of leaves to spot anthracnose. The second is to look for petals that are unaffected by water. The table "Bikini Roses" lists a selection of such varieties. Regional nurseries and chapters of the American Rose Society (ARS) give further region-specific information about varieties.

■ Roses for the Semishady Existence—Roses for Graves

Light is the most important elixir of life for the rose. When you take away access to it, flowering suffers and susceptibility to fungus disease increases. The fact that roses are usually completely leafless inside the bush demonstrates the high light needs of the queen of the flowers. However, all levels of part shade are not the same. Robust varieties will manage with four or five hours of direct sun per day as long as the semishady location is not attributable to the crown of large trees and the roses are not under the drip line (see page 143). Some roses still feel completely comfortable in the moving shadows of buildings or walls. The table "Roses for Part Shade" lists a selection of proven varieties.

Roses for Grave Sites: Grave sites have a very special meaning for those left behind. Mourners usually look for plants for them—also as an expression of their feelings—with special love and care and they maintain these plants accordingly.

Grave sites in old cemeteries very often lie in half-shady or nonsunny locations. The large Methusalehs of trees take care of this. They surely offer important, undisturbed refuges for birds, small mammals, and insects with their imposing growth forms. However, large trees also rob the light from large areas of the cemetery. Nevertheless, it is not necessary to rule out roses for decorating a grave. Varieties like 'Surrey', 'Bonica '82', 'Aachener Dom', or 'The Fairy' can work their magic in semishady situations.

When raindrops are falling on my calyx . . . 'Schneewittchen'® ('Iceberg') does not disappoint in high-rainfall areas either.

Rain-Resistant Roses

Variety	Color	Class	Flowers	Height (in [cm])
'Aachener Dom'®	pink	hybrid tea	double	24–32 (60–80)
'Banzai '83'®	yellow	hybrid tea	double	32–39 (80–100)
'Bonica '82'®	pink	bedding	double	24–32 (60–80)
'Centenaire de Lourdes'	pink	shrub	semidouble	59–79 (150–200)
'Christopher Columbus'®	orange	hybrid tea	double	24–32 (60–80)
'Edelweiss'®	white	bedding	double	16–24 (40–60)
'Escapade'®	lavender/white	bedding	semidouble	32–39 (80–100)
'Flower Carpet'®	pink	area	semidouble	24–32 (60–80)
'Ghislaine de Féligonde'	yellow	shrub	double	59–79 (150–200)
'Grand Hotel'®	red	shrub	double	59–79 (150–200)
'Heidepark'®	pink	bedding	semidouble	24–32 (60–80)
'IGA '83 München'®	pink	shrub	double	32–39 (80–100)
'La Sevillana'®	red	bedding	semidouble	24–32 (60–80)
'Lichtkönigin Lucia'®	yellow	shrub	double	39–59 (100–150)
'Pheasant'®	pink	area	double	24–32 (60–80)
'Play Rose'®	pink	bedding	double	24–32 (60–80)
'Polka '91'®	amber	shrub	double	39–59 (100–150)
'Romanze'®	pink	shrub	double	39–59 (100–150)
'Rosarium Uetersen'®	pink	climber	double	79–118 (200–300)
'Schneewittchen'® ('Iceberg')	white	shrub	double	39–59 (100–150)
'Schöne Dortmunderin'®	pink	bedding	double	24–32 (60–80)
'Silver Jubilee'®	pink	hybrid tea	double	24–32 (60–80)

If the grave site is sunny, the Lilliputians among the roses, the miniature or dwarf roses, are a special solution for the planting. They use the extremely limited area effectively and colorfully.

Rose varieties like 'Pink Symphonie' or 'Sonnenkind' take some of the mournfulness from grave sites while not looking too exuberant about it. In combination with other shrub and perennial midgets, but also with annuals, suitable grave arrangements can be created for the entire year.

■ Salt Roses—Roses for the Salt of the Earth

Locations in which the shrub is directly exposed to salt contact—be it through spray near the coast or along a main road with winter salting—are certainly not ideal for roses. In the winter stage, the injury affects the already formed buds. It is manifested through irregular sprouting in the spring. Nevertheless, foliage and shoots of certain varieties appear to be relatively

No rule without its exception: Although roses are sun lovers, some vital varieties can manage well with a limited amount of light.

'Frau Dagmar Hartopp'

'Yellow Dagmar Hastrup'®

'Polarsonne'®

'Polareis'®

'Foxi'®

insensitive to the effects of salt. Moreover, one group of roses that both tolerates direct salt contact on the above-ground parts and withstands salty soils relatively well is the *rugosa* hybrids. Coastal stands of own-root seedlings document this salt tolerance. Thus, we can draw conclusions about the suitability of the *rugosa* hybrids for similar situations, provided the plants are growing on their own roots.

On the other hand, *rugosas* react very sensitively to lime, one of the reasons that their tendency to chlorosis (yellow coloration of the young leaves) increases as lime in the soil increases. The chlorosis appears in own-root and grafted stock, which means that gardeners should be particularly careful to avoid lime soils when using *rugosa* hybrids.

■ Weekend Roses—Roses for Community Gardens

Many urban dwellers who long to grow vegetables, fruits, herbs, and flowers but lack adequate personal yard space have found refuge in community gardens. Many cities like New York, San Francisco, Cleveland, Philadelphia, and Seattle are famous for their extensive network of community gardens. Many community gardens were also started during World War I and World War II when victory gardens were promoted, indeed, critical in

'Schnee-Eule'®

Tried-and-true roses are recommended for the community garden: Little care, much free time.

Salt-Tolerant Roses

Variety	Color	Class	Flowers	Height (in. [cm])
'Ballerina'	pink/white	area	single	24–32 (60–80)
'Bingo Meidiland'®	pink	area	single	16–24 (40–60)
'Bonica '82'®	pink	bedding	double	24–32 (60–80)
'Dortmunder Kaiserhain'®	pink	area	double	32–39 (80–100)
'Flower Carpet'®	pink	area	semidouble	24–32 (60–80)
'Foxi'®	pink	*rugosa* hybrid	double	24–32 (60–80)
'Frau Dagmar Hartopp'	pink	*rugosa* hybrid	single	24–32 (60–80)
'La Sevillana'®	red	bedding	semidouble	24–32 (60–80)
'Marjorie Fair'®	red	area	single	24–32 (60–80)
'Marondo'®	pink	area	semidouble	24–32 (60–80)
'Mirato'®	pink	area	double	16–24 (40–60)
'Palmengarten Frankfurt'®	pink	area	double	24–32 (60–80)
'Pierette'®	pink	*rugosa* hybrid	double	24–32 (60–80)
'Pink Meidiland'®	pink/white	area	single	24–32 (60–80)
'Polareis'®	pink	*rugosa* hybrid	double	24–32 (60–80)
'Polarsonne'®	red	*rugosa* hybrid	double	24–32 (60–80)
'Repandia'®	pink	area	single	16–24 (40–60)
'Repens alba' *(Rosa* x *paulii)*	white	*rugosa* hybrid	single	12–16 (30–40)
'Robusta'®	red	*rugosa* hybrid	single	59–79 (150–200)
'Rugelda'®	yellow	shrub	double	59–79 (150–200)
'Schnee-Eule'®	white	*rugosa* hybrid	double	16–24 (40–60)
'Surrey'®	pink	area	semidouble	16–24 (40–60)
'The Fairy'	pink	area	double	24–32 (60–80)
'Yellow Dagmar Hastrup'®	yellow	*rugosa* hybrid	semidouble	24–32 (60–80)

helping to produce extra food for domestic and military needs.

Many community gardens begun during the war years continue today. In fact, community gardens in large metropolitan areas continue to function as a kind of modern victory garden, i.e., victory over urban decay and crime. Many of today's community gardens have been salvaged from vacant, garbage-strewn lots that sorely needed attention and beautification.

A community garden brings together a diverse group of individuals all joined by their love of growing their own plants and the love of being outdoors working the soil. America's community gardens also reflect our immigrant past. Immigrants coming to America from rural farming lives in the old country were anxious to have a patch of earth in which to grow fresh vegetables as they did in Europe. These immigrants also brought over seeds and cuttings of their favorite flowering plants, including roses, which found their way into community gardens.

Many community gardeners grow abundant plantings of ornamentals, including many hybrid tea, bedding, climbing, and shrub roses, to add beauty and charm to the plots.

If you would like to acquire a space in a community garden, inquire with your local extension service, botanical garden, or city government office about where and how to apply for gardening space.

■ The Wallflowers—Trailing Roses for Walls

Roses and masonry in combination always provide for an especially romantic ambience. Here, first of all, area and climbing roses with long, overhanging canes display their ability as enchanting wall artists that grow from the crown of the wall to its base.

Climbing roses? The idea of planting them may be confusing if one thinks of them only in their role of making walls green from bottom to top. However, limiting them to this function alone would make these multitalented roses seem one sided. With certain varieties, turning the direction of growth the other way causes

no problems at all. The location of the wall is important. Hot southern exposures are unsuitable. They promote the plants' susceptibility to attack by spider mites and powdery mildew. Only roses with a prairie character, perhaps *Rosa hugonis,* should be considered here if at all. Someone who wants to avoid experimentation will therefore choose walls that have an east-west orientation for planting with roses.

TIP The wall will not become green the first year; a little patience does not hurt. Therefore, you should choose the variety with special thought so as to avoid disappointment. Pay special attention to robustness and to the harmonizing of the colors of flowers and masonry.

In addition, any uncovered, slightly inclined stonework between the rose canes can be charmingly planted with appropriate perennials. Especially recommended are species and varieties that can manage with little soil and the little rainwater that can run down through the slight incline into the cracks. Among these are Aubrieta, moss pinks, houseleek, or even alpine edelweiss.

Roses for Community Gardens

Variety	Color	Class	Flowers	Height (in. [cm])
'Aachener Dom'®	pink	hybrid tea	double	24–32 (60–80)
'Astrid Lindgren'®	pink	shrub	double	39–59 (100–150)
'Ballerina'	pink/white	area	single	24–32 (60–80)
'Blühwunder'®	pink	bedding	semidouble	24–32 (60–80)
'Bonica '82'®	pink	bedding	double	24–32 (60–80)
'Dornröschenschloss Sababurg'®	pink	shrub	double	39–59 (100–150)
'Elina'®	yellow	hybrid tea	double	32–39 (80–100)
'Flower Carpet'®	pink	area	semidouble	24–32 (60–80)
'Fragrant Cloud'®	red	bedding	double	24–32 (60–80)
'Graham Thomas'®	yellow	shrub	double	39–59 (100–150)
'Ilse Krohn Superior'®	white	climber	double	79–118 (200–300)
'La Sevillana'®	red	bedding	semidouble	24–32 (60–80)
'Leonardo da Vinci'®	pink	bedding	double	24–32 (60–80)
'Lichtkönigin Lucia'®	yellow	shrub	double	39–59 (100–150)
'Louise Odier'	pink	shrub	double	59–79 (150–200)
'Loving Memory'®	red	hybrid tea	double	24–32 (60–80)
'New Dawn'	pearly pink	climber	double	79–118 (200–300)
'Pariser Charme'	pink	hybrid tea	double	24–32 (60–80)
'Raubritter'	pink	shrub	double	79–118 (200–300)
'Romanze'®	pink	shrub	double	39–59 (100–150)
'Rosarium Uetersen'®	pink	climber	double	79–118 (200–300)
'Schneeflocke'®	white	bedding	semidouble	16–24 (40–60)
'Schneewittchen'® ('Iceberg')	white	shrub	double	39–59 (100–150)
'Silver Jubilee'®	pink	hybrid tea	double	24–32 (60–80)
'Sunsprite'®	yellow	bedding	double	24–32 (60–80)
'Surrey'®	pink	area	semidouble	16–24 (40–60)
'The Fairy'	pink	area	double	24–32 (60–80)
'The McCartney Rose'®	pink	hybrid tea	double	24–32 (60–80)
'Westerland'®	apricot	shrub	semidouble	59–79 (150–200)

Roses for Trailing Over Walls

Variety	Color	Flowers	Cane Length (in. [cm])
'Alba Meidiland'®	white	double	39 (100)
'Apfelblüte'®	white	single	59 (150)
'Fairy Dance'®	red	double	39 (100)
'Ferdy'®	pink	double	59 (150)
'Flammentanz'®	red	double	118 (300)
'Ghislaine de Féligonde'	yellow	double	59 (150)
'Immensee'®	pink	single	118 (300)
'Lovely Fairy'®	pink	double	39 (100)
'Magic Meidiland'®	pink	double	79 (200)
'Marondo'®	pink	semidouble	59 (150)
'Max Graf'	pink	single	79 (200)
'New Dawn'	pearly pink	double	118 (300)
'Pheasant'®	pink	double	79 (200)
'Repandia'®	pink	single	99 (250)
'Rosarium Uetersen'®	pink	double	118 (300)
'Scarlet Meidiland'®	red	double	59 (150)
'Snow Ballet'®	white	double	39 (100)
'Super Dorothy'®	pink	double	118 (300)
'Super Excelsa'®	crimson pink	double	118 (300)
'Swany'®	white	double	39 (100)
'The Fairy'	pink	double	59 (150)

Attractive "wallflower": versatile 'The Fairy'.

Enjoying Roses

For centuries, rose flowers and fruit have delighted the senses, by refining food and beauty products, adorning floral arrangements in vases, and enchanting with their lovely flower perfume. With roses, many meanings are expressed at the same time.

Seeing, Smelling, Tasting Roses

A rose is always ornamental whether it be as a single flower of a native wild rose with its golden yellow stamens surrounded by hovering bees or as a luxuriously elegant hybrid tea rose. How highly the decorative value is prized and what is found beautiful is decided individually by the person viewing the rose. The spherical flower fullness of the English Roses in lush arrangements or the fragrant hybrid teas in a vase, tree rose flowers at eye level or long-stemmed roses as dried flowers—the breadth of aesthetic ways to use roses is simply inexhaustible.

The rose is not only the queen of the flowers in the garden but also in the vase.

Roses for Cutting

Every rose lover has the opportunity of having her or his own cut roses from the garden during the summer months. Most roses will give pleasure in a vase for many days—if they are cultivated and handled properly. Many gardeners have magnificent roses growing in their very own gardens but do not realize how suitable the flowers are for cutting. As so often happens in this life, it is only a matter of knowing how to cut the flowers for numerous small details determine the lasting power of a cut rose after it is picked.

Cut Rose Types: A glance at a table with a selection of cut roses shows that many come from the class of the so-called **hybrid teas.** Hybrid teas are characterized by long stems, predestined for cutting, on which grow large, elegantly shaped, well-doubled flowers. Usually one flower grows per stem. Originally, hybrid teas were produced from crossings with Chinese tea roses, but they soon surpassed them in winter hardiness and therefore superseded the Chinese tea roses.

The variety 'La France', introduced in 1867 by the Frenchman Jean Baptiste Guillot, is considered the first hybrid tea. Today, many hundreds of varieties are available in the trade. In the selection presented here, the robustness of the variety has been given top consideration.

The second interesting group of roses for cutting are the so-called cluster flowered, particularly cluster-flowered varieties with a number of flowers per stem. They consist of varieties from the bedding and area rose classes. Their advantage lies in their high degree of effectiveness. Very lush bouquets can be created with a few, even very short-stemmed flower clusters.

The very doubled shrub varieties of the **English Roses** are made-to-order for arrangements. Even one flower in an arrangement improves the arrangement fantastically. English Roses are difficult to find at floral shops, if they can be found at all. English Roses are very expensive. However, at the same time, they have the highest possible value when used in floral arrangements.

All the varieties mentioned are only a selection, which cannot be representative. Anyone who already has different rose varieties in the garden can evaluate them for cutting by the following rule of thumb. The more doubled the variety is, the longer its flowering life is, as a rule, so the more suitable it is for cutting. Unfortunately, the very fragrant varieties, whose vase life rarely lasts more than seven days, deviate from this rule.

Disbudding: If you grow roses in the garden for cutting, you want to control the development of the remaining flowers by breaking off some buds early, before they mature. In hybrid tea roses—as in thinning the fruit on a fruit tree—breaking off the side buds promotes the better development of the topmost lead bud. This topmost bud is then the only one to consume nutrients and water and consequently grows larger.

In bedding and area roses, which develop clusters of flowers, the procedure is just the opposite. The uppermost lead buds must be removed, and all the side buds are left. If this does not happen, the lead bud opens way before the other side buds. When the stem is then cut, the closed side buds never get to open, and the mass of flowers is lost.

Flower Maturity: Cutting at the proper time determines the duration of the vase life of the cut rose. Who has not rejoiced over a marvelous bunch of roses from the florist and then been angry because the expensive flowers were already hanging their heads and wilting by the very next day? The main reason for this is the fact that the cut-rose specialists sometimes cut

their roses too early in the budding process. These specialists have to be afraid that they will get less from the flower wholesalers for roses that are already in flower, that is, already beginning to open a little, which is really the right time for cutting roses.

The **appropriate time to cut** is when the outer sepals of the flower have opened and point down, while the first petals are slowly beginning to open on the upper edge of the flower. The flower bud feels soft in this stage. A flower that is cut in the hard, budded condition with closed sepals will not open in the vase and will wither ahead of time.

When to Cut: Roses selected for the vase should be cut in the early hours of the morning while nighttime coolness still prevails. Roses should never be cut during the heat of midday for arranging. For one thing, their water content is at the lowest then, and for another, their inner store of nutrients is at its greatest. When cutting during midday, an above-average amount of nutrients and carbohydrates are lost, adversely affecting the further development of the rose plant. These nutrients lost to the cut flower cannot be transported back to the older plant parts and the roots at night.

How Deep to Cut: When cutting, it is best to make sure not to cut the stem too long. Each leaf the rose loses diminishes the surface available for producing the saccharides that are vital for life. The rose carries on respiration through its leaves; the foliage with its chlorophyll content is the plant's "lungs." Only when sufficient leaf mass remains after cutting can the rose keep on growing steadily.

Therefore, only two stems per rose plant, maximum, should be cut at any one time. At least three normally developed, seven-pinna rose leaves should be left on new (this year's) growth. (*Pinna* means that the rose has pinnate leaves whose individual leaflets look like independent leaves. Botanically, however, they are collectively termed a leaf—a pinnate leaf, in fact.) Anyone who cuts into several-years'-old wood has cut way too deep and is endangering the vitality of his or her rose. A useful tool for cutting roses for the vase is the so-called presentation pruning shears, which are described in the section entitled "Times Table of Tools" (beginning on page 177).

Prolonging Vase Life: Evaporation is influenced by leaf surface, air temperature, humidity, light intensity, and air currents. The water intake is accomplished through the conduction pathways of the stem. Studies have shown that roses that were placed in the sun for fifteen minutes after cutting lasted for a considerably shorter

Roses for Cutting

Class	Variety	Color	Flowers	Height (in [cm])
hybrid tea	'Aachener Dom'®	pink	double	24–32 (60–80)
hybrid tea	'Banzai '83'®	yellow	double	32 –39 (80–100)
hybrid tea	'Barkarole'®	red	double	32 –39 (80–100)
hybrid tea	'Carina'®	pink	double	32 –39 (80–100)
hybrid tea	'Christopher Columbus'®	orange	double	24–32 (60–80)
hybrid tea	'Elina'®	yellow	double	32 –39 (80–100)
hybrid tea	'Peace'	yellow/red	double	32 –39 (80–100)
hybrid tea	'Golden Medaillon'®	yellow	double	32 –39 (80–100)
hybrid tea	'Ingrid Bergman'®	red	double	24–32 (60–80)
hybrid tea	'Landora'®	yellow	double	24–32 (60–80)
hybrid tea	'Paul Ricard'®	amber	double	24–32 (60–80)
hybrid tea	'Senator Burda'®	red	double	24–32 (60–80)
hybrid tea	'Silver Jubilee'®	pink	double	24–32 (60–80)
Sprays/Clusters of Flowers:				
bedding	'Bonica '82'®	pink	double	24–32 (60–80)
bedding	'Diadem'®	pink	double	32 –39 (80–100)
bedding	'Europas Rosengarten'®	pink	double	24–32 (60–80)
bedding	'Fragrant Cloud'®	red	double	24–32 (60–80)
bedding	'Make Up'®	pink	double	32 –39 (80–100)
bedding	'Royal Bonica'®	pink	double	24–32 (60–80)
bedding	'Rumba'®	red/yellow	double	24–32 (60–80)
bedding	'Träumerei'®	pink	double	24–32 (60–80)
bedding	'Warwick Castle'	pink	double	24–32 (60–80)
area	'Alba Meidiland'®	white	double	32 –39 (80–100)
area	'Lovely Fairy'®	pink	double	24–32 (60–80)
area	'Scarlet Meidiland'®	red	double	24–32 (60–80)
area	'The Fairy'	pink	double	24–32 (60–80)
Roses for Nostalgic bouquets:				
English Rose	'Abraham Darby'®	apricot	double	59–79 (150–200)
English Rose	'Constance Spry'	pink	double	59–79 (150–200)
shrub	'Eden Rose '85'®	pink	double	59–79 (150–200)
English Rose	'Graham Thomas'®	yellow	double	39–59 (100–150)
English Rose	'Heritage'®	pink	double	39–59 (100–150)
bedding	'Leonardo da Vinci'®	pink	double	24–32 (60–80)
English Rose	'Othello'®	red	double	39–59 (100–150)
shrub	'Polka '91'®	amber	double	39–59 (100–150)
English Rose	'The Squire'	red	double	39–59 (100–150)
English Rose	'Wife of Bath'	pink	double	32 –39 (80–100)

Disbudding: Early pinching out side buds (see arrows) using the thumb and forefinger allows the main bud of the hybrid tea to grow larger.

time than normal. This period permitted the water in the conduction pathways to flow out and caused an obstructing vacuum to develop there.

Therefore, after cutting, one should immediately (!) place the stems into a pail filled up to the halfway level with lukewarm water. Warm water facilitates the water uptake. Ideally, the pail should then be placed into a cool, shady spot, e.g., a cellar area, in which the cut roses can cool down for about four hours. This slows down the physiological breakdown processes inside the stem considerably. If there is no chance of placing the roses into water immediately, they should first be wrapped in damp newspaper and then placed out of the direct sunlight. Never pour water onto cut roses from above. The water droplets trapped inside the flowers will very quickly develop into rotten spots. The following tips apply not only to cut roses from your own garden but also to ones bought from the store.

How to Keep Cut Roses Longer

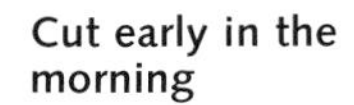

Cut early in the morning

Place flowers into water immediately

Cut stems on the diagonal

Dip stems in boiling water for 3 seconds.

All of the prickles and leaves that would be underwater are removed from the rose stem so that no rose leaves will be in the water. The best way for removing prickles has proven to be with a special hand-protecting tool, the so-called thorn stripper. Furthermore, the ends of the stems are cut with a sharp knife on a diagonal line about 1½ to 2 in. (4 to 5 cm) long so that nutrients and water can enter over the largest possible surface. Cutting stems on the diagonal in this manner ensures that the stems do not stand flat on the bottom of the vase.

The best final move in the preparation of roses for the vase is for the bunched stem ends now to be dipped briefly—for about 3 seconds—into boiling water. This is not an April Fool's joke but a method used by cut-flower specialists to extend the life of cut roses. As soon as the stems are cut, small air bubbles immediately enter the conduction pathways of the stem through the cut—the vacuum already warned about. These small air bubbles plug up the water channels. This procedure is something like an embolism in a human being, in other words when air gets into the bloodstream. By dipping the cut rose stems into boiling water for 3 seconds, the boiling water draws the bubbles out of the conduction pathways, leaving the way free again for the cut stem to be further supplied with water.

The containers and vases in which cut roses are going to be placed must be scrupulously clean. The prepared cut rose bouquet tolerates no direct sunlight and no drafts. The water in the vases should be changed daily, again to keep the conduction pathways open. Fully open roses need considerably more water than do budded stems.

Preservatives: Stores offer cut-flower preservatives in the form of tablets or powder. The gardener stirs these preservatives into the water in the vase. Their disinfectant action lengthens the life by several days of a cut rose that has been

Rosy arrangement with lady's-mantle, delphinium, sage, and syringa. Tip: Cut roses last appreciably longer if one (1) uses clean containers, (2) lets clean, lukewarm water run in slowly (without oxygen buildup), (3) does not place the arrangement in direct sun or in drafts, and (4) puts the vase in a cool place at night.

cut and prepared properly. Bacteria in the vase water, which plug the conduction pathways, are killed by the preservatives. However, these preparations do not perform miracles, and they can compensate for poor previous handling of cut roses only to a certain degree. Moreover, the directions on the package as to quantities to use should be followed exactly, for too much of the preservatives will injure the roses rather than keep them fresh. You can also make your own preservative. A home recipe calls for combining the following ingredients: 1 quart (liter) of water, 1 teaspoon of apple or other fruit vinegar, 1 tablespoon of sugar.

TIP If long-stemmed roses wither too soon, you can freshen them by leaving the roses overnight in a bathtub filled with $^{3}/_{4}$ to 1 inch (2 to 3 cm) or so of water. The drooping roses are then recut and thus renewed plugging of conduction pathways with air bubbles is prevented.

'Paul Ricard'®—its unusual anise scent is unique and amazingly reminiscent of a refreshing pastis.

Guiding Flower Production: In the outdoor culturing of flowers for cutting, there is the problem that many varieties bloom all at once. To spread the harvest over a longer period, so-called **pinching back** of young shoots has proven useful. About three or four weeks before the main blooming season—as desired—a portion of the growing canes are shortened. Canes pinched back in this way are forced to put out new growth again. Their flowers are delayed for about six to seven weeks.

End of the Cutting Season: After October 1, roses should not be cut anymore so that the rose can rest and mature.

Fragrant Roses—Roses with Soul

In many respects, scents exert powerful stimuli in our lives. However, our language provides only a few terms for the over 3,000 different scents that we can smell. Thus fragrances are, in the truest sense of the word, indescribable to us. We perceive them, but we lack the words to describe them precisely. This speechlessness is connected to the fact that the sense of smell is the only one of our senses linked directly with the emotional center in the brain, bypassing the speech center. All our other senses initiate elaborate processing responses in the speech center when we receive stimuli. Therefore, we link particular scent notes not with descriptive words but with memories. For instance, when we smell a new straw hat, we do not conjure up the word hay in our minds. Instead, the scent may awaken the memory of a country walk that led across fields that smelled of fresh hay. Smells call up recollections, but on the other hand, we are not able to recollect smells. They have a fleeting character, even in the memory. This occurs although the more than 100 million densely packed receptors in the olfactory mucous membrane of the nose receive scent molecules constantly. The scent stimuli released there travel over some of the 20 million olfactory nerve fibers directly to the limbic system of our brain.

The limbic system is developmentally the oldest part of the human brain and plays a vital role in the arousal and control

Cut Roses—Flower Maturity

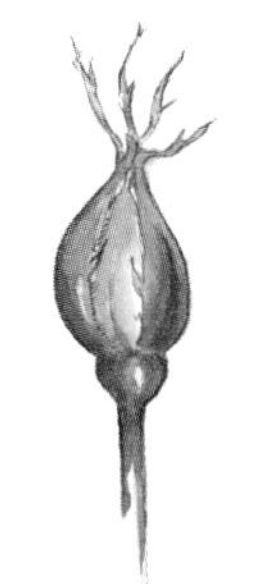

Wrong: Roses cut in the bud will not open.

Bud just before the ideal cutting time

Proper time to cut a rose in summer.

Opening bud: Good for cutting in cooler seasons.

Wrong: Opened roses have only a short vase life ahead of them.

of feelings. Among other things, it has direct influence on an endocrine gland that regulates the secretion of sexual hormones. Even before such things were established scientifically, people recognized the connection between scents and sexual stimulation. So they placed their hopes in the stimulating smell of musk perfume, whose basic ingredient came from the male glands of the musk deer. Sometimes they used the high portion of sexual pheromones in truffles to trick the searching sow into thinking a rutting boar was in her vicinity. Even Napoleon believed in the aphrodisiac effect of body odors. Each time he was getting ready to return from a campaign, he would write to Josephine, "Don't wash—I'm coming home!"

Modern researchers have established that in earlier times, women were more aware of a "stink" than men; women definitely have a better sense of smell. This probably has to do with their higher concentration of the female hormone estrogen, which activates the olfactory receptors. When the concentration of this hormone reaches its high point during ovulation, a woman's sense of smell is also most pronounced.

Rose Fragrance: What is rose fragrance? The world-famous dendrologist Gerd Krüssmann describes its chemical structure in his definitive work, *Rosen, Rosen, Rosen* as follows: "Aromas are volatile secretion products of certain glands or individual gland cells or entire organs (flowers, leaves). The fragrance is often a mixture of several kinds of smell. . . . . Aromas are unsaturated and aromatic alcohols, aldehydes, fatty acids, phenols, carbonic acid, and their esters, essential oils, and resins." Moreover, a chief component is geraniol. Thus, the cranesbill *(Geranium)* also smells a little like roses and is used in the manufacture of an inexpensive rose oil variant.

Rose analyst Krüssmann's reminder of the chemistry classes of long-gone school days is the general formula for the rose fragrance, which for many creatures, especially humans, sends out the highest possible attractive power. Anyone who gets close to a rose immediately bends over and smells its flowers. One expects scent from

Roses with Fragrance

Class	Variety	Color	Flowers	Height (in [cm])
bedding	'Fragrant Cloud'®	red	double	24–32 (60–80)
bedding	'Manou Meilland'®	pink	double	24–32 (60–80)
bedding	'Rose de Rescht'	red	double	32–39 (80–100)
bedding	'Sunsprite'®	yellow	double	24–32 (60–80)
bedding	'Träumerei'®	pink	double	24–32 (60–80)
hybrid tea	'Barkarole'®	red	double	32–39 (80–100)
hybrid tea	'Deep Secret'®	red	double	32–39 (80–100)
hybrid tea	'Duftrausch'®	lavender	double	32–39 (80–100)
hybrid tea	'Elina'®	yellow	double	32–39 (80–100)
hybrid tea	'Eroica'®	red	double	32–39 (80–100)
hybrid tea	'Fragrant Gold'®	yellow	double	24–32 (60–80)
hybrid tea	'Karl Heinz Hanisch'®	cream	double	24–32 (60–80)
hybrid tea	'Loving Memory'®	red	double	24–32 (60–80)
hybrid tea	'Papa Meilland'®	red	double	24–32 (60–80)
hybrid tea	'Pariser Charme'	pink	double	24–32 (60–80)
hybrid tea	'Paul Ricard'®	amber	double	24–32 (60–80)
hybrid tea	'Polarstern'®	white	double	32–39 (80–100)
hybrid tea	'The McCartney Rose'®	pink	double	24–32 (60–80)
area	'Immensee'®	pink	single	12–16 (30–40)
area	'Lavender Dream'®	lavender	semidouble	24–32 (60–80)
wild rose	*Rosa gallica*	pink	single	32–39 (80–100)
wild rose	*Rosa pimpinellifolia*	cream	single	32–39 (80–100)
climber	'Compassion'®	pink	double	79–118 (200–300)
climber	'Ilse Krohn Superior'®	white	double	79–118 (200–300)
climber	'Lawinia'®	pink	double	79–118 (200–300)
climber	'Morning Jewel'®	pink	semidouble	79–118 (200–300)
climber	'New Dawn'	pearly pink	double	79–118 (200–300)
climber	'Sympathie'	red	double	79–118 (200–300)
rambler	'Bobbie James'	white	semidouble	118–197 (300–500)
rambler	'Paul Noël'	pink	double	118–197 (300–500)
rambler	*Rosa rugosa*	pink	semidouble	118–197 (300–500)
rugosa hybrid	'Frau Dagmar Hartopp'	pink	single	24–32 (60–80)
rugosa hybrid	'Foxi'®	pink	double	24–32 (60–80)
rugosa hybrid	'Yellow Dagmar Hastrup'®	yellow	semidouble	24–32 (60–80)
rugosa hybrid	'Pierette'®	pink	double	24–32 (60–80)
rugosa hybrid	'Polareis'®	pink	double	24–32 (60–80)
rugosa hybrid	'Polarsonne'®	red	double	24–32 (60–80)
rugosa hybrid	'Schnee-Eule'®	white	double	24–32 (60–80)
shrub	'Abraham Darby'®	apricot	double	59–79 (150–200)
shrub	'Constance Spry'	pink	double	39–59 (100–150)
shrub	'Graham Thomas'®	yellow	double	39–59 (100–150)
shrub	'Heritage'®	pink	double	39–59 (100–150)
shrub	'Ilse Haberland'®	pink	double	39–59 (100–150)
shrub	'Kazanlik'	pink	semidouble	59–79 (150–200)
shrub	'Lichtkönigin Lucia'®	yellow	double	39–59 (100–150)
shrub	'Louise Odier'	pink	double	59–79 (150–200)
shrub	'Maigold'	yellow	double	59–79 (150–200)
shrub	'Othello'®	red	double	39–59 (100–150)
shrub	'Polka '91'®	amber	double	39–59 (100–150)
shrub	*Rosa centifolia* 'Muscosa'	pink	double	32–39 (80–100)
shrub	'Souvenir de la Malmaison'	pink	double	32–39 (80–100)
shrub	'The Squire'	red	double	39–59 (100–150)
shrub	*Rosa gallica* 'Versicolor'	pink	semidouble	39–59 (100–150)
shrub	'Westerland'®	apricot	semidouble	59–79 (150–200)
shrub	'Wife of Bath'	pink	double	32–39 (80–100)

roses. If it is not there, disappointment sets in. The Australian rosarian A. S. Thomas expressed it precisely, "A rose without fragrance is no less beautiful, but it is less attractive." Why should roses smell at all? There is basically no luxury in nature. The fragrance of certain roses must therefore have a rational, ongoing importance beyond our aesthetic perception. In the first place, it attracts insects. For bees and bumblebees, the essential oils of the often nectar-poor flowers offers a true feast. In the early-morning hours, when the rose fragrance is at its most intense, they fly around the flowers in bustling excitement. This pleasure is paid for by the pollination of the flowers. A further reason for the existence of the rose fragrance is the partially repelling effect of the essential oils on fungus, insects with evil intentions, and enemies that might chew the flowers.

Area rose 'Lavender Dream'®—not only hybrid teas but also certain varieties of other groups are fragrant.

Can rose fragrance be described? Anyone who has various fragrant roses in the garden might sometime want to set up a private fragrance seminar with friends, perhaps in conjunction with a pleasant wine tasting. This would quickly show how hard it is to try to describe a rose fragrance—and a wine bouquet. When attempting to describe the scent, begin with the characteristic so-called lead fragrance of a variety and then describe the central and bass notes, the so-called bouquet. In this regard, the French perfume specialists provide help, having coined the various terms with a fragrance that can be depicted: perhaps a fruity, animal, metallic, powdery, or grassy scent, a scent of anise, apple, or lemon. Also, in her book *Begegnung mit Rosen (Introducing Roses),* Alma de l'Aigle characterized in great detail and very imaginatively the fragrance notes of many rose varieties. She has integrated terms like *warmly smoky, flowery sweet, forest floor in fall,* and *sunbathed girl's skin* into her fragrance vocabulary. The title still provides tips to fans of scented roses, even forty years after its publication.

Fragrance strengths and fragrance notes can vary in scented roses of one and the same variety. Fragrances change with the type of soil, the location, the time of day, and the developmental stage of the plant, among other things. The Greek philosopher Theophrastus—a pupil of Aristotle, who is considered the father of botany and lived from 371 to 287 B.C—was already writing, "The intensity of color and fragrance can be different from one place to another; also rosebushes (of the same species) that grow on the same piece of ground do not smell the same." These observations concur with the modern experiences of rosarians who, with many fragrant varieties, have discovered an increase in the intensity of a fragrance from plants grown in loamy to clayish soils.

Which roses smell? The table on page 116 names a small selection of fragrant roses. The assortment does not limit itself merely to the fragrant hybrid tea roses—the classic perfume roses. It also includes wild roses, climbers, bedding, shrub—yes, even area roses with a strong fragrance. Thus, any gardener can choose the appropriate fragrant rose for his or her individual garden situation.

One more note: It is often said that an old rose means the same thing as a fragrant rose. That is a fairy tale. With the old varieties as with the newer ones, a similar proportion of them are fragrant and nonfragrant. Perhaps only a quarter of the rose varieties of the last century possessed real scent. This proportion seems higher today because it is primarily those of the old varieties with fragrance that are being sold today and those without fragrance have been dropped.

Fragrance in Rose Breeding: Of course, every rose breeder knows that fragrance is an appealing sales argument for a new variety. Unfortunately, genetics today, as at the beginning of systematic rose breeding, has established no clear boundaries. Fragrance genes are inherited recessively, which means that other genes dominate them. Often, the offspring of fragrant parents have no scent at all. However, the exact opposite also happens frequently. In addition, the genes present for scent are often coupled with particular genes of other characteristics like color or disease resistance. Lilac-colored varieties, the so-called blue roses, almost always have scent but also suffer from very high susceptibility to fungal diseases. Yellow/apricot is also considered a fragrant color. (Nevertheless, it is often insufficient in winter-hardy varieties.) Fragrance is hard to introduce into orange varieties. Red varieties often have fragrance. In spite of all their successes, rose breeders still have a long way to go to achieve new varieties that are fragrant and, at the same time, vigorous and resistant.

Rose Oil Production: Anyone who wants to read about the old methods of harvesting perfume raw materials detailed in an exciting, entertaining, and amusing way should read Patrick Süskind's fascinating detective story *Das Parfüm (The Perfume).* In it, the novel's protagonist, Jean Baptiste Grenouille, extracts an essence from the body odors of murdered maidens. When he himself is sprayed with an inadvisably high dose of this perfume, he ———. Every reader and lover of fragrance has to read the story to find out what comes next.

The word *perfume* comes from the Latin expression *per fumen,* which means *through smoke.* To stabilize the volatile scents, even in antiquity, the petals of roses were mixed with oil or fat. Even today, in the perfume capital of the world—Grasse in southeastern France, the **enfleurage** is carried out according to the principles of these old methods. For example, rose petals are strewn onto tablets of beef suet. The suet fixes the essential oils, which are later extracted with alcohol.

Newer, more effective methods of producing essential oils are extraction and distillation. For **extraction,** the dried raw materials are dampened and stacked on top of one another on sieve plates in stainless steel barrels. With the addition of the solvent hexane at room temperature, the plant cells break open. With rose petals, this procedure lasts 7 minutes; it is repeated several times. After that, the contents of the barrels are heated to 176°F (80°C), the hexane evaporates, and the so-called concrete is left behind in a vacuum. The concrete is the first step of the extraction, a waxlike mass with the essential oils. From this, the absolute—the pure essence—is retrieved by dissolving in alcohol and by a frost treatment at –31°F (–35°C).

In **distillation,** hot steam withdraws the the essential oils from the fresh flowers. The precious steam is captured, cooled in a so-called Florence flask, and condensed. The lighter, utterly pure essential oils float on the surface of the condensation water and are siphoned off. This method is—except for roses—primarily used for lavender and sandalwood.

Most of the flowers used for producing rose oil come from the so-called oil rose, *Rosa* x *damascena* 'Kazanlik' (see picture, page 200). Barely half the worldwide rose production currently comes from Ukraine, a quarter from Bulgaria, and the rest from Turkey and Morocco, among others. The harvest of the semidouble flowers takes place in the early-morning hours, starting in May. To make 2.2 pounds (1 kg) of rose oil, about 6,614 pounds (3,000 kg) of flowers are needed. Pure rose oil rewards the effort; it costs about twice as much as gold. For example, Bulgaria has stored hundreds of barrels of rose oil in foreign banks as security for its trade balance.

The Rosy Fragrance Garden

The design for a rosy fragrance garden might look like the following: In March, the winterhazel *(Corylopsis)* opens the fragrant procession, followed by the early spring *Rhododendron* x *praecox* and the daphnes (*Daphne* species and varieties). In May and June, the fragrant azaleas finish the spring, and they also provide fiery fall color with their foliage. In summer the roses take over the fragrance paradise—English Roses and fragrant bedding and hybrid tea roses beguile the nose. Behind a bench, the Apothecary Rose creates a scented backdrop. The strongly fragrant honeysuckle *Lonicera* x *heckrottii* decorates the top of the wall along with rambler roses. In the meantime, by acting as fillers, catnip *(Nepeta),* sage *(Salvia),* aubrieta *(Aubrieta),* moss pinks *(Phlox subulata),* and—last but not least—lily of the valley *(Convallaria)* cast their spells.

Old-fashioned Roses

The term *old-fashioned roses* embraces the Old, romantic, and English Roses. Really, these old-fashioned, nostalgic roses do not refer to any one individual rose class, for many different growth habits are represented, in particular a number of varieties with shrub habits. Common to all

Old and Romantic Roses

Class	Variety (Year Introduced)	Color	Fragrance	Height (in. [cm])
bedding	'Gruss an Aachen' (1909)	cream		16–24 (40–60)
bedding	'Leonardo da Vinci'® (1993)	pink		24–32 (60–80)
bedding	'Rose de Rescht' (unknown, before 1880)	red	fragrant	32–39 (80–100)
shrub	'Eden Rose '85'® (1985)	pink		59–79 (150–200)
shrub	'Ghislaine de Féligonde' (1916)	yellow	fragrant	59–79 (150–200)
shrub	'Jacques Cartier' (1868)	pink	fragrant	39–59 (100–150)
shrub	'Polka '91'® (1991)	amber	fragrant	39–59 (100–150)
shrub	'Raubritter' (1936)	pink		79–118 (200–300)
shrub	'Louise Odier' (1851)	pink	fragrant	59–79 (150–200)
shrub	*Rosa centifolia* 'Muscosa' (1796)	pink	fragrant	32–39 (80–100)
shrub	'Souvenir de la Malmaison' (1843)	pink	fragrant	32–39 (80–100)
climber	'Gloire de Dijon' (1853)	orange/cream	fragrant	79–118 (200–300)
English Rose	'Abraham Darby'® (1985)	apricot	fragrant	59–79 (150–200)
English Rose	'Constance Spry' (1960)	pink	fragrant	59–79 (150–200)
English Rose	'Graham Thomas'® (1983)	yellow	fragrant	39–59 (100–150)
English Rose	'Heritage'® (1984)	pink	fragrant	39–59 (100–150)
English Rose	'Othello'® (1986)	red	fragrant	39–59 (100–150)
English Rose	'The Squire' (1977)	red	fragrant	39–59 (100–150)
English Rose	'Warwick Castle' (1986)	pink	fragrant	24–32 (60–80)
English Rose	'Wife of Bath' (1969)	pink	fragrant	32–39 (80–100)

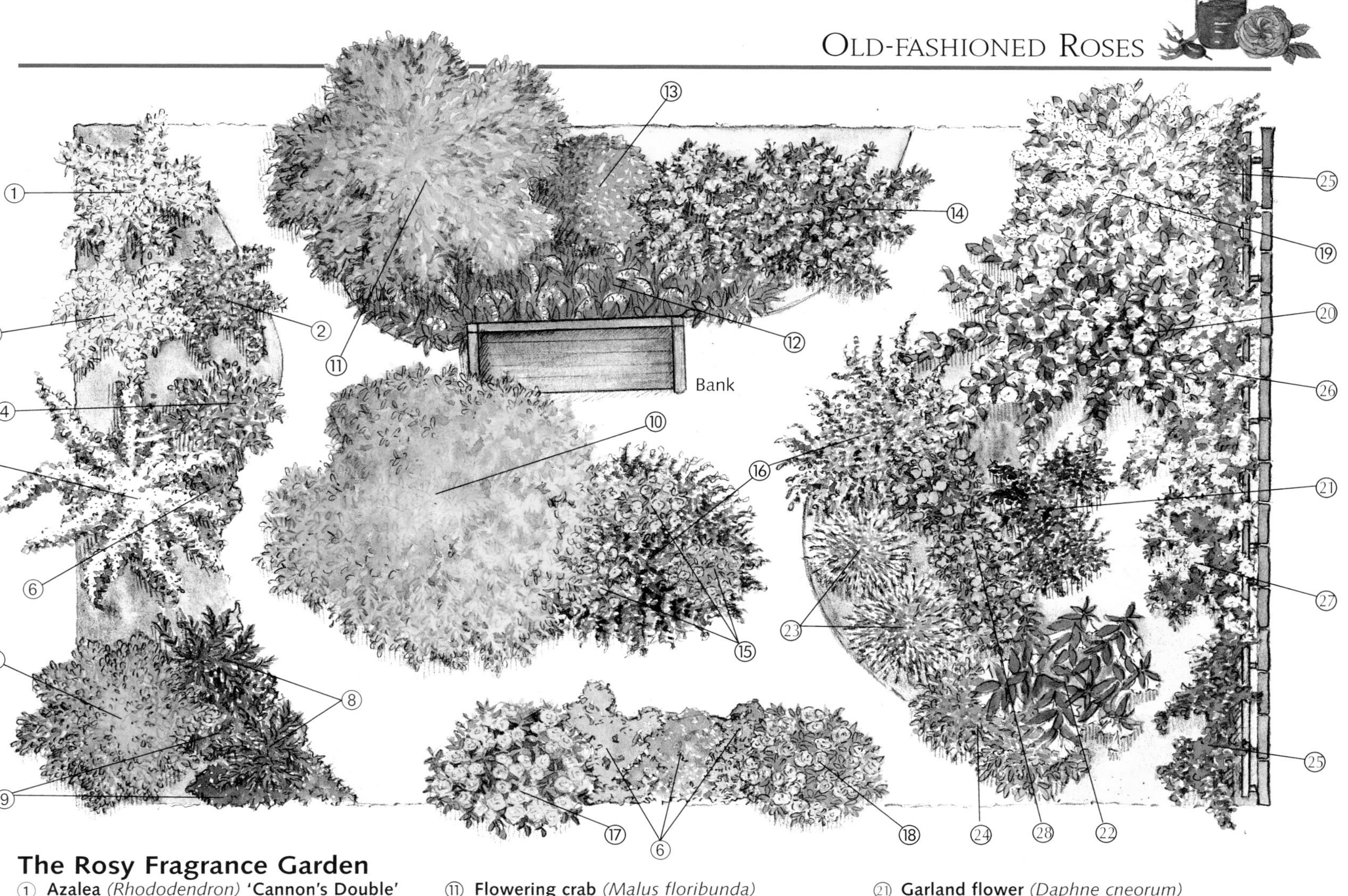

The Rosy Fragrance Garden

1. **Azalea** *(Rhododendron)* **'Cannon's Double'**
2. **Azalea** *(Rhododendron)* **'Sarina'**
3. **Azalea** *(Rhododendron)* **'Phebe'**
4. *Rhododendron* x *praecox*
5. **Bridal wreath** *(Spiraea* x *vanhouttei)*
6. **Moss pink** *(Phlox subulata* **varieties) in colors**
7. **Winterhazel** *(Corylopsis spicata)*
8. **February daphne** *(Daphne mezereum* **'Select')**
9. **Aubrieta** *(Aubrieta* **hybrids)**
10. **Black locust** *(Robinia pseudoacacia* **'Umbraculifera')**
11. **Flowering crab** *(Malus floribunda)*
12. **Lily of the valley** *(Convalleria)*
13. **Boxwood sphere** *(Buxus)*
14. **Apothecary Rose** *(Rosa gallica* **'Officinalis')**
15. **Bedding rose 'Friesia'®**
16. **Catnip** *(Nepeta* x *faassenii)*
17. **English Rose 'Graham Thomas'®**
18. **English Rose 'Abraham Darby'®**
19. **Common lilac 'Madame Lemoine'** *(Syringa vulgaris* **varieties)**
20. **Mock orange** *(Philadelphus coronarius)*
21. **Garland flower** *(Daphne cneorum)*
22. **Thorn apple, white-flowered, in tub** *(Datura)*
23. **Lavender** *(Lavender* **varieties)**
24. **Sage** *(Salvia* **varieties)**
25. **Honeysuckle** *(Lonicera* x *heckrottii)*
26. **Rambler 'Bobby James'**
27. **Rambler 'Paul Noël'**
28. **Hybrid tea 'The McCartney Rose'®**

the varieties of Old, romantic, and English Roses are the dream doubled, rosette- or balloon-shaped spheres of flowers, which magically rivet the eye. In contrast with many modern hybrid tea roses, which develop markedly pointed flowers and usually are most fascinating when they are just opening, the old-fashioned roses become more beautiful with each day as the flowers open further. In addition, many varieties exude a wonderful fragrance. Just a few blossoms of these roses are enough to transport one back to grandmother's garden.

Romantic and Old Roses: Although no other plants have been so severely manipulated by breeders in this century as the rose and marvelous new varieties with many advantages have been developed, no small number of gardeners are still drawn to the old varieties.

Certainly, the longing for grandma's rose garden or the cosy kitchen garden plays an important role in the expression "good old varieties." Also, many rose lovers find the old colors softer than the "garish," eye-blinding shades of modern varieties. In fact, pink, the main color in the realm of Old Roses, is of unsurpassable purity in some varieties. One more thing is certain: Old Roses have charm!

New rose forms are always a mirror of the spirit of the times. Old Roses, for example, reflect the period of their development with lush fullness of flower and scent. They recall the splendid ball gowns of their period, fantasies of silk and velvet, trimmed over and over with ruffles, visual high points of sweeping opera balls and splendid festivities.

But when is a rose considered old? The so-called Old Roses are classified on the basis of their year of introduction. In this respect, the term has been precisely defined. The American Rose Society declared in 1966, "A rose is an **old rose** if it belongs to a class that already existed before 1867, the year of the introduction of the variety 'La France' as the first hybrid tea."

This ruling does not satisfy anymore these days. After 'La France' and in the last twenty years, especially, roses have arisen that have the flair of older varieties—we need think only of the English Roses. Many rose lovers therefore do not embrace this strict definition and see the term *Old Roses* as a description of style rather than of period. Since the beginning of this century, rose breeders have produced a series of roses that do not in any way lag behind the Old Roses in flair, and thus, these are gathered together into one group. These new Old Roses are termed **romantic roses.** Like their forebears from earlier centuries, they enchant the viewer and try to awaken a romantic sense of life—a far cry from the current hectic state. Old and new should go together as a matter of course—almost as in human cooperation.

The gardener expects truly fantastic things of Old Roses and romantic roses—after having his or her interest awakened by articles and fantastic illustrations in many magazines and catalogs. Disease resistance, fragrance, and continuous flowering are prerequisites, especially in view of the selling price, which is often almost double the price of modern roses. Their price is high because, in spite of the renaissance of very double rose varieties, the demand—measured against that of other varieties—is relatively modest for Old Roses and romantic roses. Correspondingly, the numbers of plants produced in the nurseries—oftentimes specialty operations that deal only with this segment of the rose market—are relatively small. What is produced in only small quantities usually costs more than mass market products.

Nevertheless, Old Roses are roses just like all the others. They do not remain unharmed by the vicissitudes of a rose's life and can be infected with disease exactly like their younger varietal colleagues. Some old varieties are even very susceptible, especially to powdery mildew and spot anthracnose. As a rule of thumb, a gardener should be more careful when selecting a variety that scintillates with repeated flowering, doubled flowers, and strong fragrance.

Many Old Roses develop almost no hips. In the first place, the stamens (with the male pollen) of the flowers have been transformed into petals, as in the improbably double Centifolias; thus they no longer develop fruit. However, one thing is considered a charter claim of the Old Roses: They are probably not always the most resistant or the most ecological roses; however, their toughness, which enables them to survive in a wild garden uncared for even for decades, is unarguable. The selection in the table on page 118 primarily considers such vigorous and robust varieties.

Damask rose for beginners: 'Rose de Rescht'.

The history of Old Roses begins at the end of the Middle Ages. At that time, rose fragrance and doubled flowers reached Europe from the Near East, China, and India and marked the beginning of the first rose breeding. Since then, numerous groups of Old Roses have developed. To treat them all in detail would fill a fat book all by itself. Therefore, here only the most important classes and the proven varieties for beginners are included.

Bourbon Roses: The first Bourbon roses and the later varieties descended from them probably developed at the beginning of the nineteenth century from crossings of Damascus with China roses on the island Île de Bourbon (today La Réunion) in the Indian Ocean. French settlers bounded their fields with rose hedges and found these seedlings. Later, the rose came to France, where it was further crossed to produce the class of Bourbon roses. An important variety in this group is 'Souvenir de la Malmaison'. For many, this is the most beautiful Old Rose. 'Louise Odier', the darling of Victorian England, is one of the most certainly remontant of the Old Roses.

Damask Roses: A distinction is made between two kinds of Damask roses. The first variant, which blooms in the summer, is an offspring of *Rosa gallica* and *Rosa phoenicia.* The second, the fall Damask rose, which blooms in the summer and then again in the fall, probably arose from a crossing of *Rosa gallica* and *Rosa moschata.* Typical of the Damask varieties are the soft, gray-green foliage and the heavy, lush scent, which they have passed along to many of the rose varieties descended from them. One of the best varieties in this group is 'Rose de Rescht'. It is absolutely winter hardy, strongly fragrant, and repeat flowering and has received a respectable portion of robustness from the Gallica line. 'Rose de Rescht' is a reliable beginner's variety for all fans of the Damasks—and for those who want to become so.

Portland Roses: The Portland roses are extremely close relatives of the Damasks. This group probably resulted from a crossing of the recurrent fall Damask rose with *Rosa gallica* 'Officinalis', the Apothecary Rose. Portland roses are very good for the home garden; in particular, 'Jacques Cartier' (which is often classified with the Damasks) is worth a try. Its foliage is dense and is considered healthy; so, the **tip** for all Portland novices is to grow 'Jacques Cartier'.

Centifolias and Moss Roses: The Centifolias are considered purely and simply the embodiment of the Old Roses. They have lushly doubled flowers and a fantastic fragrance. This group of roses has the honor of its fragrance—the *Centifolia fragrance*—being a fixed element in numerous fragrance descriptions of modern varieties. However, one should know that the Centifolias become a good $6^1/_2$ feet (2 m) tall; in addition, the shrub tends to fall apart. A variety of the Centifolias, the moss rose, grows more compactly. Due to a mutation (i.e., a spontaneous change) of the flower shoot, the moss roses have a mosslike covering of dense glands on their pedicels, ovaries, and calyxes. Probably the most beautiful of these roses would be the extremely fragrant *Rosa centifolia* 'Muscosa', the monastery garden rose of the olden days, immortalized in many paintings by Flemish masters.

Noisette Roses: This class of repeat-flowering current roses goes back to Frenchman Louis Noisette. By using seedlings from his brother living in the United States, Louis began breeding these roses at the beginning of the nineteenth century. The class has low and climbing varieties.

Tea Roses: These roses are part of the Old Rose group and have very doubled flowers and extraordinary scent. A climbing rose for anyone who might like to try this class is 'Gloire de Dijon', a robust, moderately winter-hardy tea rose.

Alba Roses: Last but not least, the procession of the important classes of the Old Roses leads to the "white" rose—*Rosa* x *alba* and its forms. Even in ancient times, the Romans and Greeks were fascinated by it. It is the oldest garden rose known. The forms and varieties bred since then display very good winter hardiness, robust growth, and outstanding fragrance. The porcelain pink variety 'Maiden's Blush', a once-flowering miracle of fragrance with compact growth is recommended.

The Hobby of Old Rose Collecting

Anyone who is susceptible to beautiful old things, is a passionate collector of history, and has a garden has here a new, rich lode of discovery opening before him or her. Be careful. Great danger lurks when one becomes infected with the virus of an unusual hobby that, once it has taken hold of a person, can become an obsession.

Ideal Bourbon rose: 'Louise Odier'.

Healthy Portland rose: 'Jacques Cartier'.

Mildew-sensitive: *Rosa centifolia* **'Muscosa'.**

Fragrant *Rosa alba:* **'Maiden's Blush'.**

Old Roses are living antiques. The scouting trip through the rose classes shows what breadth the different groups have in themselves, not to mention the many, many varieties. It is impossible to maintain all of these souvenirs of the past commercially. The already-mentioned specialty nurseries for Old Roses of course try to offer an astonishingly large assortment of many hundreds of varieties within this framework. Nevertheless, certain varieties today are definitely rarieties whose existence is now known only in collectors' circles.

Therefore, particularly passionate rose collectors often propagate their varieties themselves. They want to acquire them without having to buy any more, want to keep renewing the vitality of their valuable collectors' items through reproduction, and want to obtain sought-after rose rarities as private, priceless presents for friends and other enthusiasts.

What should be kept in mind when building a collection? The selection of varieties must be left to the individual. Declaring what should constitute a standard asssortment would run counter to any idea of individuality. However, one should first poke around in the individual classes of the Old Roses. It is best to buy the proven, usually easily manageable varieties like the classics mentioned. This is a reasonable investment with which one can first gain some hands-on experience.

Getting experience in working with roses is essential. For one thing, the gardener must learn to check the available sites in the garden for their suitability to grow Old Roses. In addition, experience will teach the gardener how to evaluate better how much time he or she is going to be willing to invest in any future rose collection. One thing must be clear to any rose collector—the more rose antiques that are concentrated in a small garden area, the more maintenance will be required and the more time it will take. Maintaining a historical rose treasury is a time-consuming undertaking. Anyone who observes well-maintained rose collections with a sharp eye can get an idea of what an enormous amount of hand labor is behind all the glory displayed in a garden full of Old Roses. A true rose lover should not shy away from the amount of care, digging, pruning, tying, and leaf gathering, but should regard it as rosy pleasure gained.

'Constance Spry'—The first of David Austin's English Roses (1961), not yet repeat-flowering.

'Heritage'®—The best English Rose for beginners.

'Othello'®—An English Rose for the experienced.

'Graham Thomas'®—An English Rose with the most beautiful shrub growth habit for the medium-sized garden.

Once the first, positive experience is garnered and the urge to go on has been kindled, it is a good idea to connect with people of similar mind. The regular meetings of the local chapters of the ARS (American Rose Society) provide a forum for this. The exchange of information with other rose lovers in social clubs offers the great advantage of getting information about varieties in light of the regional weather and soil conditions. This information cannot be found in any reference book and having it avoids many disappointments in building a private rose collection.

English Roses

The life work of the English rose breeder David Austin is English Roses, the kitchen garden rose of the modern day. For over 35 years, Austin—born in 1926 and originally an agriculturalist—busied himself with specialized breeding of new rose varieties that would link the special charm and seductive rose perfume of grandmother's rose varieties with the repeat-flowering and robustness of modern forms. In addition, many of the very double Old Roses are fair-weather roses, that is they are very sensitive to rain. Their petals stick together and form ugly, gray-brown flower mummies. Austin's English Roses, on the other hand, can be classified as proof against rain, for even in wet years the flowers remain attractive.

The worldwide triumphal procession of the English Roses began in 1961 with the once-flowering variety 'Constance Spry'. Austin had crossed the modern Floribunda rose 'Dainty Maid' with the old Gallica 'Belle Isis'.

This first success already illuminated the doctrine of Austin's rose breeding. His critical, brilliant basic idea was, and is, to combine the slumbering, never really fully utilized breeding potential of Old Roses with the developmental leaps of modern breeding work. It was as if Leonardo da Vinci had had a computer and modern machines available to perfect and realize his technical fantasies that were so very far ahead of his time.

David Austin may be seen as the most successful breeder of garden roses in recent years. Probably most rose lovers in the world possess at least one Austin variety.

Naturally, one cannot recommend all the English Rose varieties sight unseen for all climate conditions. The very even Atlantic climate of the island of Britain allows varieties to thrive that would be ruled out in the latitudes of Germany. For instance, in Saxony, the almost continental-type climate with its sometimes extremely hot summers and periods of polar winter weather would not allow varieties to grow there that would grow in Britain. The selection of Austin varieties in the table on page 118 takes this into consideration and gives preference to varieties that have proven themselves in the climate zone encompassing Germany.

The term *English Roses* does not refer to a firmly defined class such as bedding or shrub roses. Austin classifies his English

The tried-and-true shrub rose 'Schneewittchen ('Iceberg')—selected as a world rose in 1983—is also problem-free and, in tree form, is good for beginners.

Roses as also bedding or shrub or climbing roses of varying heights. Not all have fragrance, and there are also once-flowering varieties—if only a few—among them. Most of the new English Roses introduced here, however, are very double, fragrant shrub roses. Austin also succeeded in creating new color nuances. The amber of a 'Graham Thomas' or the crimson of 'The Squire' is not really found among the Old Roses, so that English Roses have also broadened the color range.

Standard Roses

Standard, or tree, roses are an art form of rose culture with a long history. In earlier times, the grafting of standard roses was counted as a high garden art. Standard roses were ideal for the severe symmetry of formal, noble garden parks. They stood like tin soldiers in geometric beds, a precisely positioned guard of honor. This formal stiffness increasingly lost its charm and, as a result, the standard rose more and more disappeared from gardens.

Today, standards—like the tree forms of other shrubs—are experiencing a renaissance. There are reasons for this. The average size of gardens is constantly diminishing, but with standard roses, one can also design a small area. Balconies and terraces are increasingly larger, and standards in tubs offer excellent design possibilities for them. Not least, standard roses today are considered harmonious elements of the kitchen or potager garden, a popular garden form with great tradition and a back-to-nature ambience.

What are standard roses? It is not a matter of a single rose class. Rather, the rose and tree nurseries graft suitable varieties of all rose classes onto special stocks (*Rosa canina* 'Pfander' or *Rosa pollmeriana*) of various heights that are over three years old. The height of the graft determines the beginning of the crown of the standard. Accordingly, these roses are classified into four different categories:

16 in. (40 cm) Quarter Standards (Miniature Standards): Most miniature roses, but also some area roses, are grafted at the height of 16 in. (40 cm). The trade often offers quarter standards at Mother's Day; they fit pefectly into small pots and containers. Because of the leaves' distance from the soil surface, the threat of infection with soil-borne fungal diseases like spot anthracnose is diminished. Therefore, the dwarf roses that are considered really at risk can again be worked into the garden design as quarter standards.

24 in. (60 cm) Half Standards: Usually bedding and area roses are grafted at the height of 24 in. (60 cm). Half standards are the ideal rose trees for pots since their crowns unfurl the beauty of their flowers at eye and nose level.

36 in. (90 cm) Full Standards: As a rule, compact bedding, hybrid tea, and area roses are grafted at the level of 36 in. (90 cm). Full standards are the classic tree roses. When planted in the garden or on both sides of a front door as an entryway greeting, they are at the best height for viewing and enjoying the fragrance. Upright-growing varieties are also good for children, for as a rule the prickles of the full standard rose do not pose a risk of injuring the smallest rose lover. Real fans of full standard roses plant several standards as an avenue in the garden. These are only some examples of the multifold design possibilities that they offer for the garden and on the terrace.

55 in. (140 cm) Weeping Standards: Most climbing or arching area rose varieties are grafted at a height of 55 in. (140 cm) to create weeping standards. In particular, the soft-caned rambler varieties do honor to

Repeat-Flowering Standard Roses

16 in. (40 cm) Quarter Standards:

Class	Variety	Color	Growth Habit
area	'Swany'®	white	emphatically weeping
area	'The Fairy'	pink	weeping
miniature	'Dwarfking'®	red	upright
miniature	'Guletta'®	yellow	upright
miniature	'Orange Meillandina'®	orange red	upright
miniature	'Peach Brandy'®	apricot	upright
miniature	'Pink Symphonie'®	pink	upright
miniature	'Sonnenkind'®	yellow	upright

24 in. (60 cm) Half Standards:

Class	Variety	Color	Growth Habit
bedding	'Bonica '82'®	pink	bushy
bedding	'Fragrant Cloud'®	red	upright
bedding	'Leonardo da Vinci'®	pink	bushy
bedding	'Schneeflocke'®	white	bushy
area	'Alba Meidiland'®	white	bushy
area	'Ballerina'	pink/white	weeping
area	'Flower Carpet'®	white	bushy
area	'Lovely Fairy'®	pink	weeping
area	'Mirato'®	pink	weeping
area	'Scarlet Meidiland'®	red	weeping
area	'Sommermärchen'®	pink	bushy
area	'Surrey'®	pink	bushy
area	'Swany'®	white	emphatically weeping
area	'The Fairy'	pink	weeping

16 in. (40 cm) quarter standard

24 in. (60 cm) half standard

36 in. (90 cm) full standard

55 in. (140 cm) weeping standard (cascade standard)

Repeat-Flowering Standard Roses

36 in. (90 cm) Full Standards:

Class	Variety	Color	Growth Form
bedding	'Bonica '82'®	pink	bushy
bedding	'Diadem'®	pink	bushy
bedding	'Fragrant Cloud'®	red	upright
bedding	'Leonardo da Vinci'®	pink	bushy
bedding	'Play Rose'®	pink	bushy
bedding	'Schneeflocke'®	white	bushy
bedding	'Sunsprite'®	yellow	bushy
hybrid tea	'Aachener Dom'®	pink	upright
hybrid tea	'Duftrausch'®	lilac	upright
hybrid tea	'Elina'®	yellow	upright
hybrid tea	'Loving Memory'®	red	upright
hybrid tea	'Peace'	yellow/red	upright
hybrid tea	'Silver Jubilee'®	pink	upright
area	'Alba Meidiland'®	white	arching
area	'Flower Carpet'®	pink	arching
area	'Lovely Fairy'®	pink	arching
area	'Mirato'®	pink	arching
area	'Sommermärchen'®	pink	bushy
area	'Sommerwind'®	pink	bushy
area	'Swany'®	white	pronounced trailing
area	'The Fairy'	pink	arching
climber	'Rosarium Uetersen'®	pink	pronounced trailing
shrub	'Ghislaine de Féligonde'	yellow	pronounced trailing
shrub	'Raubritter' (once-flowering)	pink	pronounced trailing
shrub	'Schneewittchen'® ('Iceberg')	white	bushy

55 in (140 cm) Weeping Standards:

Class	Variety/Flowering Characteristics	Color
area	'Nozomi'® (once-flowering)	pearly pink
area	'Magic Meidiland'®	pink
area	'Marondo'® (once-flowering)	pink
area	'Pheasant'®	pink
area	'Scarlet Meidiland'®	red
area	'Snow Ballet'®	white
area	'The Fairy'	pink
climber	'Golden Showers'®	yellow
climber	'Ilse Krohn Superior'®	white
climber	'Lawinia'®	pink
climber	'Morning Jewel'®	pink
climber	'New Dawn'	pearly pink
climber	'Rosarium Uetersen'®	pink
climber	'Santana'®	red
climber	'Sympathie'®	red
rambler	'Flammentanz'® (once-flowering)	red
rambler	'Paul Noël' (remontant)	pink
rambler	'Super Dorothy'®	pink
rambler	'Super Excelsa'®	crimson pink
shrub	'Raubritter' (once-flowering)	pink
shrub	'Schneewittchen'® ('Iceberg')	white

Note: Quality, planting, staking, winter protection, removal of wild canes, and pruning of standard roses are described in the appropriate sections in the chapter "Practice" (starting on page 141).

this form of standard with their trailing cascades of flowers. Often the cascade standards are offered under the term weeping roses or standards—a sad term that this joyous fountain of flowers in no way deserves and that impedes their use rather than promotes it. Especially beautiful specimens can be admired in the cascade rose gardens of the Deutschen Rosarium in Dortmund.

The assorted roses listed to the left are available commercially. Varieties at 16, 24, and 36 in. (40, 60, and 90 cm) are repeat-flowering unless otherwise noted.

'Super Excelsa'® is a robust area and climbing rose that is best suited for weeping standards. In contrast to the old variety 'Excelsa' (please do not mix them up!), it is resistant to mildew and is often repeat-flowering in sunny locations. It comes from the breeding workrooms of Karl Hetzel, who has dedicated himself to breeding healthy, robust roses.

Miniature roses are rose thumblings that are offered in many colors.

Miniature Roses

Miniature roses are the Tom Thumbs among the roses. They probably originated in China. The original miniature rose *Rosa* 'Roulettii', which is also offered under the name 'Pompon de Paris', achieved world fame in the last century as "the smallest rose in the world." In 1929, the renowned plant breeder Georg Ahrends of Wuppertal brought this rose to Germany from England and immediately propagated it in large quantities. At the same time, numerous other rose breeders got to work on this variety and created further improvements.

The modern miniature roses exhibit a growth height that is barely over 12 in. (30 cm). Despite this weak vigor, the good dwarf varieties develop an enormous quantity of flowers for their size. Their foliage is charmingly small and rounds out the whole cute impression of this David of a rose.

The dwarfed growth of these varieties allows them to be used in many different ways in balcony flower boxes, in the garden and rock garden, or as cutting-propagated potted roses. In short, when space is limited, miniature roses play their trumps.

On the other hand, their decreased vigor is also accompanied by a relatively high propensity for fungal diseases of all kinds. Therefore, when miniature roses are used for area plantings, several sprayings with fungicides are necessary during the summer months.

However, this does not mean that the environmentally conscious rose lover must entirely avoid miniature roses. If the following advice is observed and the needs of the plants are complied with, the varieties named in the table can make it through the summer without a black eye. An absolutely perfect location for roses, that is with a good play of light and air (at least 8 hours of sunshine daily), must be provided, for any sins of siting are paid for by miniature roses within a very short time. The foliage of minature roses must always remain dry. The problems with fungal disease are markedly diminished if the leaves are not too close to the ground. The resting spores of spot anthracnose fungus are especially easily transferred to healthy leaves from the soil surface by splashing rain or irrigation water.

This problem can be solved by cultivating roses in troughs or balcony containers and watering from below. Furthermore, the miniature roses are then in the immediate vicinity of the rose lover. On the terrace, the many details can be appreciated close-up.

Various different forms of these roses are offered commercially. Primarily on Mother's Day, potted roses propagated by cuttings are available. They are selected to be houseplant roses and, because of that, are only conditionally suitable for the garden. These roses have the advantage of having very small root balls for planting into

Miniature Roses

Variety	Color	Flowers	Height (in. [cm])
'Dwarfking'®	red	double	16–24 (40–60)
'Guletta'®	yellow	double	12–16 (30–40)
'Orange Meillandina'®	orange/red	double	12–16 (30–40)
'Peach Brandy'	apricot	double	12–16 (30–40)
'Pink Symphonie'®	pink	double	12–16 (30–40)
'Sonnenkind'®	yellow	double	12–16 (30–40)

flower boxes. They are always worth a try, as they can usually be bought surprisingly inexpensively.

The selection of varieties listed are grafted rosebushes. The vigor of the understock allows them to grow somewhat more vigorously than their own-rooted greenhouse colleagues. They are also more robust and sufficiently woody to survive overwintering in freezing temperatures. Because of their strong root systems they do not fit into small flower containers; the container used must have a depth and breadth of at least 12 in. (30 cm).

Rock Gardens: The miniature roses listed are also suitable for sunny rock gardens. They are best planted into troughs and basins or into terrace beds.

Roses for Floral Arrangements

Creative flower arrangements with roses make it possible to have the queen of the flowers in the house year round. In the summer months, fresh roses can be turned into charming vase arrangements. In fall and winter, dried roses can be displayed in painterly still lifes or as dried petals in wonderfully smelling potpourris that attract the eye and the nose.

Floral Arrangements with Dried Roses

The rose is suited for drying as is no other flower. For years, decades, sometimes even for centuries, nothing of their romantic personality is lost; their charisma, albeit morbid, is retained. With the right methods, an astonishing approximation of the freshness and color of living roses can even be achieved.

The drying of plant parts has a long tradition. For the people of past centuries—long before photography and colored garden magazines were known—it was one of the few ways of fixing plants permanently to be able to enjoy them close-up for a longer time. Today the old art of flower preservation is being rediscovered. In particular, the very doubled fragrant roses from the hybrid tea, Old Rose, and English Rose groups are outstandingly suitable for this. However, very double bedding roses with their lush flower fullness are also popular construction materials for small, pretty, old-fashioned bouquets.

Outstanding for drying roses: An airy attic.

Drying means removing the water from the rose very slowly and very carefully. Several possible methods can be used to do this.

Air Drying: Anyone who has time and space and would like to try the most natural conservation method for roses can hang the roses in bunches upside down in a well-ventilated, drafty, dry room—perhaps in the attic. The leaves are removed from the stems before hanging and the bunch secured with a loose rubber band. Inelastic ties like string are not suitable because the stems shrink as the water evaporates and then will fall out of the bunch. By depending on the fullness of the flowers and the length of the stems, the preservation process takes ten to thirty days. Each bunch should not contain more than ten rose stems. Too thick a bunch will dry only very unevenly, and the inner stems, in particular, then cannot be exposed to enough circulating air. For the same reason, the individual bunches must be hung with adequate space between them.

Salt Drying: In craft stores, one can buy a drying salt (dessicant) to use for drying rose flowers—and, naturally, also the flowers of other plants—and retain their true colors. Ideally, the drying is done in a closed box, perhap an old, shallow tin that had cookies in it. The bottom of the box is

A dream of a dried flower arrangement with roses, grasses, and delphiniums.

To preserve them, roses are completely covered with the drying salt.

Fragant ornaments for bath- and living rooms: rose and lavender potpourris.

covered with dessicant. The flower heads, which have had their stems cut very short, are laid onto it and then sprinkled with more dessicant until they are entirely covered and nothing more of them can be seen. The tin box is closed and stored in a warm, dry place. After some days, the box is opened and the flowers removed. The dessicant is recycled by being strained and dried in the oven. It can then be reused.

How can dried roses be used? You can make a small old-fashioned bouquet of dried red roses swathed in ivy vines with ivy inflorescences or a rose presentation basket filled with dried long-stemmed roses and grasses. A mixed bouquet of dried lavender and grasses from the garden with the addition of eucalyptus leaves from the florist looks beautiful in a corner. For someone looking for something striking, a rosy clothing accessory might be recommended, for instance, a necklace to which rose petals have been piled on top of each other. Table decorations that cause guests to exclaim "Aha" can be created by groups of three flower heads. Something special is roses captured in cast acrylic for paperweights—an out of the ordinary object and an eye-catcher of the highest order. The creative possibilities have no limits. Some special things should still be noted at this point.

Fragrant Potpourris: These consist of loose flower petals or heads. They are harvested in the early morning during dry weather. For drying, the flowers are spread out next to one another—not on top of each other—on wire mesh and newspaper. The paper sucks up the superfluous water, and circulating warm, dry air provides the correct climate for drying. When sprinkled with some rose oil and stored in jars, little bags, or shallow dishes, the potpourri is placed into hallways, living rooms, baths, and cupboards. A mixture of herbs, spices, and rose petals is also good. In the eighteenth century, the peak period for potpourris, the container was warmed over the open fire in the wintertime and then the cover was lifted and the spicy aroma enjoyed.

Fragrant bouquet with dried roses, sage, star anise, cinnamon, and lavender balls.

The following recipe has proven good: 4 cups of dried rose petals from scented roses, 1 cup (240 ml) dried rosebuds, 1 cup (240 ml) lavender flowers, $\frac{1}{2}$ cup (120 ml) ground violet roots, 1 teaspoon (5 ml) each of ground allspice, cinnamon, and nutmeg, $\frac{1}{2}$ teaspoon (2.5 ml) ground cloves, and a few drops of rose oil.

Rose Balls: Anyone looking for small but fine Christmas presents or birthday surprises can transform everyday objects with a covering of dried rose petals. Christmas balls, table tennis balls, jars, little boxes, plastic flower pots, and much more—everything can be turned into a decorative accessory with

applied flower petals. A craft glue or hot glue gun is used for the pasting. Decorations with cord and ribbons add to the uniqueness of the finished object.

Little Rose Trees: For the somewhat advanced floral arranger, making little artificial rose trees is a special production. The standard is made of dried rose canes with especially attractive prickles, which are fastened together in a bunch of four or five. These are crowned with a ball or a pyramid formed of thin willow whips and decorated with flower petals. One can also use a crown of dry craft base material, such as Styrofoam, into which the smallest possible dried roses might be stuck row by row. To keep the foam base from becoming too porous from poking too many holes, three flowers at a time are wired together with a cuff of baling wire to make a unit, and these are used to make the crown. An old flower pot that already has developed a patina can be filled with sphagnum moss to serve as a base. For more stability, the pot can also be filled with plaster, which is then covered with moss after it hardens.

Glycerine for Rose Trimmings: Many leaves, when dried, are ideally suited for accessories to bouquests of dried roses. Examples of these compatible plants include beech, maple, laurel, and ivy. If they are air-dried, they crumble too quickly. To keep them flexible, therefore, they are placed into glycerine during the drying process. The clear, syruplike glycerine is available from the drugstore. Mix one part of glycerine with two parts of water. The stem ends with leaves removed and shoots with bark removed are placed into the solution to a depth of about a hand's breadth. The container is stored in a cool, dark place. As soon as the leaves show the first drops of glycerine, the preparation is finished.

Salt-dried roses: roses that never fade.

Floral Arrangements with Fresh Roses

Making bouquets and other arrangements with fresh cut roses expands the rose garden into the living areas. The question of which colors to combine with which accessories eludes discussion because everyone has his or her own preferences. One person may find the homogeneity of a bouquet of roses of the same variety and color particularly soothing and relaxing. Another individual will find the multicolored mixture of many colors and varieties far more exciting. There are no rules of thumb. The older literature however, contains descriptions of precisely which color of rose goes with which accessory for which birthday or a wedding: e.g., baby's breath and red roses, lady's-mantle and yellow roses, white daisies with pink roses.

Ultimately, the display of flowers from one's own garden depends on what's available there. It is much more important to follow the rules for cutting flowers for the vase, which are presented in detail in the section "Roses for Cutting" (see pages 112 through 123). Also extremely important is that the rose lover develop a garden of plants of many different varieties, which provide the appropriate accessory flowers. Anyone who regularly cuts flowers for the house and wants to arrange them in various ways makes herself or himself largely independent of having to buy them in summer by growing the following accessory plants (see the selection listed below). At the same time, the wealth of species in the garden is considerably increased:

Perennials: Anemones, astilbes, baby's breath, bleeding hearts, coreopsis, delphiniums, grasses, iris, lady's-mantle, lavender, marguerites, monkshood.

Vines: Clematis, ivy.

Trees and Shrubs: Colored-leaved maple varieties (*Acer negundo* varieties), blue Atlas cedar, copper beech, hydrangeas, buddleias, *Cotoneaster* 'Skogholm', deutzias, broom, pines, buckthorn, weigelia, wild rose species, flowering crabs, and Japanese cherries.

Anyone who might like to give a garden party an additional rosy framework, decorate an entryway, or quickly transform a simple container into an artful vase can use carpet tape on the desired spot (walls, wooden latticework, or similar robust materials). Then the individual can just place the roses on it very close together.

> TIP
>
> Also in the area of rose arrangements are **water roses.** These fat, very double roses are able to float on water beautifully. A flick of the wrist is enough to make shallow dishes and containers filled with water the ideal frame for water roses. Arrangements of a particular species can create an attention-getting decoration for outdoor barbecues and summer festivities.

Fresh Roses for Boutonnieres: Thin-walled glass tubes about as thick as a finger with a stickpin attached and filled with water keep a rose blossom fresh to wear in a buttonhole. Recently, these glass tubes have become available from nurseries that deal in roses.

Roses in the Kitchen (Flower Recipes)

Since it is generally known—and most recently described in Umberto Eco's monastery detective story *The Name of the Rose*—that the monks in medieval cloisters were not careless of costs, it will also be no surprise that so many monastery cooks seasoned meat and sweet dishes with fresh rose petals. By growing roses, the canny monks killed two flies with one blow. On the one hand, they could enjoy the fragrant flowers of the roses in the cloister garden. On the other hand, the monastery kitchen and apothecary made use of the rose flowers and fruits.

Recently, many rose lovers have rediscovered the culinary rose. The mild flavor of the rose corresponds to its delicate fragrance. Pectins, tannins, essential oils, and flavone are some of the constituents that make the rose a unique ingredient. Roses enrich the taste of many dishes in an inimitable way and manner.

As a rule, those fragrant roses are processed. These can be English Roses, Centifolias, or the new generation of perfume roses. Certain rules hold true for all of them: **under no circumstances** may they have been treated with any chemicals, their origins **must** be known, and, ideally, they are from one's own garden.

Natural beauty for eyes and palate: Rose potpourri and rose water.

A fruity natural buffet: Hips of various rose species.

Recipes

Rose Water

Ingredients:

4 handfuls of very fragrant rose petals

2 cups ($^1/_2$ l) water

Preparation:
Cut off the light tips of the petals (these can be bitter). Heat the water, and pour it over two handfuls of petals. Cover entirely, and let steep for 2 days. Strain and pour the water over the remaining rose petals. Let steep some more, and then put the finished rose water into bottles.

Tip: The rose water can be frozen. Use for marzipans, glazes, sherbets, and so on.

Rose Syrup

Ingredients:

2 handfuls of very fragrant rose petals
1 lemon
2 cups ($^{1}/_{2}$ l) water
sugar

Preparation:
Cut off the light tips of the rose petals (these can be bitter). Bring the petals and the water to a boil together, turn off heat, keep covered and let steep well for 15 minutes, and then strain. For more intense flavor, repeat this procedure with fresh rose petals. Add a 1:1 (or less if desired) ration of sugar, then bring the rose extract to the boil once again and put into bottles.

Tip: The rose syrup can be frozen. Use for fruit salads, ice cream, glazes, and so on.

Rose Vinegar

Ingredients:

1 handful of very fragrant rose petals
1 handful of raspberries
2 cups ($^{1}/_{2}$ l) of white wine vinegar

Preparation:
Cut off the light tips of the petals (these can be bitter). Carefully put the petals and the raspberries into a bottle that has been washed out with hot water and dried well (in a warm oven) beforehand. Add the vinegar. Let steep in a warm, sunny place for 2–3 weeks (no longer), and then filter. Store the finished vinegar in a cool, dark place.

Tip: For presentation, put the vinegar into a decorative bottle with 2–3 fresh rose petals.

Rose Ice Cubes

Ingredients:

Rose flowers in the desired quantity (especially suitable: the small flowers of the variety 'Mozart')
Water

Preparation:
Fill an ice tray with water and put in the flowers. Freeze.

Tip: These are extraordinarily decorative, not only in drinks, but also when serving a special bottle of champagne or white wine.

Homemade rose vinegar.

Rose Liqueur

Ingredients:

$4^{1}/_{2}$ oz. (125 g) fragrant rose petals
2 cups ($^{1}/_{2}$ l) water
2 cups ($^{1}/_{2}$ l) brandy or cognac
$8^{3}/_{4}$ oz. (250 g) sugar
$^{1}/_{2}$ tsp. (2 ml) cinnamon (according to taste)

Preparation:
Cut off the light tips of the petals (these can be bitter). Pour the water over the petals, and let stand, covered, for 2 days. Strain through a fine sieve, and fill up with brandy. Add the sugar and cinnamon, and let stand in a closed container for another 14 days, shaking well now and again. Then filter the liqueur and put into bottles.

Rose Punch I

Ingredients:

Fragrant petals of about 10 roses (quantity according to the desired intensity of flavor)
$3^{1}/_{2}$ oz. (100 g) cube sugar
1 small glass cognac or Grand Marnier (according to taste)
2 bottles dry white wine (or 1 bottle white wine, 1 bottle mineral water)
1 bottle champagne

Preparation:
Cut off the light tips of the petals (these can be bitter). Put the petals and the sugar into a bowl, pour the cognac and a half bottle of wine over it, cover, put into the refrigerator, and let steep for about 1 hour. Strain, and shortly before serving, add the chilled wine and champagne. Let some fresh rose petals float in the bowl for decoration.

Rose Punch II

Ingredients:

Fragrant petals of 25 roses
$5^{1}/_{4}$ oz. (150 g) sugar
$^{1}/_{2}$ vanilla bean
stick cinnamon
juice of 3 oranges
$1^{1}/_{2}$ fl. oz. (4 cl) rose liqueur
4 cups (1 l) red or white wine
1 bottle champagne
1 bottle mineral water

Preparation:
Cut off the light tips of the petals (these can be bitter). Put the petals into a bowl. Add the sugar, vanilla bean, cinnamon. Pour the orange juice, rose liqueur, and slightly warmed wine over it. Let the whole thing steep for 6 hours. Strain, and before serving, fill up with chilled champagne and mineral water. Serve the punch cool and decorated with fresh rose petals.

Rose Tea I

Ingredients:

1 handful fresh or dried fragrant rose petals
2 cups ($^{1}/_{2}$ l) water

Preparation:
Cut off the light tips of the petals (these can be bitter), and pour the boiling water over them. Let steep for 10 minutes. Sweeten to taste.

Rose Tea II

Ingredients:

1 handful of dried, fragrant rose petals
$3^{1}/_{2}$ oz. (100 g) good black tea

Preparation:
Mix the petals with the black tea, and use this as usual. Store the rest of the tea mixture in an airtight container.

Rose Jelly

Ingredients:

Petals of about 15 strongly scented roses
2.2 lb. (1 kg) sugar w/pectin
3 cups (3/4 l) dry white wine
Rose water, optional

Preparation:
Cut off the light tips of the petals (these can be bitter). Heat the petals in the wine, let draw for 10–15 minutes, and strain. Add the preserving sugar to the cooled liquid. Bring to a rolling boil and cook, stirring, for 4 minutes. Add some rose water as preferred. Put the jelly into hot jars, and close them immediately.

Rose Jam

Ingredients:

Fresh petals of 40 fragrant roses
8 3/4 oz. (250 g) sugar
2.2 lb. (1 kg) sugar w/pectin
Juice of 5 lemons

Preparation:
Cut off the light tips of the petals (these can be bitter). Mix the petals with the sugar, and put through a meat grinder. Bring this rose sauce to a boil with the preserving sugar and lemon juice, and let boil for 7 minutes, stirring constantly. Put the jam into hot jars, and seal immediately.

Rose-Blackberry Jam

Ingredients:

Petals of 20 fragrant roses
4 1/2 oz. (125 g) sugar
26 1/2 oz. (750 g) blackberries
17 3/4 oz. (500 g) honey
0.35 oz. (20 g) powdered gelatin

Preparation:
Cut off the light tips of the petals (these can be bitter). Mix the petals with the sugar, and put through a meat grinder. Mash the blackberries, and bring to a boil together with the rose sauce and the honey. Dissolve the gelatin according to the package directions, and let all boil together, stirring, for 2–3 minutes. Put the jam into hot jars, and seal immediately.

Fast and simple to make: Stimulating rose tea.

Rose-Apple Jam

Ingredients:

17 3/4 oz. (500 g) fresh petals of fragrant roses
4 1/2 lb. (2 kg) apples
17 3/4 oz. (500 g) sugar
Juice and peel of an untreated lemon
1 cup (1/4 l) water

Preparation:
Cut off the light tips of the petals (these can be bitter). Peel the apples, and cut into small pieces. Put the apples, petals, sugar, lemon juice, and lemon peel into a pan with the water. Let stand, covered, for a half hour. Then bring the whole thing to a boil, and let cook until the apples are soft. Press all through a sieve or puree with a food processor. Bring to a boil once more, stirring constantly, allowing it to thicken. Put the jam into hot jars, and seal immediately.

Rose-Champagne Jam

Ingredients

Petals of 40 fragrant roses
8 3/4 oz. (250 g) sugar
2.2 lb. (1 kg) sugar w/pectin
1 packet of vanilla sugar
1 1/2 cups (3/8 l) champagne (or possibly rosé)
1/2 cup (1/8 l) rosé

Preparation:
Cut off the light tips of the petals (these can be bitter). Mix the petals with the sugar, and put through the meat grinder. Bring this rose sauce to a boil with the preserving sugar, vanilla sugar, champagne, and wine. Let boil 5 minutes, stirring constantly. Put the jam into hot jars, and seal immediately.

Rose Cream

Ingredients:

1/2 cup rose jelly (see recipe above left)
2–3 cups (350 g) raspberries, strawberries (cut up if large), fresh currants
7 oz. (200 g) cream
1 packet of stiff, whipped cream

Preparation:
Stir the jelly, slightly warmed, into the berries, and fold in the stiffly whipped cream.

Rose Torte

Ingredients:

1 baked biscuit or batter shell (bought or homemade)
Fresh strawberries, raspberries, blackberries, and so on.
Rose jelly or jam
Cream
Candied rose petals

Preparation:
Carefully mix the fruit and jelly. Beat the cream until stiff, and fold in. Spread the entire mixture into the shell, and decorate with candied rose petals or single beautiful berries.

Deep-Fried Rose Petals

Ingredients:

25 petals of fragrant roses
2 3/4 oz. (75 g) flour
1 egg
4 Tbs. (60 ml) white wine
1 Tbs. (15 ml) milk
1 pinch salt
1 tsp (5 ml) sugar
Oil for frying
1/2 cup (1/8 l) rose syrup (see page 131)
4 Tbs. (60 ml) crème fraîche
Powdered sugar

Preparation:
Cut off the light tips of the petals (these can be bitter). Make a smooth dough of the flour, egg, wine, and milk, and season to taste with salt and pepper. Heat the oil (the temperature is right when small bubbles rise around an inserted wooden skewer). Roll 20 petals, all told, first in rose syrup and then in the frying batter. One after the other, fry them until light brown in the hot oil. Let drain on paper towel. Cut the remaining five petals into little pieces, and heat them in the rose syrup. Put the fried petals onto plates with a spoonful of crème fraîche. Serve with rose syrup and dusted with powdered sugar.

Tip: Decorate with fresh petals.

Rose Sherbet I

Ingredients:

1 cup (1/4 l) rose syrup (see page 131)
1 cup (1/4 l) rosé

Preparation:
Mix syrup and wine, and let stand in the freezer. The sherbet should be loosened a little with a fork before serving.

Rose Sherbet II

Ingredients:

1–2 handfuls of fragrant rose petals
4 1/4 oz. (120 g) powdered sugar
12 oz. (350 ml) water
Peel and juice of 2 untreated lemons
2 tsp. (10 ml) rose water (see page 130)
1 egg white

Preparation:
Cut off the light tips of the petals (these can be bitter). Dissolve the powdered sugar in water, and bring to a boil with the lemon peel. Let simmer 5 minutes, take off the stove, add the rose petals, and let all cool. Then strain and stir in the lemon juice. Season to taste with rose water. Let the mixture freeze partially. Beat the egg white stiff and fold into the sherbet. Now allow the sherbet to freeze entirely, and before serving, loosen somewhat with a fork.

Tip: Offer fresh fruit drizzled with rose syrup or rose sugar and cookies with the sherbet. Decorate with candied rose petals.

Boned Chicken Breast with Roses

Ingredients:

Petals of 4 roses
1/2 cup (1/8 l) vegetable broth
2 Tbs. (30 ml) light soy sauce
2 Tbs (30 ml) sake
1–2 Tbs. (15–30 ml) rose vinegar (see page 131)
1 Tbs (15 ml) pink pepper
8 3/4 oz. (250 g) button mushrooms
Oil
Salt
Black pepper

Preparation:
Cut off the light tips of the petals (these can be bitter). Cut the chicken breast into narrow strips. Clean the mushrooms, and slice them lengthwise. Heat the oil in a wok, and fry the chicken breast strips over high heat, stirring. Add the mushrooms, and continue cooking for 1–2 minutes. Add the petals, broth, soy sauce, wine, vinegar, and coarsely ground pink pepper, season with salt and freshly ground pepper, and serve immediately.

Tip: As a side dish, serve transparent noodles or brown rice.

Vitamin Roses (Rose Hip Recipes)

Rose hips begin developing as soon as a rose has been pollinated. The hips, which become noticeable after the petals have fallen off, contain the seeds of the rose. Rose hip colors, varying from bright red to deep purple, add extra interest and beauty to a garden—especially in the winter.

For repeat-blooming roses, it is best to remove any developing hips since, as the hips grow, they make the plant slow down and decrease flower production. For once-blooming roses, just allow the hips to develop and enjoy the show.

For centuries the fruit of roses has been used for food. Without our ancestors knowing it, the hips were an important, extraordinarily high-yielding source of vitamin C—especially in the winter months when no fresh fruits or vegetables were available. Roses with hips can therefore be called vitamin roses without any hesitation.

The vitamin C content of rose hips fluctuates very widely and is directly related to their ripeness. When picked at the right moment, hips leave even lemons in the dust in this regard.

Hips are best harvested when the fruit is fully colored but still hard and crisp. When harvested too early, hips still do not have the maturity of taste. When picked too late, soft, overripe hips have already begun the breakdown of the nutrient substances through the continued addition of oxygen.

The hips should be processed right after harvesting since they begin to deteriorate immediately.

Even cooking for too long produces vitamin loss. If the fruit is thoroughly chopped first, the cooking time is decidedly shortened, and thus the vitamins are preserved. In this way, jam made from rose hips becomes the richest supplier of vitamins that we know.

Before processing, rose hips must be thoroughly deseeded. The seeds taste sour and furry and are just not a joy to the palate.

Many rose species and varieties produce hips. The table on page 135 shows only a small selection, which can be easily broadened. Especially rich are the hips of the 'Apple Rose' *(Rosa villosa)*—so called because of its apple-shaped hips— and the following vitamin roses:

Hips—like crabapples and chestnuts (marron)—are suitable for preparing many delicious recipes.

***Rosa rugosa* Hybrids:** The *rugosa* hybrids are offspring or hybrids of *Rosa rugosa.* This species comes from China and is very widespread, particularly in East Asia. Today it is considered a naturalized wild rose in Germany. It flowers pink to bright red.

The growing desire for roses with resistant foliage have caused the demand for *Rosa rugosa* and its offspring to increase constantly.

In humid northern European coastal areas, they offer definite advantages with their minimal requirements for environmental situations. The high tolerance of *rugosa* hybrids propagated on their own roots for salty soils is unusual. Unfortunately, alkaline soils with a high pH value limit their planting.

The hips of the *rugosas* reach an unusual size and provide much valuable fruit flesh.

Syrphid flies use the flowers of the *rugosa* hybrids as favorite alighting stations. Their larvae have a great importance as natural enemies of the aphid in the framework of biocontrol of insect pests.

Rugosa hybrids are very frost hardy and robust, although they do have the disadvantage that the flowers of various varieties are susceptible to rain damage. The petals stick together and form ugly brown mummies, which are not self-cleaning.

To prevent the roses from becoming leafless too early, the varieties should be cut back each year by about one-third to one-half.

Children, and those who—as Erich Kästner said—have not laid aside their childhood like an old hat and then forgotten it, harvest the hips of the *rugosas,* remove the seeds, and plague their friends and relatives with the furry seeds. After all, the *rugosa* seeds really do make outstanding **itching powder!**

Pillnitzer Vitamin Rose 'Pi-Ro 3': Not all roses have the same high level vitamin C content. Also, a high quantity of usable fruit flesh does not necessarily correlate with a large vitamin content. Therefore, in places such as Russia and the former East Germany, people have for many years sought strenuously to select forms of vitamin roses that combine the two factors, quantity and quality. They have found the group of Pillnitzer vitamin roses, from which the selection 'Pi-Ro 3' comes very close to the desired ideal. In free-growing, decorative hedges, this shrub rose, which is produced from the native variety *Rosa pendulina,* displays a lush burst of flowers in the spring and an abundance of hips in the fall.

For someone who wants to grow hips in grand style with the highest food value (which within three years of planting yields an average vitamin C content of 1,150 mg per 100 g of seeded fruit flesh), the relatively prickle-free canes will also be an additional cause for rejoicing since they facilitate harvesting.

Rosa jundzillii: Also known as *Rosa marginata,* this rose is a secret tip among lovers of rose hips. It develops very beautiful brilliant red, round hips that are very rich in vitamin content. Someone who wants to be informed about the exact vitamin content of the hips of many wild roses should read the book *Wild- und Gartenrosen* (*Wild and Garden Roses*) by S. G. Saakov. The notes, going back to 1930, of this Russian professor who studied vitamin C are unique in their fullness on the intensive investigation of wild rose species in the former U.S.S.R.

Vitamin Roses

Class	Variety	Color	Height (in. [cm])
bedding	'Bonica '82'®	pink	24–32 (60–80)
bedding	'Escapade'®	lilac/white	32–39 (80–100)
bedding	'Heidepark'®	pink	24–32 (60–80)
bedding	'La Sevillana'®	red	24–32 (60–80)
bedding	'Märchenland'	pink	24–32 (60–80)
bedding	'Matilda'®	pink	16–24 (40–60)
bedding	'Play Rose'®	pink	24–32 (60–80)
bedding	'Ricarda'®	pink	24–32 (60–80)
bedding	'The Queen Elizabeth Rose'®	pink	32–39 (80–100)
area	'Apfelblüte'®	white	32–39 (80–100)
area	'Ballerina'	pink/white	24–32 (60–80)
area	'Pink Meidiland'®	pink/white	24–32 (60–80)
area	'Red Meidiland'®	red	24–32 (60–80)
area	'Royal Bassino'®	red	16–24 (40–60)
wild rose	*Rosa arvensis*	white	32–39 (80–100)
wild rose	*Rosa gallica*	pink	32–39 (80–100)
wild rose	*Rosa eglanteria* [Formerly *R. rubiginosa*]	pink	79–118 (200–300)
wild rose	*Rosa villosa*	pink	59–79 (150–200)
climber	'Dortmund'®	red	79–118 (200–300)
climber	'New Dawn'	pearly pink	79–118 (200–300)
rugosa hybrid	'Frau Dagmar Hartopp'	pink	24–32 (60–80)
rugosa hybrid	'Foxi'®	pink	24–32 (60–80)
rugosa hybrid	'Polareis'®	pink	24–32 (60–80)
rugosa hybrid	'Polarsonne'®	red	24–32 (60–80)
rugosa hybrid	'Schnee-Eule'®	white	16–24 (40–60)
shrub	'Bourgogne'®	pink	59–79 (150–200)
shrub	'IGA '83 München'®	pink	32–39 (80–100)
shrub	'Pi-Ro 3'	pink	59–79 (150–200)
shrub	*Rosa hugonis*	yellow	79–118 (200–300)
shrub	*Rosa moyesii* (grafted)	red	79–118 (200–300)
shrub	*Rosa sericea f. pteracantha*	white	79–118 (200–300)
shrub	*Rosa sweginzowii* 'Macrocarpa'	pink	79–118 (200–300)
shrub	'Sharlachglut'	red	59–79 (150–200)
shrub	'Schneewittchen'® ('Iceberg')	white	79–118 (200–300)

Hip Recipes

The high vitamin C content of the hips makes the fruit interesting as a food. The body needs vitamin C regularly but cannot store it. Therefore, the body must constantly be able to ingest it.

Vitamin C prevents feverish colds, influences the function of the adrenal glands, and directly controls the production of important hormones.

Finally, it is essential for wound healing in humans.

In this book, the culinary pleasures that hips have to offer are the point of focus. Naturally, the centuries-old uses of the rich vitamin content of hips when consumed as a medication, e.g., as tea, should be and are discussed. That information appears in the section entitled "The Medicinal Rose" (see pages 138 and 139).

Fresh food: Pit the hips and take a bite. Enjoy!

The history of preparations of hips for human consumption reaches back for several thousand years. In Switzerland, scientists have found hip seeds in large quantities along with other food remains in the cave dwellings of Stone Age inhabitants.

Rose Hips as Fresh Fruit

When you consume rose hips as fresh fruit, first wash the hips and remove the seeds. Chew the fruit thoroughly to release all the nutrients. Some especially digestible hips to eat fresh are those of *Rosa sericea f. pteracantha,* which have a sweet-sour taste.

Rose Hip Paste (Basic Ingredients for the Following Recipes)

Ingredients:

Hips

1 glass white wine or water

Preparation I:
Pour the water or wine over trimmed, washed, seeded hips and place into a cool, dark place for 4 to 6 days. Stir now and then. When the hips are soft, work through a sieve.

Preparation II:
Mince the trimmed, washed, seeded hips in the food processor and boil with the wine and the water for 20 to 30 minutes, stirring frequently. Work the hips through a sieve.

Rose Hip Syrup

Ingredients:

17¾ oz. (500 g) hips

17¾ oz. (500 g) sugar

4 cups (1 l) water

Preparation:
Mince the trimmed, washed, and seeded hips well, bring to a boil in the water, and let simmer for 20 minutes. Work through a very fine sieve or cotton dish towel. Slowly reheat the juice with the sugar, stirring constantly. Pour the hot syrup into jars, and seal immediately.

Tip: Use this like rose syrup for flavoring fruit salads, ice cream, and so on. You can use the syrup to make a summer drink when chilled champagne is added. The syrup can be frozen.

The fruit meat is obtained by pushing the precooked hips through a strainer.

The differing proportions of fruit flesh are clearly seen: Plenty with *Rosa rugosa* **(above), scanty with** *Rosa canina.*

Rose Hip Liqueur

Ingredients:

17¾ oz. (500 g) hips (harvested after the first frost)

5¼ oz. (150 g) rock candy

1 bottle kirsch

Preparation:
Mash the trimmed, washed, and seeded hips (or puree in the food processor), and put into a container with rock candy and kirsch. Close tightly, and let steep for several weeks. Then pour the liqueur through a fine sieve and put into bottles. The hip liqueur is mature enough to enjoy after 4 months of storage.

Rose Hip Wine

Ingredients:

8¾ lb. (4 kg) hips

4½ lb. (2 kg) sugar

5 qt. (5 l) water

Preparation:
Put the trimmed, washed, seeded, and halved hips into a demijohn or a large bottle. Dissolve the sugar in water, and pour over the hips. Close the container (careful—the fermentation produces pressure!), and place into a warm spot. Let ferment for 6 months, shaking now and then. Then filter the wine through a cloth, and put into bottles. Let rest in the cellar for several weeks longer.

Rose Hip Jam

Ingredients:

17¾ oz. (500 g) hips

2 cups (½ l) water

Sugar

Peel of an untreated lemon

Preparation:
Pour the water over the trimmed, washed, and seeded rose hips, and let stand for 1 day. Heat the now tender fruit with the soaking water, let cook until they are soft, and then work through a fine sieve. Weigh the resulting fruit paste, add the same quantity of sugar and the lemon peel, heat the entire mixture once more, and boil for 10 minutes, stirring. Put the jam into hot jars and seal immediately.

Rose Hip Honey

Ingredients:

Hips in the desired quantity

Water

Sugar

Preparation:
Cover the trimmed, washed, and seeded hips with water, bring to a boil, and cook until soft. Work through a fine sieve. Mix the fruit paste with the cooking water, measure, and for every quart of fluid (1 kg) add a scant 2¼ pounds of sugar. Heat the mixture, and cook together until it has become as creamy as honey. Put into hot jars, and seal immediately.

Rose Hip Chutney

Ingredients:

7 oz. (200 g) hips

7 oz. (200 g) apples

7 oz. (200 g) zucchini

7 oz. (200 g) tomatoes

7 oz. (200 g) onions

3½ oz. (100 g) golden raisins

3 cloves garlic

14 oz. (400 g) brown sugar

1 cup (¼ l) wine vinegar

Ground cinnamon

Ground allspice

Cayenne pepper

Preparation:
Trim, wash, and seed the rose hips; peel the apple, onion, and garlic cloves; scald and peel the tomatoes; trim and wash the zucchini. Dice all the ingredients until fine; press the garlic. Slowly heat with the remaining ingredients in a large pot, stirring constantly. Let simmer over low heat until the chutney has a creamy consistency. Put into jars while the chutney is still hot, and seal immediately. Store in a cool place, and let mature for several weeks.

Rose Hip Kisses

Ingredients:

- *2 Tbs. (30 ml) hip paste (see page 136)*
- *$17^{3}/_{4}$ oz. (500 g) sugar*
- *Peel of half an untreated lemon*
- *1 egg white*

Preparation:
Stir the hip paste with half the sugar and the finely chopped lemon peel until foamy. Beat the egg white very stiff, and fold in with the rest of the sugar. With two spoons, place little mounds on a baking sheet and let dry in the oven at very low heat (about 250 °F [125 °C]).

Rose Hip Macaroons

Ingredients:

- *2 Tbs. (30 ml) rose hip paste (see page 136)*
- *4 egg whites*
- *$8^{3}/_{4}$ oz. (250 g) confectioners' sugar*
- *$13^{1}/_{4}$ oz. (375 g) ground almonds*
- *Wafer molds, optional*

Preparation:
Beat the egg whites until very stiff, and carefully fold in the sifted confectioners' sugar. Knead the hip paste, almonds, and 2 tablespoons (30 ml) of the beaten egg whites into a dough. With two spoons, place a small mound of dough into each mold or directly onto a baking sheet lined with baking paper, press a little hollow into the center of each one, and fill with beaten egg whites. Bake the macaroons for about 35 minutes in a slow oven (250–300 °F [125–150 °C]) until light brown.

Rose Hip Compote

Ingredients:

- *$17^{3}/_{4}$ oz. (500 g) hips*
- *$17^{1}/_{4}$ oz. (500 g) sugar*
- *Cinnamon*
- *Peel of an untreated lemon*
- *Water*

Preparation:
Bring to a boil the trimmed, washed, and seeded rose hips with the sugar, cinnamon, the lemon peel cut into fine strips, and a little water. Let simmer until the fruit is soft. Lift out with a skimmer, and put into jars. Let the syrup continue cooking until it has become viscous. Pour over the hips, and seal the jars.

Tip: Serve this as a light dish with yeast dumplings or baked noodles. Serve as a dessert with cream or vanilla ice cream.

Rose Hip Sauce

Ingredients:

- *6 Tbs. (90 ml) rose hip paste (see page 136)*
- *2 cups ($^{1}/_{2}$ l) water*
- *1 Tbs. (15 ml) cornstarch*
- *Red wine*
- *Lemon juice*
- *Sugar*
- *Salt*
- *Mustard*

Preparation:
Bring the rose hip paste to a boil with the water, and thicken with the cornstarch dissolved in a little bit of water. Season the sauce until it tastes spicy by using wine, lemon juice, and spices.

Tip: Serve with fondue, broiled meat, or roasts.

Rose Hip Salad

Ingredients:

- *$8^{3}/_{4}$ oz. (250 g) dried rose hips*
- *2 cups ($^{1}/_{2}$ l) water*
- *$4^{1}/_{2}$ oz. (125 g) sugar*
- *5 Tbs. (75 ml) lemon juice*
- *3 Tbs. (45 ml) currant juice*
- *Some mustard seeds*

Preparation:
Soften the hips in the water overnight. Then bring them both to a boil, and let them simmer until only 2 to 4 tablespoons of fluid are left. Now mix in the sugar, and drizzle with lemon juice and currant juice. Sprinkle with crushed mustard seeds, and let cool.

Tip: Serve with dark roasted meats, use like cranberries.

Rose hip jam is a choice, nutritious spread for connoisseurs and gourmets.

Roses Without Hips: Occasionally one wants roses that do not form hips and thus also do not attract birds. They are used in situations in which birds can produce acute danger, such as airports, or places in which the danger is for the creatures themselves, e.g., from other animals (cats, and so on), as well as for technical constructions. Examples of nonfruiting rose varieties are 'Alba Meidiland", 'Ferdy', 'Pheasant', 'Magic Meidiland', 'Palmengarten Frankfurt', and 'The Fairy'.

The Medicinal Rose

Since antiquity, wild roses have been used for medicinal purposes. In A.D. 77, Pliny the Elder reported that thirty-two diseases might be healed by using the power of the rose. People prepared the various medicaments from rose roots, flowers, fruits, and leaves, which made the early doctors able to treat their patients independent of the season. Not only in spring and summer when the roses bloomed and bore leaves could one make use of their power—medications produced from various rose elements were available throughout the year.

Countless medicinal recipes containing the ingredient *rose*—alone or mixed with up to twenty herb species—substantiate the function of the rose as an important support of early medicine. People used it to try to cure stomachaches, headaches, toothaches, and the pain of wounds, to relieve insomnia, even to "purify" the mind in cases of mental illnesses. Fresh rose petals were placed onto swollen eyes and burns to cool them. In the Middle East, Arab doctors ascribed a primarily constipating effect to the rose. The monks in Europe increasingly refined the methods of production of juices, syrup, honey, tinctures, and salves; numerous chapters in the old monastery books bear witness to this. In addition, the Christian symbol of the rose increased the value of the plant.

An excerpt from the treatise *Herbolario Vulgare,* published in Venice in 1522, impressively documents the omnipresence of the rose as medicine: "Rose sugar . . . helps when the intestine is overtaxed, . . . also against vomiting. Rose syrup is prescribed for melancholics and enfeebled cholerics. For inflamed liver, rub the liver and for headache the forehead and temples with oil of roses. Ulcers in the mouth are helped with rose honey mixed with rose water. For oppression of the heart and heartache, one drinks rose water and one also uses same to wash the face."

Later, a Frenchman, Pierre Joseph Buchoz (1731–1807)—an all-around scribbler who is striking less for his profound thought than for his spectacular medical prose—even recommendeds, "Women of Provence, for hysterical attacks, drink a drink of three ounces of rose water and the same amount of orange-blossom water in which a cube of sugar has been dissolved over a small flame." Appropriately, rose water was often added to unpleasant-tasting or evil-smelling medicines to improve their taste and scent.

Obviously, hardly an ailment was not curable through the healing powers of the rose. This also explains why a good 600 years ago, in Provins, a town near Paris, the cultivation of large areas of the semidouble Apothecary Rose, *Rosa gallica* 'Officinalis' was begun. The variety name derives from the term for the medieval apothecary work room, the *officina.* In the beginning, the roses grown in Provins were mainly used for medicines. Gradually, however, the production of first-class rose oils began to take on importance. On both sides of the main street stood a noteworthy number of drugstores and apothecaries, which sent their products all over the world. This lasted until the middle of the 19th century—in 1860, for example, thirty-six tons of rose blossoms were exported to America from Provins.

Of the many old recipes, the ones having to do with rose hips are the ones most often still in use in our time. Numerous rose hip recipes have already been presented on the foregoing pages. How to prepare some teas from roses is now presented in this text.

Rose Hip Tea: The hips for tea are gathered in the fall, halved and deseeded, then dried and stored in a dry place. To prepare the tea, put them into cold water. Let them cook for about 10 minutes, and then let them steep for 15 minutes. Use about 1 heaping teaspoon (5 ml) of hips to 1 cup (1/4 l) of water.

Of course, the hips contain very little vitamin C once they are dried. However, they do contain a large palette of other important ingredients. As Healer-Preacher Kneipp praisingly reported, these have a healing effect on bladder and kidney ailments, rheumatism, and gout. The tea is also excellent for use as a pleasant-tasting preventive measure during the winter cold season. To increase the content of vitamin C in the tea, you can stir in a shot of lemon juice. Mixing in linden flowers—at the same time as the hips—strengthens the cold-preventing effect of this homemade tea.

Seed Tea: An effective, agreeable tea tasting of vanilla can also be brewed from seeds of the rose hip. To make this tea, the furry seeds are dried in the oven at very low heat until the villi fall off. Two teaspoons (10 ml) of crushed seeds are used per cup (1/4 l) of water. As with the dried hips, they are boiled for about 10 minutes. The tea is sweetened with a shot of honey.

In folk medicine, the seed tea is considered a special health tea. It is taken for coughs and colds. Side effects due to rose hip teas occur very seldom, certainly.

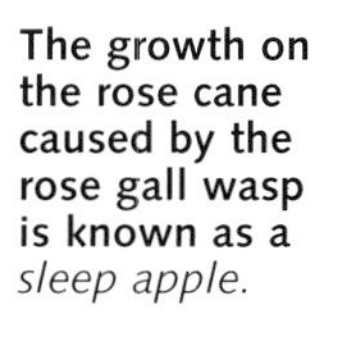

The growth on the rose cane caused by the rose gall wasp is known as a *sleep apple.*

However, a source has reported that allergies have appeared in some patients after long-term use of the seed tea.

Sleep Apples: While not a commonly occurring problem, the rose gall wasp causes a mosslike, clearly visible outgrowth, the so-called sleep apple, which is striking because of its hairy projections (see picture on page 138). The sleep apple, which can be up to 2 inches (5 cm) in size and was termed *fungus rosae* in the Middle Ages, used to be placed under children's pillows to make them sleep. These sleep apples appear most commonly on species roses.

Roses for Body Care

Roses have a very long tradition as beauty products. Especially because of their alluring fragrance, roses have been prized as cosmetics for thousands of years. People used rose water to refine all sorts of things, and sprinkled doorsills, pieces of clothing, and also the hair with this water. Today, the addition of valuable rose oil ennobles salves, creams, lotions, and shampoos.

Cosmetic Rose Water: One can buy rose water, but one can also make it. Fresh petals of especially fragrant, unsprayed roses are put into a dish, and boiling mineral water is poured over them. Finally, pure or very high-percentage alcohol is added to them. The ratio is about one part alcohol to ten parts water. The whole mixture is allowed to steep and is then strained into sterilized bottles. Before the bottles are closed, a few fresh rose petals are added to the rose water.

Rose Facial Water: An especially fine and rare recipe for a special rose facial water requires one to mix equal parts of rose blossoms, orange blossoms, and witch hazel water with pure alcohol. The ingredients are obtainable from the pharmacy or—like the rose water—can be made at home.

Rose Beauty Water: Barley is cooked soft in water and, at the end, the barley water is filtered off. Mix this with equal parts of rose water. About 0.88 ounces (25 g) of pure alcohol is mixed in for every 3½ ounces (100 g) water. The beauty water is stored in tightly corked bottles.

Rose water for body care.

Rose Hand Cream: This cream makes roughened hands smooth again. To make it, one needs 2 tablespoons (30 ml) of fresh rose petals, 4 tablespoons (60 ml) of almond oil, 8 tablespoons (120 ml) of lanolin, 4 tablespoons (60 ml) of glycerine, and several drops of rose oil. All the ingredients can be bought. The production of rose oil is described in the section "Fragrant Roses—Roses with Soul," which begins on page 115. The moisturizing cream is produced as follows. Pour hot mineral water over the petals; let the mixture draw for awhile and finally cool off. Carefully put the glycerine, lanolin, and almond oil into a glass standing in a water bath to melt. Now the mass is thoroughly and evenly stirred together with patted-dry, finely chopped petals as the rose oil is added.

Rose Bath Oil: A great many rose-scented bath oils are available commercially, but their effect and fragrance is not always satisfactory. Anyone who wishes to be certain that the bath is going to be luxurious and relaxing should use the following recipe. Mix three parts glycerine with one part rose oil. Put the mixture into a decorative bottle, which is then stored, corked, in the bathroom. When preparing the bath, take a spoonful of the valuable essence, and hold it under the hot water as the water runs into the tub. After the bath, sprinkle the body with cooling rose water. Cleopatra was said to have enjoyed her royal bath in this fashion.

Rose Bath Packets: The Romans in the highest levels of society prized fragrance packets filled with rose petals, which they carried in their clothing. These packets can be made of muslin or cotton. They can be thrown into the water to create an aroma. The petals are dried and prepared as explained in the section "Roses for Floral Arrangements" under the heading "Fragrant Potpourris" (see page 128). The further addition of other herbs such as lavender or rosemary increases the fragrant effect.

Rose Steam Bath: The relaxing treatment of the facial skin with rose steam is simple and effective. People with normal or oily skin can use this pleasant, fragrant facial sauna; those with sensitive skin had better avoid it.

One simply puts one or two handfuls of fragrant, untreated flower petals into a dish with hot water, lets them steep briefly, and then holds one's face over the rising steam, being careful that it is not too hot.

If a hand towel is placed over one's head and the bowl during the steam bath, the rosy steam has an even more intensive effect on the facial skin. The pores open, the skin is cleaned and refreshed.

The addition of a few drops of rose oil can increase the pleasant effect of the rose steam bath even more.

Finally the face is patted with rose facial water to cool it.

PRACTICE

At the time of purchase, the rose is at first only a piece of wood with great expectations fastened on it. To ensure that these will be fulfilled and the rose can live up to its reputation as a supershrub, it is necessary to consider some things about planting and maintenance.

The Rose Site— A Place in the Sun

Roses need sun and air. The gardener who follows this formula has already taken the first step toward fulfilling his or her rosy garden dreams. The choice of an appropriate site for the desired rose variety—or, to put it better, the choice of the appropriate variety for the available site—substantially determines the success with the rose as a design element.

Simply said, the gardener asks himself or herself several questions. How can I tell whether the sites existing in my garden are suitable for a rose planting that will give me much pleasure and little trouble over the years ahead? What must I be aware of? What rules should I follow? As is so often the case, no patent recipes apply to choosing and gardening with roses. This is because the **siting equation** is composed of several factors that are interdependent. The particular factors in the formula for success with rose plantings are light, soil, climate, and robustness of variety. If one of these values changes, it influences them all. Another value may not act as compensation to counter it.

Light

Roses love sun. Of course, the sun requirements and heat tolerance differ among the individual species and varieties. However, despite all species-specific differences, roses remain children of the sun. Their particular requirements depend, among other things, on genetics, the size of the leaf, and the general siting conditions.

Genetics: If one considers the extensive distribution area of the genus *Rosa* and its many species, it becomes clear that very different light requirements could have entered the gene pool of a cultivated variety through the hybridized wild species. For example, the wild roses from an almost polar habitat are attuned to a short summer period with long days. Rose breeders have taken these genetic givens into consideration in their breeding work.

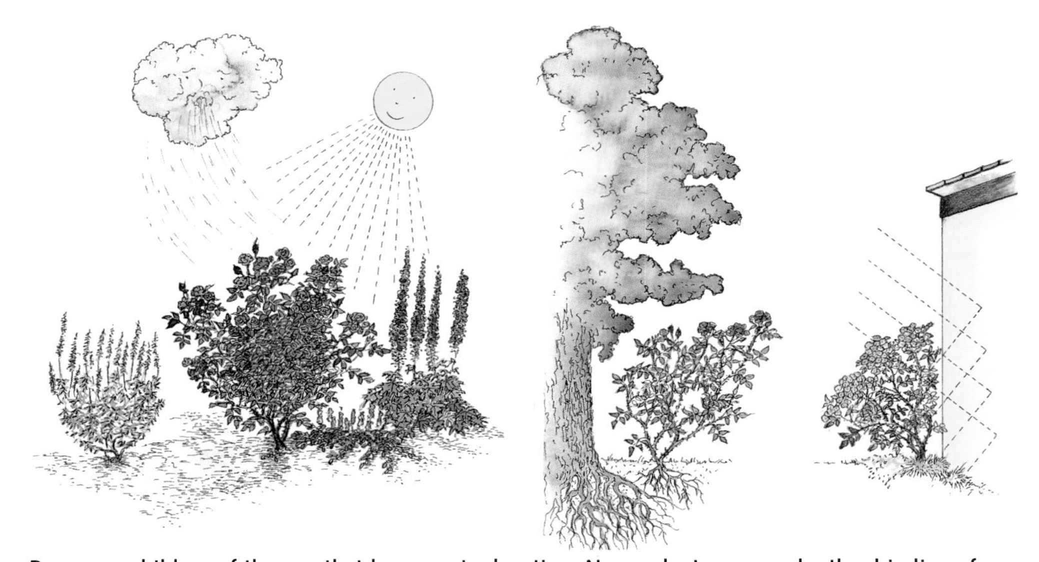

Roses are children of the sun that love an airy location. Never plant roses under the drip line of large trees. Reflected sun in front of south walls can lead to leaf scald.

During the selection processes throughout many years, rose breeders investigated whether a new variety corresponds to the local climate. All the roses described in this book are chosen with this in mind.

Leaf Size: Roses are sun worshipers, but not all varieties tolerate dog day heat the same way. In general, the small-leaved varieties, which can be found in the group of area roses, for example, display greater heat tolerance and also thrive in exposed southern locations.

Site Conditions: Sunny sites offer light, the important flower and leaf fuel, in generous quantities. Furthermore, wet leaves dry off faster, which again considerably diminishes the risk of fungal disease attack. However, caution is suggested in planting roses in front of **hot south walls.** Normally, the rays of the sun strike only the top sides of the plant's leaves. However, heat-trapping walls and plastered or cemented soil all around reflect the sun's rays, so that a high dose of light strikes the undersides of the leaves as well. In spite of good sun tolerance, the leaves can be regularly scalded as a result, especially if there is no cooling, light-absorptive upper-soil surface. Such sites are also problematic because of their very dry air. These create ideal preconditions for the spread of spider mites, also known as red spider, which stress the heat-plagued plants even more.

Soil

Roses love deeply cultivated, sandy-loamy soils with a sufficient portion of humus. Many books about soils describe the characteristics of pure sand, clay, and humus soils. However, these pure soils are rarely encountered in the garden. Much more often, it is knowledge of mixed soil forms and their components that is of importance in evaluating the physical characteristics of a soil.

A good rose soil is always a successful team effort by humus, loam, and sand in which no component assumes extreme prominence. The spade test shows the tendency in one's own garden soil. If the soil can be turned over without any difficulty, you are dealing with a rather light, sandy soil. If digging quickly becomes a sweat-producing sports exercise, you are dealing with a heavier soil with a high clay or loam content.

Culturing roses in pronouncedly sandy soils with poor water and nutrient storage demands a great expenditure of effort in watering and maintenance. Gardening roses in this sort of soil is practical only after thoroughly improving the soil.

Humus: Humus has a dark color and consists of organic material. Since it is produced by dying organisms, it has the capacity, together with the clay portions of the soil as partner, to retain nutrients

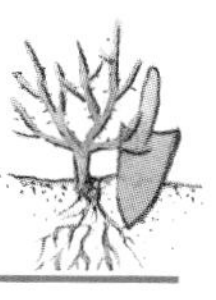

important for the roses. Humus also gives up these nutrients gradually. In humus-rich soils, there is an active soil life. These soils are good storers of moisture and warm easily. The humus content in a rather sandy soil should be about 8 percent; heavier loamy soils, on the other hand, manage with a content of 3 percent. The humus content can be increased by working garden compost into the soil before planting.

Loam: A high loam content with its accompanying clay particles provides a great inner surface for the soil so it can bind many nutrients—a basic prerequisite for vigorous roses. Too high a clay content diminishes the water drainage, however, and the aeration of the soil. It becomes a damp, cold, hard-to-work clump, which becomes crusted in dry spells and as hard as cement.

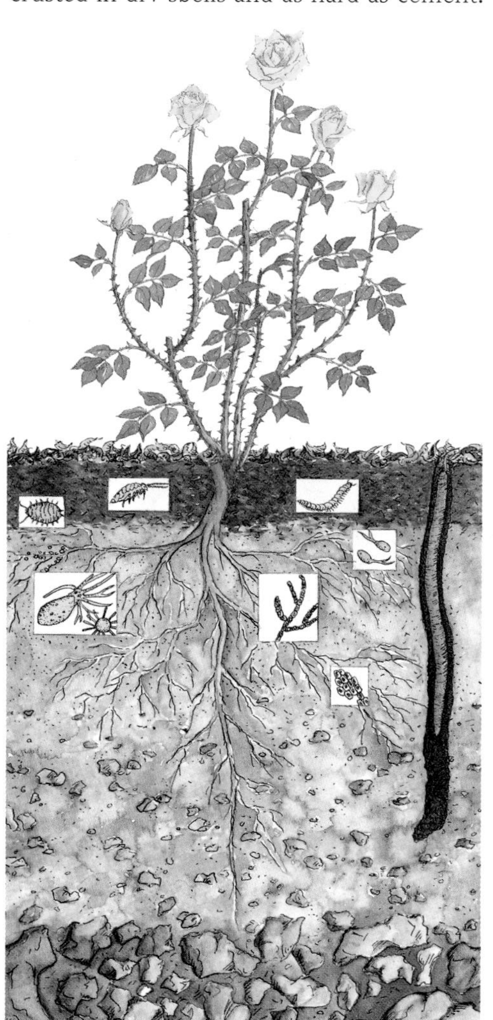

Since they have deep roots, roses prefer deeply cultivated, active soil.

Sand: With its coarse granularity, sand benefits soil, and the roses grown in it, in various ways. Sand loosens loamy soils and those with a high clay content. It also provides for better aeration and facilitates water drainage.

Deep Cultivation: Roses root deeply. This means that they anchor themselves in the soil with the aid of a deep-growing tap root that sometimes grows as long as several yards (meters). If this root encounters any insurmountable obstructions in the upper soil layers, the plant suffers and displays only meager growth. The soil depth should especially be checked in gardens planted near new construction. At these sites, the topsoil has sometimes been poured onto compacted construction debris. A soil layer that roots can penetrate of least 24–32 inches (60–80 cm) is absolutely necessary for successful rose planting. A basic application of fertilizer before the garden is finally installed supports soil life activity.

Macroclimates, Local, and Microclimates

The climate determining the characteristic course of the weather in an area during the year consists of temperature, humidity, air quality, day length, length of the summer, air flows, and intensity of sun radiation. It cannot be influenced on any large scale. Three different aspects of climate—macroclimate, local climate, and microclimate—will be explained next.

Macroclimate: If one thinks of the Mediterranean Palatinate with its mild winters around Freinshelm and compares this German Tuscany with the arctic cold of a winter in Saxony, one must be aware that a uniform, macroclimate does not exist from the point of view of roses. Therefore, nationwide species recommendations are always relative and mistakes are not ruled out.

Local Climate: Because no single, consistent climate exists in all the regions of Germany, the goal of classifying a variety should be to describe the necessary average climate of a locality. Sometimes enormous regional differences arise because of landscape-imprinted characteristics, perhaps lakes or mountain chains, but also because of striking local differences. So, for example, in the Black Forest, the high altitudes of the mountain fields and the habitually warmer valley altitudes around Freiburg are only a few miles (kilometers) apart as the crow flies. Nevertheless, both have entirely different local climates. Different roses thrive in these two areas. Advice on the choice of plants in such situations should be obtained from local garden centers and nurseries.

Microclimate: This term denotes the climate in the smallest garden area. While the macroclimate and local climates cannot be controlled by the hand of humans, the gardener can contribute very much to a rose microclimate in his or her own green refuge—especially with the knowledge of which microclimates are not very pleasing to roses.

Such a microclimate with a negative effect on roses has already been presented in the section "Light," in the discussion of **hot south walls** (see page 142). Unusually high temperatures can also develop on southern hilly sites. Only roses that are not heat sensitive are appropriate for such a situation.

Another microclimate that is associated with risks for rose plantings is found in tight, **unventilated corners,** in which there cannot be sufficient air exchange. These niches are favorite landing places for fungi, especially powdery mildew. In addition, pests like aphids, which show themselves to be dedicated rose lovers, often thrive on roses planted in unventilated corners.

Under the **drip line** of old, large-crowned leafy trees exists another microclimate with bad consequences for the rose. When under the dripping of the tree crown, the foliage of the rosebush cannot dry off quickly enough because of the increased humidity. This environment thus offers an ideal breeding ground for powdery mildew and spot anthracnose fungus. Foggy areas also promote fungal disease with their extremely high humidity. However, these locations are not to be put into the same category as areas with heavy downpours and classified as fundamentally unsuitable for roses. One can definitely find suitable "rain roses" (see page 104), provided, however, that the variety selected also receives breezes so that the foliage can dry quickly.

The microclimate of a semishady site is not considered to be particularly tolerable for a rose either, as a rule. However, by considering a selection of robust hybrids, one can possibly plant roses here (see page 104).

Finally, hollows where cold air collects offer two potential problems for roses. First, one encounters an increased risk for wintering over. Second, these hollows pose a high danger of late frost for the new rose growth. The young, soft canes can tolerate a few degrees below freezing for only a very short time. This also goes for the new growth of especially frost-hardy roses such as the *rugosa* hybrids.

This survey shows that roses do not like any climatic extremes. Notes on the varieties that will nevertheless tolerate specific sites are found in the section entitled "The Rose for Special Situations" (see pages 99 through 109).

Rose Sickness (Specific Replant Disease)

Rose sickness is a complex, not yet completely explained phenomenon. It can occur when roses are replanted in an area that has already been used for growing roses or with other representatives of the large rose family (Rosaceae)—for example, apples and pears. Thus the rose is not actually sick. Instead, the problem is accurately termed *replant disease* or *specific replant disease,* as it is known. When this occurs, the many metabolic effects of bacteria, nematodes, and substances given off by the roots of the previous Rosaceae affect the new planting of roses. This results in the freshly planted roses showing poor growth. Simply, one might say that roses do not feel very well in the "garbage" of their predecessors.

Old, deeply rooted roses can grow best for many decades in the same spot without "soil exhaustion" occurring. One need only think of the more than 600-year-old rose at Hildesheim Cathedral. This fact makes it clear that the deeply sunk roots of old plants grow toward the nutrients and,

Failing roses on soil with rose replanting disease.

of course, always into ever more remote layers that are still "unrosed." In the wild, roses do not as a rule grow into areas that are already filled with the roots of earlier rose plants. Replanting with the human imprint was not foreseen in the developmental history of the rose.

What can be done about rose sickness? In the first place, there is no need to be dramatic about the phenomenon. Even in tired soil, a successful rose planting can be achieved with the help of a soil exchange. To do this, the soil must be removed to a depth of 20 inches (50 cm) and replaced with new, "unrosed" soil. It is a tiresome procedure, but the gardener is then on the safe side.

Recent experiments at the Deutsche Rosarium in Dortmund have suggested a less sweat-producing alternative for dealing with specific replant disease—if dealing with the problem becomes necessary. All the accessible roots must be pulled out with meticulous care when uprooting the old rosebush or picked out of the soil afterward as much as possible. The soil must then be improved by adding horse manure that was composted for a year and by using unsprayed straw for mulch. Use a wheelbarrow load of horse manure for every 43 square feet (4 m^2) of land planted. (Use caution: residues of weed killers can negatively affect roses!) Several months later, roses of vigorous, robust varieties like 'Surrey' and 'Lavender Dream' should be planted. At that time, a small amount of well-rotted manure should be added to the planting hole. When researchers at the Deutsche Rosarium used this procedure, they observed during the years immediately following that no diminution of vigor resulted from specific replant disease.

Whether any rose sickness will occur can be determined by using a test. Growers of cut roses can use the following procedure. They must plant several cuttings of a pot rose variety, which basically roots very easily. One cutting should be placed into an untreated test sample from the old rose bed. Another cutting should be placed into a treated test sample from that same bed. This treatment consists of steaming the soil at 212 °F (100 °C). If the cutting grown in the steamed soil clearly grows better than the one grown in the untreated soil, the gardeners will have to reckon with rose sickness when planting new roses in the particular bed.

TIP The phenomenon of specific replant disease does not occur to the same degree in all soils. Special care is needed for *light soils poor in humus,* in particular. In soils that are rich in humus, this problem, if it exists at all, is noticeable only in a considerably diminished form.

About Planting

Dealers in roses make a living from their wares. Understandably, each one gets the best price he or she can for the roses. Therefore, in the colorful catalogs, dealers show each variety from its most beautiful side. These pictures do not fail in their effect, as experience shows. They seduce the viewer into spontaneous rose purchases. Many rose novices acquire their first roses this way—and are disappointed when the variety does not live up to their expectations if it does not harmonize with other plants in its color and growth habit, for example, or becomes at all diseased. The novice then blames the variety.

Buying Roses

When buying roses, there is really no such thing as a "wrong" rose—however, often the site available and the chosen rose do not go together. Since it is easier to fit the choice of rose to the given site than the other way around, the following questions about the siting conditions are extremely important to answer. What light conditions prevail? Must more robust varieties be selected because the site is not optimal? Is it a long-distance effect that is wanted or will the plants be up close to the viewers? What color are the flowers of the plants surrounding it? How big might the rose become?

After it is clear which type of rose and what colors can be considered, one can again bury oneself in the catalogs and/or books. The long winter evenings are good for this when it is unpleasantly cold, dark, and melancholy. A carefully considered selection repays the effort.

Sources of Supply: When the choice of variety is clear or at least narrowed, the question of where to purchase arises. Basically, there are two ways to buy: either by mail order or by buying directly on the spot.

Mail Order: Nurseries sell by catalog and list their varieties. The known growers are traditional firms, which as a rule send quality plants and supply responsibly. Although offers and catalogs are sent all year long, the supplying of plants is limited to the usual planting times in fall and spring. Anyone who is interested in particular new introductions should make his or her wishes known early and make an advance reservation since new varieties—and any special rarities—are not available in large numbers and are quickly spoken for. Anyone who does not worry about telephone bills can also turn directly to the supplier and telephone with specific questions about a variety.

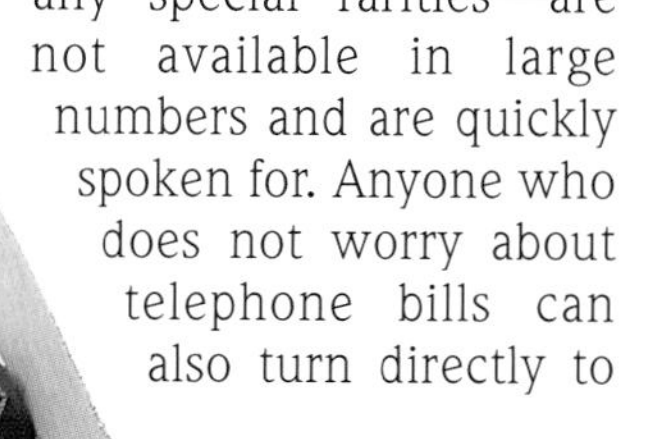

Clean packaging of mail-order roses.

Plants sent by mail often reach the buyer at a time that is unsuitable. **Short-term storage** of the unopened package in a cool but frost-free place—perhaps in a garage—is possible without causing any problems. Should the roses that were ordered arrive during an extended period of freezing, the package should be allowed to thaw slowly in a frost-free place.

Buying Locally: If the choice of variety is not yet clear and intensive consultation is desired, it makes sense to take a walk around a tree nursery, rose nursery, or garden center. Also, home builders' supply stores with connected well-supplied rose departments are a good place to buy. It is critical to examine apparently cheap examples of nameless roses in supermarkets that are offered on sale on the rummage counter with little dignity and no information. Without knowledge of the exact variety, planting such roses becomes like playing the lottery, with numerous duds included.

Direct from the Grower: An additional advantage of buying from a nursery that grows their roses themselves is that during the summer, in July and August, one can look at the desired variety *in natura* in the fields. This can be fun for the whole family and can be linked with an excursion to the country. The impression the plant makes in the field can often say more than many pictures and also shows the flowers in their natural colors. In addition, regionally grown plants are adapted to the prevailing climate and soil conditions. Besides, the varieties offered are guaranteed and, as a rule, are tailored to local conditions.

How They Are Sold

The trade offers roses in a wide range of forms. Thus, it is unimportant which kind of rose is involved, whether it is sold as a bush or a standard.

Fundamentally, all roses sold are harvested in the fall. Depending on the form of the offer, however, these plants come into trade at various appropriate points in time. For the best planting times for the different planting forms, see pages 147 and 148.

Bare-Root Plants: The sale of roses as bare-root plants is the original and most traditional nursery form. Bare-root roses are sleeping beauties, which are in a stage of total winter dormancy. The customer can see the quality of a plant at a glance in this unpackaged way of selling. However, during and after buying in the fall (from the middle of October) or in early spring, one must make absolutely sure to protect bare plants—but especially the fine roots that have no bark—from direct sunlight and drying. Even a brief exposure of the roots to brilliant sun or strong drafts (e.g., in the nursery delivery truck) can lead to irreversible damage from drying out and seriously endanger the vigor of the plant.

Bare-root rose with packing removed.

Prepacked Roses: In this type of packaging, the roots are packed in a moisture-releasing natural material, such as moss or potting soil, which is supposed to keep them from drying out. Roses prepared in this way come in a plastic bag or box with a large colored tag that contains abundant plant information for the customer. This type of packaging is commonly found in garden centers or the garden departments of home builders' supply stores. As long as the plants are fresh, this is a tidy package for self-service sale. The above-ground canes of the plants are often waxed to prevent excessive evaporation. After planting, the rose grows through the wax, so the gardener does not have to remove it. Also, removing this wax should never be attempted by force: the danger of

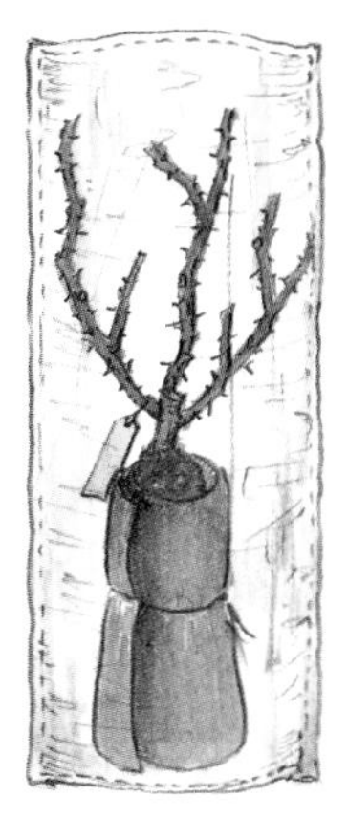

Prepacked roses (also called packaged roses), should not have sprouted at the time of purchase.

injuring the rosebush is simply too great. In addition, great care should be taken when opening the root packing, since the roots have sometimes already grown young, fragile hair roots.

Roses with Balled Roots: A still relatively new mode of packaging has several advantages for the dealer and the gardener. Balled roots means that the roots of the rose are enclosed in a ball of soil, which is held together with a carton or a net. The net is covered with a moisture-retaining plastic bag in addition, which is removed before planting. Net and carton rot in the soil and thus are planted with the rose so that the already active roots are not disturbed by the planting procedure and do not become retarded in their development. The balled rose continues to grow without interruption, even if it is already in leaf and putting out new growth. This is a form that is transportable, decreases generation of trash, and saves resources. At the same time, using roses with balled roots permits the gardener to work with a growing plant until far into the spring.

Roses sold with the root ball may already be in leaf at the time of purchase.

Container Roses: These have the unbeatable advantage of allowing the buyer to see the chosen variety in bloom and in its true color and form before buying. Besides, it can be planted the entire year round except in freezing temperatures. The main time for buying with the greatest variety of choices is generally in the months of May, June, and July, when the time for planting bare-root roses has passed. Since the introduction of roses in containers, nothing more is standing in the way of impulsively redesigning the garden and the terrace during the summer months.

Container roses for summer planting.

The volume of the plastic container ranges between 2 and 5 quarts (2 and 5 l) as a rule. Because of the higher costs of cultivation and transportation, container roses are more expensive than bare-root roses.

Quality

The quality of a rose plant cannot be determined by the form in which it is offered for sale and must be judged afresh in each individual case. The quality criteria described below are good for all garden roses whether they are sold bare-rooted, prepacked, or in pots.

The External Quality: Generalizing about plants is very difficult. Therefore, the establishment of guidelines for measuring the external quality of a plant is the start of an ideal that tries to establish a helpful support for differentiating between good and bad products. The Union of German Nurseries—the professonal association of nurseries with its office in Pinneberg bei Hamburg—has formulated for its members quality stipulations for all nursery plants, including roses. These stipulations are recognized all over Europe by other professional organizations and accepted by all divisions of the trade. Grafted roses are separated into two quality grades: quality grade A and quality grade B.

The picture tag gives information about variety and planting.

Bush roses of **quality grade A** must display at least three sturdy canes. At least two of these must grow directly out of the bud union. The third cane may begin up to 2 inches (5 cm) above the bud union. These bushes must also have a well-branched root system.

Bush roses of **quality grade B** must also be sold with a well-branched root system. They must have at least two canes growing directly out of the bud union.

The number of canes says nothing about the inner quality and vigor of a rose. In years with bad harvest conditions, there can be an above-average quantity of grade B roses for sale. In a good specialty dealership, observing the quality stipulations for roses and selling only plants that are

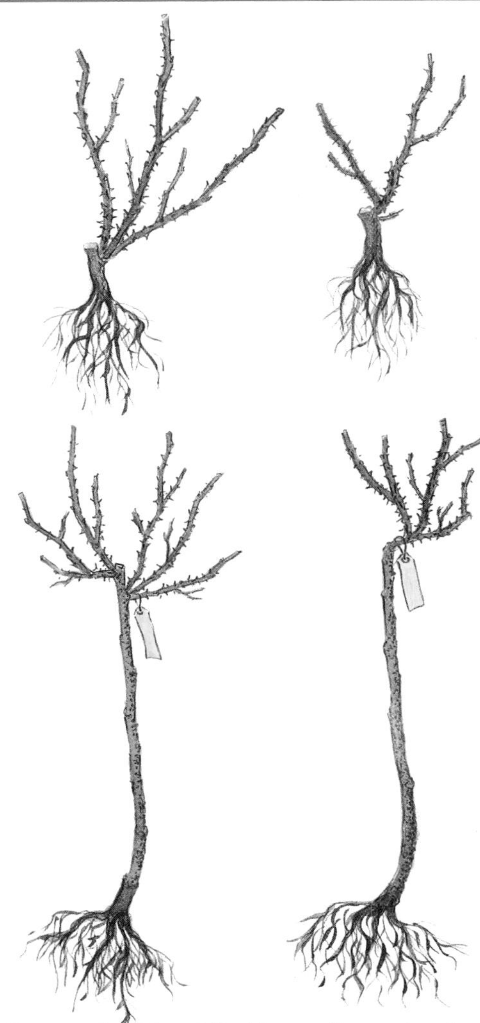

Rose grades #1 and #2. The grading system is based on the number of large, strong canes the plant contains. Grade #1 requires a minimum of three strong canes. Grade #2 must have two strong canes. Grade #3 indicates a plant with only a few small stems growing from the bud union. Only budded or bud-grafted roses are graded.

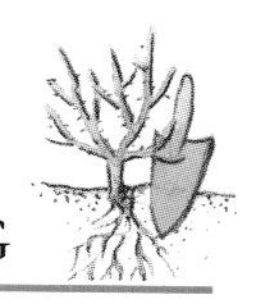

undamaged and properly handled should be expected without question.

Quality stipulations also exist for **standard roses.** Besides the number of canes, these also concern the number of bud unions existing. The experienced rose buyer makes sure that the crown consists of sturdy canes that arise from at least two bud unions (quality class A). Crowns with only one bud union unfortunately tend to one-sided growth, which can be very hard to correct later.

The Inner Quality: The inner quality of a rose cannot be measured and builds on the environmentally correct cultivation by the grower and the appropriate treatment of the plants by the dealer. Correctly cultivated roses display firm wood and smooth, sturdy canes. The roots are not transparent, and a slight scratching with a fingernail reveals a white interior. One should keep one's distance from roses that have bark spots on their canes. The roses are signaling drying injury with shriveled bark; this is caused by improper handling along the way to the point of sale.

Special Examination Criteria . . .

. . . For Prepackaged Roses: The so-called supermarket rose must not be any worse, basically, than the bare-root rose provided that we are dealing with freshly lifted plants that are sold and planted quickly. With a longer sojourn in the plastic, perhaps lasting for weeks, however, roses suffer severely, especially if they are on display in heated, very warm areas. The result is premature development of the bud inside the package; the rose develops very long, light-colored canes. When planted into the garden, such very prematurely developing plants are susceptible to frost and have unsatisfactory root development. Buying one of these leftover packaged roses is therefore inadvisable.

. . . For Container Roses: A good container rose has a pot with a minimum volume of 2 quarts (2 l) and shows a well-developed root system. Roses that have not grown in pots and have just been potted in the container shortly before sale do not justify the higher price of a container rose. They can be recognized by the insufficiently root-permeated, loose ball. As a rule, they grow only haltingly after planting.

After a year, the root ball planted with the rose is rooted throughout without any difficulty.

Do not try to remove the protective wax coating!

Planting Times

Roses can be planted year-round, except during frost periods when the soil is frozen, since they are always available in the market in one or another of the forms described. However, each form has its own particular best planting time. The important differences are explained below. In addition, other aspects, like the location of the site or the type of soil, can also have an influence on the planting time. A further note: often a bare-root rose in its leafless state looks like a lifeless stick of wood with thorns and makes one forget that it is a living plant that needs to be treated carefully. One should know that each transplanting represents a drastic turning point in the life of a rose. The more the requirements of the rose are met at the beginning, the more rarely it will disappoint later.

Fall Planting: The best time for planting bare-root and prepacked roses is between the middle of October and the middle of November. In this period, the garden soil is still warm enough, and the freshly planted roses start right when developing new root growth. Thus, the warmth remaining in the soil from the summer can be used to secure the rose an ideal starting position.

Before the middle of October—or even as early as September—it is less advisable to plant bare-root or prepacked roses, for the plants used for this must have been harvested too early. Their canes are usually not mature enough. Insufficiently matured roses quickly fall victim to frost the following winter.

Spring Planting: Bare-root and prepacked roses can also be planted in early spring as soon as the soil has thawed. If roses are to be planted in particularly cold situations or in extremely heavy soils, spring planting is generally preferred to fall planting.

Most nurseries winter over the plants to be sold. This sometimes occurs in gigantic climate-controlled chambers that are reminiscent of refrigerators. Doing so prevents premature new growth of the rosebush. The humidity within these chambers can be optimally controlled. This prevents the roots and canes from drying out. Roses held back in this way can be planted until May. However, their optimal handling (hilling) afterward is absolutely necessary, especially if the daytime temperatures are already reaching summer values. Remember the following rule. The later in the spring bare-root and prepacked roses are planted, the more difficulty they have getting started. Anyone who wants to avoid pitfalls should fall back on balled-root roses (net or carton) for late planting dates in spring.

Summer Planting: The entry of the container roses into the sales areas of garden centers and nurseries has given the rose lover a completely new planting season for roses: summer. With proper planting and regular watering, even in extreme summer temperatures, container roses settle in and grow on in the garden without any risk. When anchored in a firm root ball, the fine, vitally necessary root hairs remain uninjured and undisturbed at planting.

Soil Preparation

The soil is a complex organism with countless life forms inhabiting it (see drawing, page 143). Without this soil life, the life of plants and, as a consequence, of humans and animals would not be possible. Thus, the soil should be optimally prepared, especially for a rose planting that will be there for many years.

The Rose-Loving Earthworm: The inestimable value of the earthworm as useful collaborator in the garden was recognized by Charles Darwin more than 100 years ago. Through his sensational publication *Formation of Vegetable Mould, Through the Action of Worms, with Observations on Their Habits,* he freed the earthworm from its dominant image at that time as a harmful gnawer of roots. To be sure, the earthworm does also eat parts of plants but only if they are already dead. With the help of its digestive system and by using clay minerals, it turns roots into the best humus.

Approximately 200 earthworms plow a square yard (square meter) area. There are

The greater the earthworm population, the better the roses grow.

Green fertilizer for the rose-bed-to-be: *Tagetes erecta* **(marigolds).**

deep- and surface-grubbing species and also garden compost worms. Earthworms not only provide for more humus in the soil, they also loosen it with their underground tunneling. This induces aeration and water flow, and it highly promotes the thriving of roses. The number of earthworm passageways gives information about the density of the population. A thousand tunnels per square yard (square meter) is considered the ideal for garden soils. Anyone who wishes to may establish the number by scraping off an area of 20 × 20 × 2 in. (50 × 50 × 5 cm) of soil. The number of tunnels found multiplied by four gives the total number of earthworms per 11 square feet (1 m^2).

Green Fertilizing with Tagetes erecta: Sowing a green fertilizer in April, before roses are planted in the fall, improves the soil structure and—by increasing the humus—the fertility of the soil. In addition, green fertilizers are important food plants for bees and other beneficial insects. Especially suitable green fertilizer plants for the preculture of roses are *Tagetes erecta* (marigolds). According to the results of recent experiments that were performed on North German sandy-humus soil, this and only this *Tagetes* species drives away nematodes that are inimical to roses, in particular those of the dangerous genus *Pratylenchus.* However, the prerequisite for this environmentally friendly, groundwater protecting nematode repellent is that the surface be utterly weed free before sowing. This is so very important because, otherwise, the nematodes can find refuge on the roots of the weeds. The North German results emphatically advised against sowing *Phacelia,* as is often recommended in the literature, since *Phacelia tanacetifolia* (Fiddleneck) is also classified as a host plant for nematodes. This information is especially true for areas on which there were also other representatives of the rose family, such as apples, pears, pyracantha, and many more, before the green fertilizing.

Garden Compost: Garden compost is a preferred and attainable soil improvement material. Through its use, organic materials are returned to the biocycle. Besides, garden compost promotes an organized, controlled growth of the rose since the nitrogen elements contained in it are very slowly released according to the soil temperature. Thus, these nitrogen elements flow in harmony with the plant growth. At the time of planting, about 30 percent garden compost can be mixed in with the filling soil. The importance of garden compost as a provider of nutrients in already-established rose plantings is discussed under the heading "Fertilization" in the section "Garden Compost" (see page 161). The use of compost makes the addition of the valuable natural raw material peat—which has nothing to offer roses anyway—superfluous.

Weeding: Before the actual planting of roses, all the rooted weeds must be removed from the bed. Use a spading fork to loosen the roots in the damp soil. The work is really tiring and you may feel tempted to skimp on it. However, the weeds that remain will certainly be a much greater plague in the rose bed later than is their timely, painfully careful removal ahead of time.

Planting

Before planting, the above-ground canes of bare-root and prepacked hybrid tea roses are **cut back** to about 8 in. (20 cm) (pruning length). If thicker roots are injured or bent, they can be cut back to the point of injury. However, under no circumstances should the fine roots be cut, because the plants are able to feed only through them. The more intact root hairs a rose has, the more easily and faster it will grow and, therefore, put out new growth. Afterward, the rose is plunged into a water bath to cover all its canes and roots for two to three hours whether it is planted in fall or in spring.

The **planting hole** should be about a handbreadth larger in height, depth, and width than the ball or the circumference of the roots. The soil that has been removed is piled to one side of the hole. It can be mixed with garden compost.

Neither peat nor inorganic fertilizers should be used when planting a rose; just a handful of horn meal (about $1\frac{3}{4}$ ounces [50 g]) or some grains of slow-release fertilizer, which give up their nutrients slowly and evenly, is added with the soil.

The bottom of the planting hole is thoroughly loosened with the spading fork.

The rose is placed—or, if there is someone else to help, held—vertically in the hole. The roots should be able to dangle freely in the air without having to bend.

It is important to make sure that the **bud union** is 2 in. (5 cm) below the soil surface. In very clayish, cohesive soils, the bud union should be only $1\frac{1}{4}$ inches (3 cm) deep. This suppresses the tendency of

Planting Roses Correctly

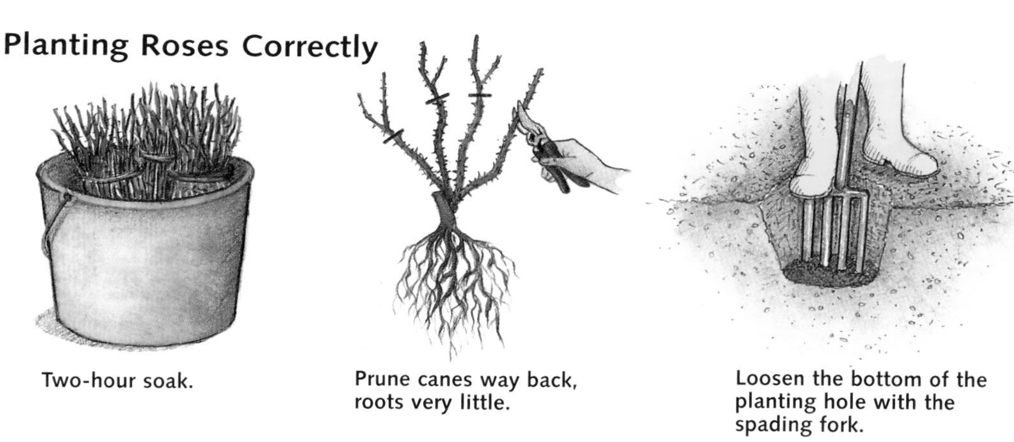

Two-hour soak.

Prune canes way back, roots very little.

Loosen the bottom of the planting hole with the spading fork.

Set the bud union 2 inches (5 cm) deep; the roots should be able to dangle free.

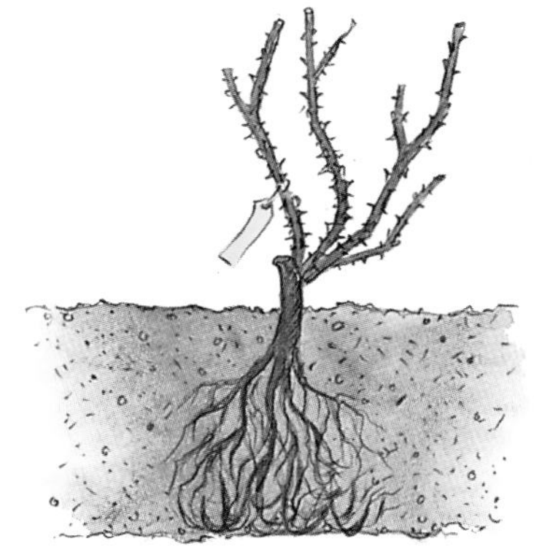

Wrong: The bud union is above the soil surface; the roots are bent in the planting hole.

The hole is filled in while gently rocking the rosebush.

Thoroughly water by using the hose or watering can.

Gently but firmly tamp down soil with the heel.

Always hill with soil!

Left: Go easy—cut back only roots that are overlong or injured!
Right: A stick laid across the top of the planting hole helps determine the proper planting depth.

the understock to send out suckers. Frost protection for the rosebush is increased.

A stick laid across the hole can help to make sure the plant is upright.

The hole is now filled up with the amended soil and tamped lightly around the roots with the heel. The hole is then filled with water with the watering can without the spray head or with the garden hose. Creating a small wall of soil around the newly planted bush keeps the muddy water from flowing away.

Finally, soil is shoveled in high enough so that only the tips of the canes of the new plant are showing. Roses must always be hilled with spring planting. Hilling protects the rose from drying winds, sun, and frost. In addition, the evaporation from the canes is reduced until the the plant can take in nutrients by using the newly anchored roots. The hills should not be removed until 8 weeks after spring planting—for fall planting, they should be removed at the end of March.

Carefully remove the pot before planting container roses.

Special Planting Tips For

▶ ***Balled Roses:*** Plastic packaging must be removed before planting. Net balling or balls in cartons may, as a rule, be planted as is. Note the detailed directions on the package about this.

▶ ***Container Roses:*** The plants are taken out of the pots. The balls are sunk into a pail of water until no more bubbles arise. Balls of container roses consist of a loose, humus-containing, nutrient-rich planting soil.

For the roots to grow into the heavier soil of the garden after being removed out of this paradise for growing, they must be well loosened. Otherwise, the roots will take the path of least resistance and stay in the area of the original ball without really anchoring in the garden. This would, of course, be detrimental for the development of an independent water and nutrient supply.

Very heavily rooted balls should be loosened at the bottom of the hole and any matted roots broken apart.

▶ ***Standard Roses:*** Planting is carried out as for grafted rosebushes, but it is not usually necessary to worry about a bud union at the base. In addition, a stake the height of the standard is always required. It should be secured with coconut, jute, or raffia in at least two places and preferably in three. The string must never cut in and strangle the trunk. In the fall, the stake and the ties should be checked to be sure they are stable and secure.

The planting holes for full standard roses can be somewhat larger than those for bush roses. However, full standards should not be "shortened" by especially deep planting holes. Doing so may, of course, be in line with aesthetic concerns in certain situations, but it injures the rose.

All freshly planted roses, whether bush or standard, whether in a container or bare root, need sufficient water in the first weeks after planting. If the weather is dry, this must be provided by the gardener—it must also be provided to fall plantings.

Note The amount of space needed for bending over at the snag during subsequent overwintering must be taken into consideration. Thus, the alignment of the snag must be also considered at planting time. More on this under "Winter Protection" (see pages 169 and 170).

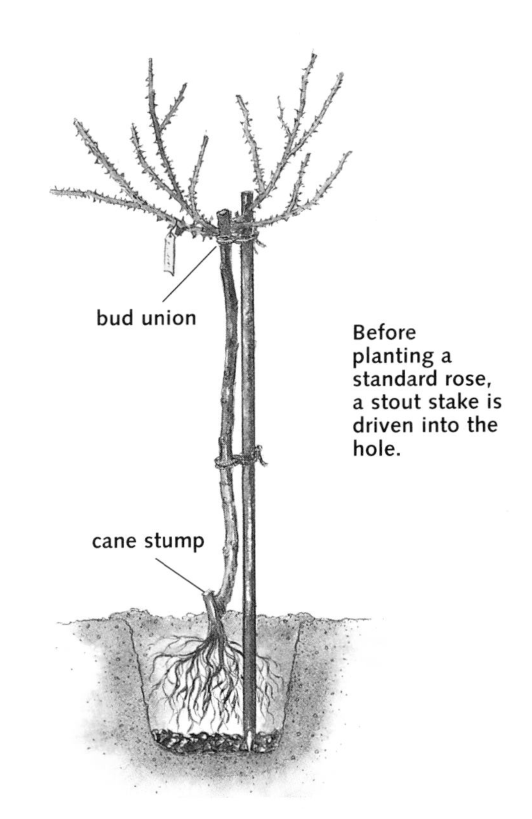

Before planting a standard rose, a stout stake is driven into the hole.

The Care of the Rose

The Calendar—A Month-by-Month Overview of the Most Important Jobs

January	February	March
–read up on rose literature –study rose catalogs –carefully knock large quantities of snow from climbing and shrub roses	–transfer garden compost –develop garden plans –protect roses from winds and any possible intense winter sun by using additional branches	–remove brush protection when snow melts or damp weather arrives –remove hilling at the end of March, distribute hilling soil around the roses –prepare soil for new plantings –undertake new plantings and transplantings –on a cloudy day, uncover soil from crowns of tree roses or remove brush –stake and tie tree roses –if special rose fertilizer is to be used, apply it now –propagation: get started with sowing seeds, take cuttings and start them
April	**May—the spring roses kick off the season!**	**June—rose summer at full speed**
–prune roses –thoroughly clean up all cuttings –loosen soil, remove weeds –mulch after pruning, e.g., with bark mulch –propagation: unhill your own grafted roses, remove any wild shoots –sow any plants intended for green fertilizer for fall rose plantings	–do not cultivate mulched beds, keep weeded –water if necessary –if slow-release fertilizer is to be used, apply it now –remove suckers at the base	–order new rose varieties –visit rose gardens (until September) –soil loosening, weeding –remove suckers –use pest controls and fungicides with only heavy infestation –water if necessary and refertilize –do not prune once-flowering roses if hips are desired
July—roses, roses, roses . . .	**August—in favorable conditions, repeat flowering**	**September—fall flowering everywhere now**
–order new rose varieties –regularly remove spent flowers from repeat-flowering varieties, water –do not apply any more nitrogen fertilizers –tie up long new canes of climbing roses –loosen soil, remove any encrustations from watering by using the grubber or hoe –summer pruning: begin grafting, take cuttings and root	–remove spent flowers, water if necessary –summer pruning for once-flowering roses –remove suckers at base	–prepare plant locations for fall planting, make sure to remove weeds, roots and all –restrain the companion plants to the roses –fertilize with potassium fertilizer with warm weather and continued growth –stop cultivating soil so that no further nitrogen release takes place
October—roses and no end	**November—frost gets the last roses**	**December**
–pick off leaves with fungal disease and burn –start new planting after the middle of the month –propagation: harvest ripened hips, stratify, undertake fall sowing	–provide brush for winter protection –continue new plantings –slightly prune shrub roses to prevent snow breakage –with bush roses shorten only very long canes –transplant old rosebushes –lay down standard roses –propagation: overwinter rooted cuttings in a frost-free, bright situation –begin hilling at the end of the month	–expand garden plans –if garden compost is to be used, add it now –if horn meal is to be used, spread it now

A Course in Rose Pruning

For many gardeners, how to purne roses is a mystery. There is so much advice given in literature and in catalogs on this subject that in the course of the lectures the overview is frequently lost. Rose pruning, after all, is no science, although many discussions give this impression.

Many of the pruning methods practiced today have their beginnings in Victorian England. If one is to believe the English sources, even at that time some methods were not oriented primarily toward the well-being of the rose but developed as busywork procedures for keeping the countless numbers of gardeners employed during the off-season.

Left: Severe pruning—few but large canes. Right: Gentle pruning—many but thinner canes.

Pruning Basics: Certain rose groups unarguably need regular pruning. For instance, only by being pruned regularly can bush roses be rejuvenated. Light can reach the lower, dormant buds so that new canes develop at the base. Without pruning, the plants develop into regular "coat racks" with decreasing flowering and size of flowers. **For modern, repeat-flowering varieties the rule is the more young wood, the more flowers.** Also, the wild rose—the forebear of all cultivated varieties—is used to regular rejuvenation in its natural habitat along the edges of southern forests by means of the browsing of animals. In other groups, such as the *rugosa* hybrids, many canes die back disproportionately as a species-specific characteristic after the third or fourth year. If these canes are not removed and thus the development of new growth not stimulated, many of these varieties will quickly age and die out.

A careful pruning also improves the light conditions within a rosebush so that more and better-developed leaves can be produced. Often the important function of the leaves as lungs of the plant is forgotten. In conjunction with a massive fertilization, more building materials (assimilates) can be produced for the rose over a larger leaf area and be made available for plant development with proper pruning.

Scope of Pruning: Argument continues about the extent of pruning roses. In this connection, one can observe that the advocates of moderately pruning their roses often live in a vineyard climate, while those who advocate more extreme pruning often have to deal with freezing situations. There the canes of the roses freeze back so regularly that pruning is unavoidable. However, the rose lovers in milder climates are spared this. Awareness of the very different climate zones throughout the United States thus illuminates the controversial discussion of pruning problems.

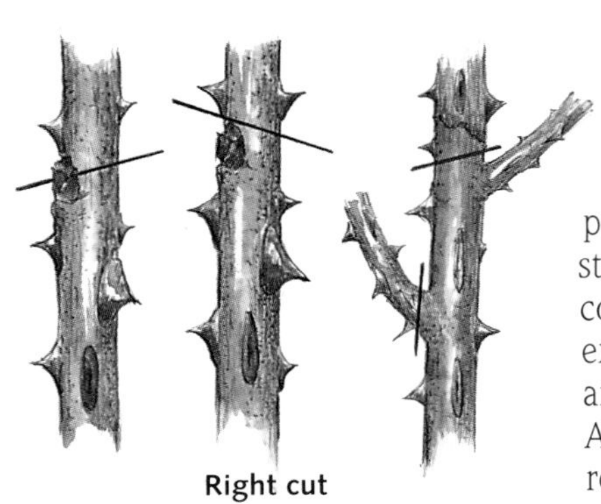

Wrong cut

Right cut

Growth Principles: For this reason, a general rule for pruning cannot be stated. However, the consequences of the extent of the pruning are unambiguous. Anyone who cuts roses back hard and leaves only a few buds, the so-called eyes, will have fewer but longer and sturdier canes. The other way around holds true, too. A light pruning produces more numerous but shorter and rather weak new canes.

> **TIP** In short, weak pruning produces new growth that is weak, and heavy pruning produces new growth that is sturdy.

Making the Cut: Anyone who prunes roses should be sure to make the cut properly. Roses should be pruned on the diagonal about $^2/_{10}$ in. (5 mm) above the eye and, of course, so that the eye is situated at the topmost point of the shortened cane. The cut should not be made at too sharp an angle, so as to keep the surface of the wound as small as possible. Also, no large stubs should be left. The pruning shears used should be sharp. They should not crush the cane but leave a smooth cut behind them.

Spring Pruning: The proper time for pruning roses in the latitude of Germany is spring. The optimal time varies with the individual regions. The milder the climate, the earlier one can begin pruning. In more severe climates, one should wait until the end of April. The flowering of the forsythia is a good timing trigger for spring pruning.

In any case, work in the rose garden should begin only when there is no longer a threat of severe frost. Even if the feeling of spring foments impatience, bear in mind that the new growth of the roses can be severely injured in late frost areas, especially if the morning sun shines intensely on the canes after a clear, cold night.

There are **ground rules** for pruning roses in the spring. First, all canes showing injury from diseases, wounds, or frost—usually brownish in color—are cut back to the healthy wood, which still shows greenish white inside. Stumps of canes that did not put out growth the previous year are also cut back rigorously to the base. This also goes for all thin, weak canes. In doing so, one should not be shy or anxious about wielding the shears. The production of sturdy, healthy new canes with good flower production can be expected only from correspondingly strong, thick wood.

Freezing of roses that have been cut back too hard in fall can lead to the death of the plant because of the loss of vital substance. Spring (when forsythia blooms) is the only proper time for pruning roses.

The once-performed **fall pruning** is rarely done today. In fall, one shortens only very vigorous, long canes to prevent damage from wind breakage or heavy snows.

There is no uniform pattern that can be followed for pruning all rose groups. Each group has its own requirements. Besides, each rose gardener can take into consideration the use to which the plants are to be put and decide, for example, whether there are to be hips or greater flower production.

One more observation: Any large pruning is, in the truest sense of the word, an incisive event in the life of the rose. Depending on the species and variety, it should be accompanied by flanking measures, e.g., by sufficient application of a balanced food necessary for the production of strong, resistant new growth.

Pruning Debris: After pruning roses, all pruning debris should be removed from the beds and borders. Pieces of canes are the ideal nurturing ground for bacteria and fungi, which can attack the canes and roots of the living rose. One recognizes the experienced rose gardener by the cleaned-up rose bed at the end of pruning. Debris from pruning is generally infected with the resting spores of dreaded fungal diseases like spot anthracnose and therefore does not belong on the compost heap. It must be destroyed.

Labeling: When pruning shrub and climbing roses, one must basically distinguish between once-flowering and repeat-flowering varieties. The introduction of repeat-flowering roses was more than just the success of an eagerly welcomed step in hybridization. It caused rules of pruning to become more complex. Therefore, the beginning rose grower should make a note about the flowering rhythm when buying a variety. He or she should attach a label noting it or enter this information into a planting plan. It is also important for a gardener who regularly cares for private gardens and wants to do the work professionally.

Pruning of Miniature, Bedding, and Hybrid Tea Roses (Bush Roses)

In the spring, after hilling is removed, miniature, bedding, and hybrid tea roses receive the hardest annual pruning of all the rose groups. First, all diseased and dead canes are removed. Thus, one gets a good overall view of the healthy parts of the plant. The height at which pruning begins varies according to the growth rules and the desired development. This height ranges from a handbreadth (perhaps with miniature roses) to 8 in. (20 cm) above the ground.

"Summer pruning of roses stimulates new growth the way stealing eggs from a nest stimulates new egg laying." *(Helmut Maethe)*

Place to cut for summer pruning

Some hybrid teas tend to develop only a few new basic canes and to grow loose and scanty. So-called **pinching back** (halving the new growth; see page 115) in the middle of May induces new growth to branch more and thus promotes bushiness in varieties that grow awkwardly. After the first flush of flowers, the miniature, bedding, and hybrid tea roses can be cut back **(summer pruning).** However, this pruning should not be too hard; the proper cut is made over the first fully formed leaf group. Too deep a cut down the cane costs valuable foliage, and with it, assimilation area. A deep cut also massively impairs the production of new flowers, especially in hot weather. The more leaves (rose lungs) that are left, the more speedily the variety will make new growth. Also consider that the corresponding root sections of a rose die back if too much leaf mass is removed in the summer. The important balance between leaves and roots is thus disturbed during the growing phase.

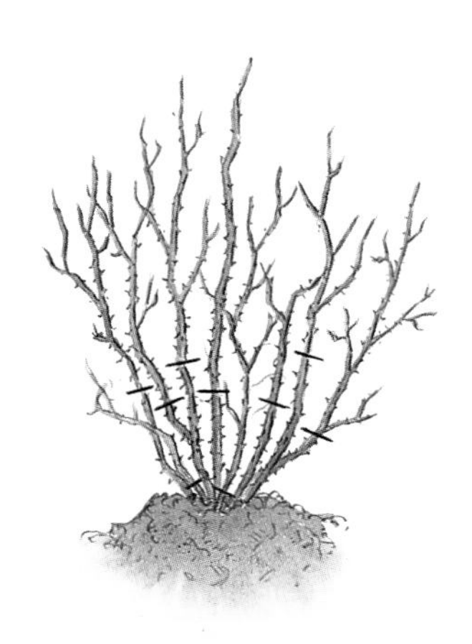

Bedding, hybrid tea, and miniature roses are pruned back to four to five eyes (buds).

Pruning Once-Flowering (Nonrecurrent) Climbers

In Goethe's day, repeat-flowering climbing roses were still unknown. His garden house in Weimar was decorated with nonrecurrent ones. So for him, the rose season was over by the beginning of July. Had he then pruned his once-flowering roses in the spring, the glory of the flowers would have been lost for that year. This would have occurred because once-flowering climbers do not bloom on the current year's long canes. Only the side canes that grow from the current long canes bear the flowers the following year.

Thus, many a nursery dealer or some professional staff in the garden centers must calm angry customers in the fall who swear that they have been sold a "blind" rose that did not produce one single flower the spring after it was planted. Something even worse is possible. If this customer is not correctly advised and again prunes back the first-year canes in the coming year, his or her next flowering failure will be preprogrammed.

Young, nonrecurrent climbers are not pruned at all but only tied up firmly. If they are to grow up into trees with thin foliage, the canes are carefully laid in the right direction or tied to previously installed frames and trellises. Some patience is required, for it takes several years to produce a flowering frenzy.

Older plants are cut back after flowering. The withered flowers sit on small side shoots, which are cut back to two to three eyes. This spurs the bush to develop new growth, which has until fall to mature—an important requirement for it to come through the winter unharmed.

If rosebushes are intended to provide forage for the bird world, the plants need to be decorated with hips in the autumn. Therefore, the flowers should not be cut off in the summer since no others will then develop. Some gardeners will gladly accept the decrease in flowers the following year in exchange.

The withered flowers of the very double varieties tend to develop into rotten, ugly mummies after long periods of rain. These flowers should be removed for aesthetic reasons.

Independent of whether the climber should fruit or not, the new, first-year canes should be left in any case. They are the framework for the flowers the following year.

Pruning Ramblers: Rambler roses with the appropriate amount of space are not regularly pruned at all. However, if they are, they are only thinned, that is, the old, several-year-old canes are removed at their bases. In the Rosarium Dortmund, where annual pruning is even omitted, a special work-saving pruning method for rambler roses has developed into regular rejuvenation. It is also interesting for the home garden, especially if the amount of space available for the very vigorously growing climbers is limited. In rotation every four years, the main canes are cut back to the ground after flowering and only this year's young canes are left—as with blackberry culture.

Pruning Repeat-Flowering (Recurrent) Climbers

Recurrent climbers flower on this year's wood as well as on last year's and several-years-old canes. As with the bush roses, all dead and injured canes are removed in the spring. Weak growth is cut back to the point of issue, sturdy side canes are shortened to three to five eyes after flowering. After the first flowers have finished, one can cut back to a well-developed eye **(summer pruning)** since the repeat-flowering roses form new flowers within six to seven weeks during the summer. These—depending on the variety—can produce hips later.

Overlong side shoots and new long canes are firmly fastened to a climbing support. They are the basis for the next year's flowers.

During pruning, untying the long canes and tying them up again after the pruning is done can sometimes be necessary. Therefore, care should be taken in tying the canes—using a material that will not strangle—to make the ties easy to loosen again.

Pruning Nonrecurrent Shrub Roses

Once-flowering shrub roses are park, moss, and wild roses. These develop flowering side canes on old wood, thus on one- and several-years-old canes.

This group is pruned only very rarely; some connoisseurs completely reject any pruning at all. Only diseased, winter-killed, and dead wood should be removed. Canes that are too close together may be thinned. If canes that must be removed have reached the size of branches, use a saw or heavy-duty tree loppers. If one takes back the canes annually in the spring, as in the bush roses, flowering in June/July will not occur. If any pruning is done at all, the spent side canes are cut back to two to

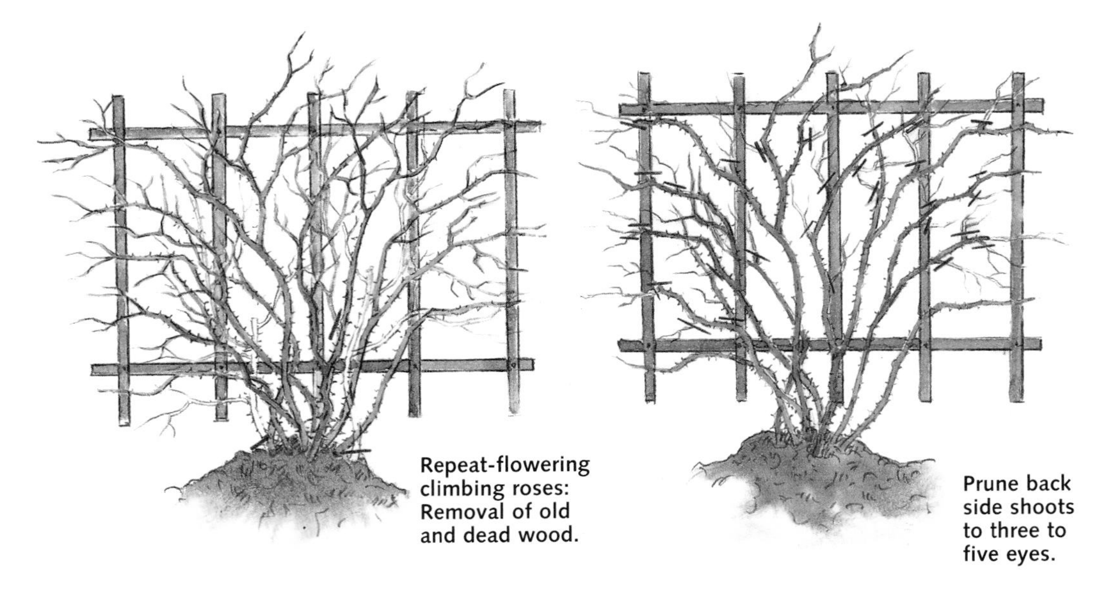

Repeat-flowering climbing roses: Removal of old and dead wood.

Prune back side shoots to three to five eyes.

three buds in summer. Short canes develop that will bloom the following year. This **summer pruning** keeps the giant shrubs in shape and permits their culture even in locations with limited space.

Pruning Repeat-Flowering (Recurrent) Shrub Roses

In contrast with the nonrecurrent shrub roses, the repeat-flowering varieties develop their flower shoots not only as side shoots on older canes but also on this year's wood. The modern varieties are, as a rule, such early or eager growers without pruning that the removal of old canes and occasional thinning is enough. Spent flowers from the first flowering can be cut back to the first strong eye **(summer pruning).** Those of subsequent flowerings, like those of the recurrent climbers, can be cut back after six to seven weeks. Summer pruning does not prevent formation of hips in the fruiting varieties, it just retards them.

If because of winterkill or declining flower production canes ever do need to be cut back, the lead canes are pruned only lightly, whereas the side shoots, on the other hand, are cut back harder. In rotation, all the old, strong ground canes of recurrent shrub roses are removed every four to five years—like many other flowering shrubs, too—right back to the base at the soil surface.

Pruning Old and English Roses

The Old Roses from great-grandmother's garden have returned to our gardens in such numbers that questions about how to prune them correctly have arisen. Pruning directions from the tradition-rich rose country, England, as one can often read in the colorful garden magazines, do not always help with this. They work for English climatic conditions and therefore cannot automatically be applied to the care of central European rose plantings.

The majority of Old Roses only bloom on one-year-old and several-years-old wood. In spring, only the old, diseased, and dead canes are removed. A rejuvenation pruning brings an overaged rose back into form again.

Many Old Roses bloom preferentially on the ends of the longer canes. Instead of cutting them, bending these and anchoring them to the ground seems more reasonable. This forces the plant to develop side wood and thus noticeably increases the flower production on parts of the cane that would otherwise be bare.

English Roses: The nonrecurrent English Roses are pruned like the Old Roses. The repeat-flowering varieties are pruned like modern recurrent shrub roses. A hard pruning is thus often possible, but the typical habit of the variety should dictate. Maintain the appropriate form of the shrub.

Side canes of repeat-flowering climbing and shrub roses . . .

. . . are cut back in the summer after the first flowering . . .

. . . and develop new flowering shoots after about six weeks.

Repeat-flowering roses can go for years without pruning.

If necessary, they can be thinned and a portion of the structural canes shortened.

Pruning Area Roses

Sometimes these groundcover roses cover very large areas. A radical pruning need be undertaken only every three to four years. If recurrent varieties are maintained as specimens or in small groups, they are pruned like bedding roses, although somewhat more moderately, to about 12 in. (30 cm) high.

If the roses are not grafted but grown on their own roots, as are being planted these days in public areas in large numbers, these plants can be cut back rigorously. To do so, use a timesaving method utilized for other groundcover shrubs. Prune these roses to a uniform height of up to 12 in. (30 cm) by using hedge clippers or even a mowing bar.

Pruning Standards

The crowns of quarter-, half-, and full standards are pruned—like bedding roses—back to about 8 in. (20 cm). Care must be taken to maintain an even, but not symmetrically round, pruned crown shape. After pruning, the outline of the crown may comfortably be looser and open. However, this outline should still appear to be natural.

The **weeping** or **cascade roses,** as a rule, consist of very vigorous climbers grafted onto the tree trunk. These are cut back according to the pruning rules for this group—depending on whether they are once- or repeat-flowering. As a rule, this leads to only light pruning, removing overaged canes but preserving wood that is several years old. The wood allowed to remain serves as a basis for flowers. In addition, here the nonrecurrent varieties are not pruned in spring but cut back only right after flowering.

Sometimes the gardener should bear in mind when pruning **weeping roses** that the plant juices must work their way through a relatively narrow trunk. An explosive burst of new output that comes about as a consequence of a very severe pruning results in a limited second flowering.

Suckers

Roses are grafted on to understocks, the so-called wildings. It can always happen that suckers may grow out of the understock—or on standard roses, from the trunk. The sucker must be removed all the way back to the point of growth. Under no circumstances should a stub with dormant eyes be left. These will only produce the next suckers, which will again rob the energy of the hybrid tea variety. Sometimes some soil must be removed to get at the base of the sucker; this is then replaced after the sucker is cut out.

Suckers (above), with their small-leafleted pinnate leaves, overgrow a grafted 'Frau Dagmar Hartopp'.

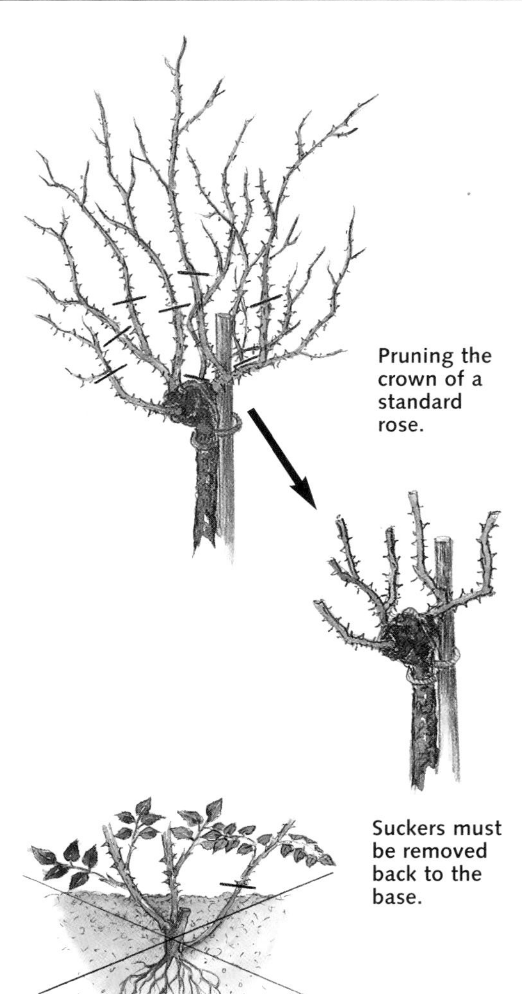

Pruning the crown of a standard rose.

Suckers must be removed back to the base.

First the sucker is unearthed entirely and then . . .

. . . cut or cleanly torn off at the point of growth.

■ The Water Factor

The characteristics of modern, up-to-date roses are lush, healthy foliage and repeated flowering. Both positive characteristics are based on the vigor of a variety. To promote it, the plants must have water available in sufficient quantity, of course. In order to provide enough water, the quantity must be measured by eye and knowledge of the facts. For roses, insufficient water promotes susceptibility to fungal disease, for example. On the other hand, sogginess makes rose roots actually suffocate from inadequate air circulation. The wrong way to water, in the very worst case an evening shower on the leaves, brings even the robust varieties to their knees and turns them into entertainment centers for fungi.

The younger a rose planting is, the more attention must be paid to watering carefully and in a way that is correct for roses. Well-rooted, older plants—and other flowering shrubs—are able, under normal conditions for the site, to regulate their own water metabolism. However, young rose plants need the gardener's supervision until they have developed deep roots and have reached appropriately moist soil layers. Therefore, the following remarks apply primarily to roses that are still in the process of developing roots.

Winter Moisture: A good store of soil moisture during the winter allows the roses to get a good start in the spring. Nutrients can dissolve only in soil that is sufficiently most, thus making these nutrients available for the plant to use. The roots are already active starting in February—depending on the winter situation—that is, long before the aboveground parts of the plant are showing new growth. Nutrients reach the plant through fine root hairs of the roses and are transported inside through the conduction pathways. The rose needs a sufficient quantity of water to carry on this intraplant transportation as well.

Spring Dry Spells: A spell of dry weather during the start-up phase in spring can have a negative effect on roses. With appropriate soil cultivation and soil care, such dry phases can be bridged more easily.

Roses prefer a loose soil. Soil surfaces that have crusted because of sun or rain are best loosened with a grubber or hoe, worked shallowly and not sunk to a depth of more than 4 in. (10 cm). Under no circumstances should there be deep digging between rosebushes. In addition, mulching of roses has shown to be helpful (see pages 157 and 158).

Summer Dry Spells: Roses need sufficient water and nutrients to develop flowers. Repeat-flowering varieties, especially, must be well provided with water after the first flush of flowers in summer in order to develop new flowers. One can determine whether enough moisture is present in the soil by poking a little hole in the soil with one's fingers. If the soil is dry at a depth of about 4 in. (10 cm), a thorough watering is necessary.

TIP The basic rule is: better to water seldom and thoroughly than often and skimpily.

Watering thoroughly means using the hose according to the weather conditions for an hour or more. As a rule, watering for too short a time dampens only the soil surface, and hardly any moisture reaches the deeper layers.

In heavy soils, the intervals between waterings even during hot spells can be up to three weeks. However, in light soils, they may be just eight days.

Especially endangered by drought are climbers and shrub roses that are situated in front of a house wall with a south exposure. The rain hardly gets to the roots, especially if there is an overhang that interferes with the natural evaporation cycle. In situations like that, the rose without sufficient water weakens and is quickly infested with pests like spider mites (red spider).

Never water rose foliage from above. The best option is perforated soaker hoses from which the water gently drips into the soil.

Additional watering should be stopped in September. The wood of the rose should mature properly and be able to go into winter well supported.

Important: This book has pointed out a number of times that water should never be poured or sprayed over rose foliage. The repetitions are intentional. This rule, in particular, is important for healthy roses, and it is disregarded over and over. Water-happy evening work with the hose raining down from above can turn the garden into a "fungus hell" for roses. This should be especially borne in mind, then, if perhaps a lawn that borders the roses is being watered. Damp foliage is the number one cause of the explosive spread of rose fungal diseases.

When to Water? Ideally, watering should be done in the early morning hours. Then leaves that may have been dampened can quickly dry off in the day's sun. The hose is laid in the bed, and the water runs out slowly—under low pressure—and seeps in. Even better, of course, is the use of a perforated soaker hose. The water drips in evenly from its many holes to where it belongs without making detours in the soil. Under no circumstances should the water spray out of the holes under too high a pressure and thus shower off the foliage.

During protracted dry spells in the summer if no water is provided, sometimes large portions of the fertilizer applied in the spring can remain in the soil undissolved. When the heavy rains start again in September, these nutrients are dissolved in a very short time. As a result, the roses begin to grow and, of course, continue into the late fall. Their wood has no time to ripen and becomes easy prey for wintry frosts, especially if the frosts appear suddenly and hit an active plant. In this case, the rootstock may even die.

If the roses are situated in a notably dry or rainy place, particularly suitable varieties can be used in such situations. Special selections are introduced in the section "The Rose for Special Situations" (pages 99 through 109).

Ground Cover Through Mulching: The term mulch derives from the Low German word *mölsch,* which means something like "soft, at the beginning of decay" and refers to the covering of the soil with organic

Anyone who wants to ruin his or her roses for sure should continually water the tops of the leaves.

material. Mulching has been with us for many centuries.

Not only do mulches improve the soil quality but they also spare the gardener continual weeding. If the weeds nevertheless become a problem, they are easier to pull out (weeding) because of the high soil humidity. In addition, the water-saving effect on the rose plants is not to be underestimated. An experiment by the Lehr- und Versuchanstalt (Teaching and Experimental Institute) in Kassel unequivocally established the positive effect of mulch on the soil moisture. All sixty-five rose species in the experiment demonstrated better development with an application of mulch than without it, and this occurred with a smaller expenditure of care and more balanced soil moisture. The soil structure was also shown to be very good and it was unnecessary—as an additional side effect, so to speak—to apply herbicides as weed killers.

The gardener can choose from a variety of **mulching materials.** These include horse and cow manure, unfinished compost, grass cuttings, straw, sawdust, postharvest cornstalks, shredded paper, and bark products.

The application of various mulching materials for rose plantings is a controversial subject among rose specialists. Whether using garden compost of one's own making or bark mulch from the garden center, all mulching materials exert a strong influence on the nutrient makeup of the soil. The effects are quite different and sometimes even injurious to roses. In particular, one should not mulch with fresh, ground wood chips since these break down into free agents that are toxic to roses. Bark products are used extensively for mulching roses. The Institute's results show that of this group, bark mulch is of special interest to the rose grower.

Bark Mulch: By the term *bark mulch* we mean chopped-up bark, which is applied as a soil covering without any further additions. In garden centers and nurseries, a standarized, properly stored material is offered as quality-guaranteed bark mulch. The application of this quality bark mulch has already demonstrated its value with other ornamental shrubs. So the rose gardener would be reasonable if he or she supposes that this positive experience will also carry over to the shrub rose.

Important: The growth of the rose is only uninfluenced by the spreading of bark mulch if nitrogen fertilizer has been spread beforehand. The strongest nitrogen fixing by the mulching material takes place in the first few months after it is applied. If compensatory fertilization is not provided for the roses, it leads to severe symptoms of deficiency. These occur primarily in fresh rose plantings and are recognizable by light yellow leaves. To prevent this, additional fertilizer applications of horn meal or horn chips of about $4^1/_2$ ounces (130 g) per square yard (square meter) have proven helpful. An added application of fertilizer on top of the mulch layer makes the mulch break down faster and is therefore not advisable.

Applying Mulch: Before mulch is applied, the soil should be painstakingly cleared of weeds, especially roots and perennial weeds. Otherwise, they will find ideal conditions for spreading under the mulch layer because of the higher soil warmth and humidity. Depending on the material, the mulch layer is applied to a height of about $1^1/_2$ in. (4 cm) deep. This ideally occurs in spring after pruning and proper fertilization. An overwhelmingly thick layer of mulch, especially near woods and meadows, lures voles and turns a positive mulching effect into the opposite.

Note

Anyone who has not been able to garner his own experience in dealing with mulching materials should not undertake any experiments with newly planted roses in the first year after they are planted.

Mulch is particularly advantageous in areas that are dry in summer.

As already mentioned, fertilization is determined by the mulch material used. This is important. Fertilization must compensate for any nutrient deficiencies and avoid providing excesses of nutrients.

Fertilization

Fertilization has become controversial in our society in recent years, although the proper nourishment of plants as well as of humans has always been of the greatest importance. However, in spite of having very complete knowledge about the effectiveness of the individual nutrients, and based on the complexity of the opposing influences, our knowledge about them can never be complete. The one-sided use of nutrients in agriculture has certainly led to increasingly higher yields. It has, however, also fostered a just as one-sided intensive agriculture that has too little regarded the interplay of plants and the environment. The results are undebatable. Along with the increase in yields there has also been an increase of the nutrient content in the groundwater and surface water.

Meanwhile, a change in thinking has taken place. Not only in agriculture has the awareness of responsibility for the environment clearly strengthened—garden lovers and hobby gardeners are also working more consciously with nature and not against it. Therefore, when it comes to the fertilizing of roses, in particular, there exist numerous reasonable ways to begin. Understanding basic relationships facilitates handling the different fertilizers. This does not imply studying the scientific details but only better recognizing the effects of one's own treatments and putting them into an overall natural context. The following explanations should contribute to this understanding.

Nutrients: Nutrients are essential for the growth of roses. Therefore enough of these heavy laborers must be available for them to use. Only roses that are well nourished in a balanced fashion develop enough flowers and maintain their disease resistance.

The nutrients needed can be determined by means of a **soil test.** The soil-testing kits available from garden suppliers permit

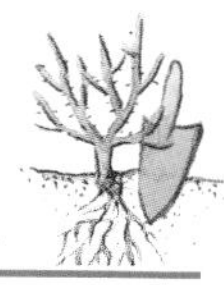

The more often roses are pruned, the more feeding they need.

determination of the content of nitrogen, phosphorus, and potassium in the soil. Anyone who wants a complete, precise soil analysis should send his or her sample to an accredited soil lab. For the most comprehensive and accurate soil test, contact your local county extension agency to determine where and how to send your soil sample for testing. A telephone book's blue government pages should list extension service phone numbers. The **principal nutrients** are nitrogen, phosphorus, potassium, calcium, and magnesium.

Nitrogen (N) is required by roses for their long growing season. Deficiencies of nitrogen are indicated by light-green leaves. The roses remain small and develop weak, thin canes that have few flowers on them. A remedy is a quickly absorbed nitrogen fertilizer, such as a liquid fertilizer. Excess nitrogen leads to the development of soft, succulent, very juicy cane tissue that does not ripen by fall and is very susceptible to frost injury. Therefore, after the first of July of any year, one should no longer apply any nitrogen fertilizer and should stop working the soil in September so that no further nitrogen is mobilized.

Phosphorus (P) is an essential component of the cell nucleus. Therefore, it is heavily involved in the protein synthesis in the plant and is of great importance, especially for the development of sex cells, flowers, and fruit. The requirement for phosphorus in roses is not large, however, and in well-prepared garden soil is sufficiently taken care of without need for further addition of fertilizer. On the contrary, many soils are overfertilized because of the year-long addition of phosphates in complete fertilizers or as Thomas meal. Phosphorus deficiency expresses itself through smaller, bluish green leaves, which display a purple-bronze coloration along the edges of the leaves. The flower buds open late, the foliage falls off prematurely, and the few fruits (hips) are colored bronze-red.

> **Note** Bone and blood meal are organic fertilizers with high phosphorous content.

Potassium (K) regulates the water balance and influences the metabolism of the rose. Fertilization with potassium ($1^3/_4$ ounces per 11 square feet [1 m^2]) at the end of August/beginning of September promotes the maturing of the wood and thus lowers the danger of winterkill of the rose canes. In addition, a sufficient provision of potassium increases the plant's resistance to spot anthracnose. Potassium deficiency shows in the form of unsatisfactory leaf development, in which dark-green or brownish gray spots appear along the edges of the leaves. Furthermore, extremely dark-green leaf color often can indicate too high a supply of nitrogen but also indicates too little potassium. Lime promotes the uptake of potassium.

Magnesium (Mg) is a component of the green of the leaves and moderates many processes in the metabolism of the rose. Magnesium deficiency is recognizable by a mosaic-like yellow coloration of the leaves. The leaves later become brownish and fall off prematurely. The yellow spots spread out from the main nerve of the leaflets. In extreme deficiency, a light scratching in of Epsom salts helps.

The pH value of the soil can be determined by the use of a quick test.

Calcium (Ca), also simply referred to as lime, is a building block and driving force for tissue growth. Besides, it regulates the pH value of the soil. Other essential nutrients can be taken up by the plant only on the foundation of a sufficient supply of calcium.

The **pH value** expresses the concentration of soil acidity. Very acid soils are indicated by very low pH values, around 4 to 5 or even lower. For roses, a value of 6.5 in the neutral to alkaline range in sandy-loamy soils is considered optimal. Quick tests obtainable commercially give information about the level of the pH in one's own garden. When the pH value is too low, i.e., too acid a soil reaction, the pH can be raised by the addition of lime.

Iron (Fe), a companion nutrient, is usually present in the soil in sufficient quantities but may often be in a form that is not available to the rose because of the low pH value (thus acidic soil conditions) or soil compaction. In addition, too high a pH value traps iron so that the plant cannot take up the mineral. Iron deficiency makes itself known through **chlorosis,** a yellow coloration of the leaves and fruit with dark-green coloring of the leaf nerves at the same time. Avoiding soggy soil by using effective drainage and good soil aeration work against iron deficiency. The addition of iron fertilizers, for example Sequestren or Fetrilon, overcomes an acute iron deficiency and makes chlorotic leaves green very quickly. However, these fertilizers do not offer the solutions to any causative problems.

Rosa rugosa varieties react to high lime content (= high pH) with chlorosis.

Fertilizers: The nutrients are made available to the roses as fertilizers. These fertilizers can be inorganic or organic in nature or a mixture of both. It is all the same to the plant no matter in what form the rose receives the needed nutrients. Fertilizers with high organic components do, however, improve the soil structure. In general, fertilizers should never be sprinkled over the plant's leaves or flowers.

Inorganic Fertilizers (Short-Acting Fertilizers): These substances are highly effective nutrient concentrates, which are easily dissolved in water and are constantly activated by appropriate soil moisture. If they are applied to the rose in too great quantities and are therefore not taken up immediately, they are quickly washed into the soil layers and are lost to the rose and contaminate the groundwater. Inorganic fertilizers have none of the soil-improving characteristics. They are seen as nutrient-heavy artillery when it is necessary to treat deficiency symptoms quickly. Some fertilizers that have been developed just for roses include Rose-Tone and Miracle Gro for Roses. However, using a brand name fertilizer made specifically for roses is not critical. Look for a water-soluble, well-balanced fertilizer like a 20-20-20 for fast-acting feeding.

If regular application is recommended at all, it is only for rosebushes that have already started growing well, perhaps shrub roses with appropriately wide-spreading root systems. Doses of 1¾ to 3 ounces (50 to 80 g) per 11 square feet (1 m^2) are sufficient for most garden soils. The responsible gardener follows the manufacturer's directions on the package. Inorganic fertilizers should be so distributed that the harmony of the soil life is not disturbed by a shocking increase in the soil salt content. Late applications after July 1 can negatively influence the maturing of the wood and make canes more subject to winter damage (see also page 169 about the subject).

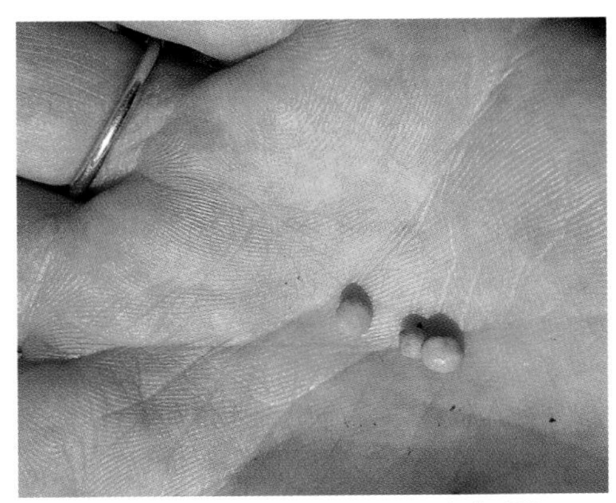
Slow-release fertilizers are inorganic fertilizers that give up their nutrients deliberately.

Inorganic fertilizers should never be used on newly planted roses. In well-prepared beds, enough nutrients are available for them so that an unnecessary addition of salts only injures the stabilization of the roots—and the life of the soil.

Slow-Release Fertilizers: The term *slow-release fertilizers* refers to inorganic fertilizers whose grains are surrounded by a semipermeable resinous shell. Thanks to this shell, the grains with the nutrients inside them are temperature-dependent. In practice, this means that more nutrient salts are released at higher soil temperatures, fewer to no nutrients are released at lower temperatures. Since plant growth is also temperature-sensitive, the quantity of nutrients given up is ideal for the growth of the woody plant. Particularly in winter, when the plant is not taking up nutrients, this type of preparation prevents the nutrients from washing out of the soil.

Slow-release fertilizer plugs are recommended for older and larger shrub and climbing roses.

The "co-thinking" slow-release fertilizers also have their price, however, and—because of the expensive production process—sell for more than regular inorganic fertilizers. However, the higher price should be regarded in the light of the effectiveness of these fertilizers. As a rule, the duration of their effect is longer than one growing season. Depending on the manufacturer and the composition they are effective for 5, 6, or 9 months and longer.

Raw garden compost makes a good rose fertilizer.

Since the time noted on the package specifies a soil temperature of 70 °F (21 °C) and, in general, outdoor temperatures are clearly much lower for long periods of time, one can still rely on the manufacturer's timing and multiply by two or three in order to get the actual period of effectiveness of the slow-release fertilizer in a rose planting.

The manufacturer recommends about 1 ounce (25 g) of fertilizer per rose plant (follow the manufacturer's directions), which is applied no earlier than the middle of May in direct proximity to the plant. Slightly scratching the grains in ensures close contact with the moist soil and the swift start of nutrient uptake. An ideal form for second applications of fertilizer to older specimen roses and larger shrubs and climbers is fertilizer grains in plugs. The earth at the base of the plant is slightly loosened with the spading fork, sparing the roots, and the plugs are pressed into the root area.

Organic Fertilizers: Organic fertilizers release their nutrients slowly. Only after they are broken down by the soil microorganisms are they available for the plants to use. This means that applying organic fertilizers requires advance planning so that the roses are provided with sufficient nutrients when they need them.

Horn Chips: Horn chips are ideal organic plant- and soil-conserving nitrogen providers. They are best spread in late fall, about 3 ounces (80 g) (for young plants) to 4¼ ounces (120 g) (for older plants) per 11 square feet (1 m^2). In spring, the gardener should use a second application of the same strength.

Horse Manure, Dried Cow Manure: The most traditional organic fertilizer is well-rotted horse manure. In previous times, it was mulch, fertilizer, and soil improver in one. Even today, it is far superior to any synthetic chemical product with this multifunctional capacity. Nevertheless, because of its high nutrient content, one should not use it on rose plantings, which are usually on very well-provided garden soils. Only in professional institutions, greenhouses supplying high-demand markets, or outdoor cutting rose fields does it still find use as a natural expansion of conventional fertilizers. In addition, realistically, it must be pointed out that the actual production of true stable manure in fewer rural areas is not unproblematic for the gardener. Nevertheless, those who, despite everything, do not wish to give up manure can find dried cow manure available from a number of manufacturers at garden supply stores.

Garden Compost: Old Rose Master Wilhelm Kordes already wrote about the subject of garden compost many decades ago: "A means of saving much money and at the same time giving his roses the best fertilizer in the world is a well-maintained compost heap." This quotation is just as apt today. It is only relative insofar as the modern household compost heap, because of the small size of the garden and thus not so abundant garden waste, is often is not enough for optimal rose feeding by itself. However, the available garden compost applied at the right time works its positive effect on the soil activity and the soil structure fully and completely.

The organic house and garden waste are collected for the compost heap. Rose leaves and branches must not be included since their undestroyed resting spores, especially that of spot anthracnose, outlive the composting and later promote further infection in the rose bed. Aside from that, the usual garden compost metabolism takes place, with earthworms and microorganisms as important helpers.

Even a small compost heap produces a large quantity of valuable garden compost.

Garden compost is added to the rose bed only when it has been well stored and is mature. This rule is especially true when dealing with new rose plantings.

Unstrained, semimature compost may be applied only when it is used as a **mulching material.** It is then spread 1 to 2 inches (3 to 5 cm) thick and sprinkled with 3½ ounces (100 g) of stone meal per 11 square feet (1 m^2). Finished garden compost should be used in concentrations of 2 to 4 pounds (1 to 2 kg) per 11 square feet (1 m^2). It should be mixed with a lime source so as to act as a good rose fertilizer and soil improver. For the advantages of mixing garden compost with the filling soil at planting, see page 149.

Rose Fertilizers: Multinutrient fertilizers, which are offered in the trade as rose fertilizers, contain the necessary organic and mineral nutrient components in the right balance of quantities. These are tailored to the requirements of the plants.

Note

Gardeners who already know at planting time that they want to add large quantities of garden compost to the beds should consider the soil-raising effect of regular additions of compost. After several years, the thickness of the soil layer over the bud union can increase from 1 in. (5 cm) to just about double that. Then roots develop above the bud union, which diminishes the influence of the understock. The result—weakly growing hybrid tea rose varieties that depend on their rootstock for their vigor can exhibit symptoms of chlorosis. Thus, the bud union should be set slightly higher than otherwise, that is, only about 1 in. (2 or 3 cm) deep. This is especially true for very clayish, cohesive soils.

Pest and Disease Control

Everything is relative. Albert Einstein, the inaugurator of the theory of relativity and the revolutionizer of physics, was certainly not thinking of roses when he formulated this famous sentence. However, he was also exactly on target in the case of the robustness of roses. Their robustness is relative, because it depends on a number of factors that are only partly under the influence of human beings.

Every Year Is Different: Numerous studies have clearly indicated that the severity of infection with the most common fungal diseases of roses—black spot and mildew—varies widely from year to year. During some years, a considerably stronger infection is registered than in the years previous or following. One variety in the same location under the same conditions of care is, of course, affected.

An important reason for this lies in the general condition of the weather, i.e., the influence of the macroclimate. In damp, humid weather, for instance, the spores have the best germinating conditions, and experts then speak of real mildew weather. If one considers the development of plant life over several decades from this standpoint, it can be seen that roses in especially sunny years have fewer problems with fungal infections than in years with a rainy summer. This phenomenon is also certainly linked with the fact that the inner resistance of the sun-loving rose is strengthened by plenty of fuel provided by sunlight.

Variety Differences: For the modern rose breeder, sturdiness has become an important, if not the most important, selection criterion. The strength scale of the variety varies widely. Because of this, experienced breeders know that a variety must be carefully observed for several years before any conclusions can be drawn about its sturdiness.

However, any such ultimate evaluation also remains relative. Among other things, a plant's sturdiness always depends on the age of the variety—some varieties appear especially robust in their juvenile stage but after the third or fourth year they fall off and must then be classified as prone to infection. The sturdiness of a rose depends on the place where selection occurs—a variety sometimes behaves quite differently in southern France than it does in northern Germany. Sturdiness also depends on the weather during the selection phase: successive years without severe cold, for example, do not permit testing for winter hardiness.

Age of Variety: Diseases and pests alter and are continually developing. They are constantly selecting, through the process of survival, for new stock and species forms that are a little more resistant than their predecessors are. This process can be noticed after only a few years because of the high speed of reproduction of the parasites. This explains the phenomenon of why older rose varieties that today are known as disease prone were, at the time of their introduction decades ago, considered robust. It is not the variety that is more susceptible. Instead, the conditions around them have changed.

Choice of Location: The behavior of a rose variety is decidedly influenced by the conditions around it. If the location is not right for roses and, for instance, is hot and airless or wet and shady, even a robust variety will go to its knees. The gardener can thus markedly promote the sturdiness of his or her roses by choosing the right site and can palpably minimize the effects of any attack.

Density: The larger the area planted with roses, the more robust the varieties have to be. The more variety of species in a planting, the easier it is to maintain. Anyone who wants to turn his or her garden into a place that is exclusively for roses had better be clear about the high maintenance requirement for such a collection. Any monoculture offers species-specific diseases and pests heavenly opportunities for spreading.

Maintenance: Even a variety that is robust in itself needs enough food and soil moisture, especially with frequent pruning. Hungry and thirsty plants are particularly prone to attack by disease and pests.

Do not leave rose foliage with undestroyed resting spores of spot anthracnose lying on the ground. The gardener who does so is offering the offending fungus an ideal place to flourish and cause new infections in the following year.

Subjectivity of Terms: After all, what is a pest? We call the aphids parasites because they perch on the rose leaves, rudely feast on our darlings, and thus greatly disturb our aesthetic perception. The ladybug, on the other hand, eats the aphids for her dinner. Because it is useful to us and to the roses in this manner, we like the ladybug and call the little beetle a beneficial insect. Nevertheless, it is an insect whose existence depends on the aphid.

Our conceptualization is subjective and not correct according to the self-contained natural cycle to which all living creatures are subject. Even the rose cannot only take but must also give something.

Numerous fungus and insect species are bound to the presence of the rose for their existence. They need the rose to survive and then again serve as basic foodstuff for other creatures. When regarded from this viewpoint, we should not describe the rose as especially prone to attack but as especially ecological. Its strong power of attraction for many creatures, including humans, underlines its high **ecological value.**

Result: Resistance is always a snapshot of an instant—over the long term there is no such thing as an absolutely robust rose. There cannot be one because of the macroclimatic and microclimatic conditions already mentioned. Such an absolutely robust, disease-free, pest-free rose would also not be desirable, for it would then have the ecological value of a plastic rose—that is, none at all.

Therefore, it is better for us to speak of **relatively healthy** roses, which can also of course sometimes catch a little cold but can overcome even an infection because of their varietal vigor and also because of their own strength. Only when the infection threatens to cause serious damage should one reach for the fungicides and pesticides.

All-American Rose Selections

The All-American Rose Selections is a nonprofit organization coordinated by a group of commercial rose growers and breeders. The results of their rose trials are aimed at giving consumers some guidance for selecting roses. The roses have been tested and evaluated.

Since 1940, only about 5 percent of all roses submitted for trial evaluation in the AARS' twenty-five test gardens located throughout the country are found worthy enough to merit becoming AARS winners.

Evaluators take data about the roses for two years. Results are compiled from all the gardens, which are located throughout the country. The rose test gardens are established at various sites like public parks, universities, and botanical gardens and also at growers' own premises. The categories roses are judged in are as follows: novelty value, bud form, flower shape, color when young, mature color finish, fragrance, stem/cluster, growth habit, vigor/renewal, foliage, disease resistance, flowering effect, and overall appearance and effect. Many of America's most popular roses like 'Queen Elizabeth', 'Peace', and 'Carefree Delight'™ have been AARS Winners.

AARS Winners inclue the following roses: 1999: 'Betty Boop'™, 'Fourth of July', 'Kaleidoscope'™, 'Candelabra'™; 1998: 'Fame', 'Opening Night', 'First Light', 'Sunset Celebration'; 1997: 'Scentimental, Artistry'™, 'Timeless'™; 1996: 'Mt. Hood'™, 'Carefree Delight'™, 'St. Patrick'™, 'Livin' Easy'™; 1995: 'Brass Band', 'MACivy'; 1994: 'Secret'™, 'Midas Touch'™, 'Caribbean'; 1993: 'Rio Samba', 'Child's Play'™, 'Solitude'™, 'Sweet Inspiration'; 1992: 'All That Jazz'™, 'Pride'n Joy'™, 'Brigadoon'™; 1991: 'Shining Hour'™, 'Sheer Elegance'™, 'Perfect Moment'™, 'Carefree Wonder'™; 1990: 'Pleasure'; 1989: 'Class Act', 'Debut'™, 'New Beginning'™, 'Tournament of Roses'; 1988: 'Amber Queen'®, 'Mikado'™, 'Prima Donna'™; 1987: 'Bonica'®, 'New Year'®, 'Sheer Bliss'; 1986: 'Broadway'™, 'Touch of Class'™, 'Voodoo'™; 1985: 'Showbiz'; 1984: 'Impatient', 'Intrigue Olympiad'™; 1983: 'Sun Flare', 'Sweet Surrender'; 1982: 'Brandy'™, 'French Lace', 'Mon Cheri'™, 'Shreveport'™; 1981: 'Bing Crosby', 'Marina'®; 1980: 'Love, Honor'™, 'Cherish'; 1979: 'Friendship'®, 'Paradise'™, 'Sundowner'; 1978: 'Charisma', 'Color Magic'; 1977: 'Double Delight'™, 'First Edition', 'Prominent'®; 1976: 'America'™, 'Cathedral', 'Seashell', 'Yankee Doodle'®; 1975: 'Arizona', 'Oregold', 'Rose Parade'; 1974: 'Bahia', 'Bonbon', 'Perfume Delight'; 1973: 'Electron'®, 'Gypsy', 'Medallion'®; 1972: 'Apollo'®, 'Portrait'; 1971: 'Aquarius', 'Command Performance', 'Redgold'; 1970: 'First Prize'; 1969: 'Angel Face', 'Comanche', 'Gene Boerner', 'Pascali'®; 1968: 'Europeana'®, 'Miss All-American Beauty', 'Scarlet Knight'; 1967: 'Bewitched', 'Gay Princess', 'Lucky Lady', 'Roman Holiday'; 1966: 'American Heritage'®, 'Apricot Nectar', 'Matterhorn'®; 1965: 'Camelot', 'Mister Lincoln'®; 1964: 'Granada', 'Saratoga'; 1963: 'Royal Highness', 'Tropicana'; 1962: 'Christian Dior', 'Golden Slippers', 'John S. Armstrong', 'King's Ransom'®; 1961: 'Duet', 'Pink Parfait'; 1960: 'Fire King', 'Garden Party'®, 'Sarabande'®; 1959: 'Ivory Fashion', 'Starfire'; 1958: 'Fusilier', 'Gold Cup', 'White Knight'; 1957: 'Golden Showers'®, 'White Bouquet'; 1956: 'Circus'; 1955: 'Jiminy Cricket', 'Queen Elizabeth'®, 'Tiffany'; 1954: 'Lilibet', 'Mojave'; 1953: 'Chrysler Imperial', 'Ma Perkins'; 1952: 'Fred Howard', 'Helen Traubel', 'Vogue'; 1951: No winners in '51; 1950: 'Capistrano', 'Fashion', 'Mission Bells', 'Sutter's Gold'; 1949: 'Forty-Niner', 'Tallyho'; 1948: 'Diamond Jubilee', 'High Noon', 'Nocturne', 'Pinkie', 'San Fernando', 'Taffeta'; 1947: 'Rubaiyat'; 1946: 'Peace'; 1945: 'Floradora', 'Horace McFarland', 'Mirandy'; 1944: 'Fred Edmunds', 'Katherine T. Marshall', 'Lowell Thomas'; 1943: 'Grande Duchesse Charlotte', 'Mary Margaret McBride'; 1942: 'Heart's Desire'; 1941: 'Apricot Queen', 'California', 'Charlotte Armstrong'; 1940: 'Dickson's Red', 'Flash', 'The Chief', 'World's Fair'.

'Magic Meidiland'.

'Elmshorn' (1950), 'Flammentanz' (1952), 'Dortmund' (1954), 'Dirigent' (1958), 'Schneewittchen' (1966) ('Iceberg' in the United States; but awarded as German rose), 'Fragrant Cloud' (1964), 'Sympathie' (1966), 'Carina' (1966), 'Lichtkönigin Lucia' (1968), 'Bischofsstadt Paderborn' (1968), 'Eroica' (1969), 'Edelweiss' (1970), 'Fontaine' (1971), 'Pussta' (1972), 'Sunsprite' (1973), 'Escapade' (1973), 'Westerland' (1974), 'Montana' (1974), 'Happy Wanderer' (1975), 'Morning Jewel' (1975), 'Compassion' (1976), 'Chorus' (1977), 'Grand Hotel' (1977), 'Deep Secret' (1978), 'La Sevillana' (1979), 'Marjorie Fair' (1980), 'Robusta' (1980), 'IGA '83 München' (1982), 'Aachener Dom' (1983), 'Bonica '82' (1982), 'Rosenresli' (1984), 'Banzai '83' (1985), 'Goldener Sommer '83' (1985), 'Romanze' (1986), 'Repandia' (1986), 'Pink Meidiland' (1987), 'Surrey' (1987), 'Lavender Dream' (1987), 'Dolly' (1987), 'Elina' (1987), 'Rödinghausen' (1988), 'Vogelpark Walsrode' (1989), 'Play Rose' (1989), 'Ricarda' (1989), 'Marondo' (1989), 'Flower Carpet' (1990), 'Apfelblüte' (1991), 'Wildfang' (1991), 'Super Excelsa' (1991), 'Schneeflocke' (1991), 'Pierette' (1992), 'Palmengarten Frankfurt' (1992), 'Schöne Dortmunderin' (1992), 'Rugelda' (1992), 'Mirato' (1993), 'Foxi' (1993), 'Armada' (1993), 'Dortmunder Kaiserhain' (1994), 'Bingo Meidiland' (1994), 'Blühwunder' (1994), 'Magic Meidiland' (1995)

ADR Certificate

Germany has a very vigorous and extensive rose trial garden network. The Allgemeine Deutsche Rosenneuheitenprüfung (ADR) certificate is awarded to a rose after it has been evaluated for three to four years in ten different places throughout Germany and deemed exceptional in a number of categories including disease resistance and cold hardiness. So far, more than 1,500 roses have gone through trials with less than 100 having been awarded the ADR certificate. The varieties listed in the box to the left have received the ADR. The year it was awarded is also listed.

Powdery mildew covers the rose leaf with a floury deposit.

Important Rose Diseases

Powdery Mildew

Signs: A floury white, removable deposit is seen predominantly on the upper sides of young leaves as well as on the calyxes and particularly on the tips of canes. The affected leaves crinkle up with a severe attack and turn reddish.
Occurrence: Powdery mildew appears even in early summer on new growth when the weather reaches temperatures over 68 °F (20 °C) and humidity becomes higher than 90 percent as often occurs in wine climates. In areas that tend to have heavy rainfalls, the thorough wetting of the leaves suppresses powdery mildew, certainly, but promotes spot anthracnose.
Prevention: Choose robust varieties. Feed roses optimally and water correctly, avoid nitrogen fertlizers. Avoid poorly ventilated locations. Collect affected leaves and destroy them—never, ever compost them. Wash weekly with a strong stream of water.
Treatment with Commercial Preparations: Fungicides generally recommended for powdery mildew include: triadimefon (Bayleton, Strike); triforine (Funginex®), thiophanate-methyl (Cleary's 3336, Domain),

Typical signs of spot anthracnose, the most dreaded rose fungus.

propiconazole (Banner), and sulfurous fungicides. When using these, follow all labels carefully. Be sure to spray both top and bottom leaf surfaces. The person who applies the pesticide has the responsibility, by law, to read and follow product directions. Changing labels and product registration may make some products illegal to use. Always check the product label, or call your local county extension agency or the Environmental Protection Agency for

Rust is recognizable by its rusty-orange pustules on the undersides of the leaves.

clarification. **The author and publisher assume no liability resulting from the use of these products mentioned above.**
Home Remedies: You can use a horsetail infusion or stinging nettle brewed with baking soda (1 tablespoon (15 ml) per gallon of water and a few drops of mild soap).

False or Downy Mildew

Signs: This appears as whitish gray moldy areas on the undersides of leaves—in contrast with powdery mildew. Dark spots are visible on the upper sides of leaves; affected leaves wither and fall off. The false mildew can easily be confused with spot anthracnose, but in contrast with this, it begins on young leaves and attacks the rose from top to bottom.
Occurrence: This disease appears especially during severe temperature changes in late summer and fall. In the evenings, after fast cooling of the air, condensed water collects on the leaves, which creates ideal nurturing conditions for this fungus.
Prevention: Use sunny, not too close positioning of plants, which permits rapid drying of the leaves. Vigorous, healthy roses are more resistant. Cut out affected canes and destroy along with affected leaves—never, ever compost them.
Treating with Commercial Preparations: If used at all, spray with a fungicide containing copper, manganese, or zinc. Do not forget the undersides of leaves when spraying.

Spot Anthracnose

Damage: Star-shaped, violet-brown to black spots with raylike projections visible on the upper sides of leaves are the signs of spot anthracnose. The leaves become yellow and fall off. Sometimes the canes also exhibit spots. The spot anthracnose permanently weakens the rose in a number of respects, affecting readiness to flower and winter hardiness, for example. These reasons make spot anthracnose one of the most insidious fungal diseases that can befall roses. It is easily confused with false mildew. In contrast, it begins on the leaves in the lower regions and attacks the rose from bottom to top so that the plant grows bare at the bottom.
Occurrence: Spot anthracnose usually appears in late summer and fall. In rainy summers, it is also observable in June, especially in regions that are already high in rainfall.
Prevention: Plant robust varieties, although hardly any rose variety is absolutely resistant to spot anthracnose; choose light, airy locations. Do not plant too close. Above all, prevent protracted moisture on the leaves since the fungus' spores need only about seven hours of sufficient moisture to germinate. Never plant roses in the shade or in damp sites. Since the focus of the infection develops close to the soil, one should be extremely careful to gather up fallen foliage and destroy it—never compost it. Make sure to provide sufficient potassium. This disease is rarely fatal and is controlled best by controlling black spot.
Home Remedy: Use horsetail infusion.

Rose Rust

Signs: After the new growth appears in the spring, rose rust appears as orange, very powdery spore deposits, about the size of the head of a straight pin, on the undersides of the leaves. In fall, the pustules are dark brown. An attack is especially likely to occur in very clayish sites.
Occurrence: Rose rust occurs at different intensities in different years. Sometimes the attacks seem to stop altogether after several plague years.
Prevention: High humidity favors an infection, so choose a site that is good for roses in which the leaves can dry quickly. Loosen very clayish soil by adding sand. With insufficient calcium and potassium, roses grafted onto the wild understock of *Rosa laxa* are more susceptible than others. Never compost fallen foliage but gather it up and destroy it.
Treatment with Commercial Preparations: See the information about fungicides for powdery mildew.
Important: Do not forget the undersides of

leaves when spraying, and also do not spray the affected roses from too close.
Home Remedies: You can use bracken, wormwood, or horsetail infusions.

Other, Generally Less Often Encountered Fungal Diseases

- Botrytis Disease: The fungus *Botrytis* causes rotten spots on petals and buds. Remove affected buds at the first sign.
- Canker: In particular, immature, soft canes exhibit brownish red spots after mild winters, which can easily be confused with winter injury. Remove diseased canes, provide for good maturing of wood of the rose, provide sufficient air circulation in winter, remove winter covering as early as possible. Use copper sprays.

Ants multiply their aphid colonies by keeping the aphids' natural enemies away from them. Therefore, deny ants access to roses.

- Valsa Disease: Canes affected with this disease wither, sometimes the entire rosebush even dies—therefore the disease is also called **cane death.** Remove diseased canes.
- Black Spot: Areas of black fungus, not to be confused with spot anthracnose, develop on the leaves. The fungus colonizes on the sugar-containing excretions from aphids and scale, which drop from other shrubs onto rose foliage. If an attack occurs, spray new leaves with a fungicide. Remove and destroy infected parts. Remove fallen debris. Water at the soil line and early in the day.

Very rarely **viral diseases** (e.g., rose mosaic virus) and **bacterial diseases** (tumor growths) may occur in roses. In these cases, treatment promises little success. The entire plant should be removed and destroyed.

If you are visiting the Portland area, stop by one of the country's most famous rose trial gardens, the Portland International Rose Trial Garden. This garden boasts ten splendid acres of roses grown for evaluation purposes. The garden, started in 1917, enjoys the title of the longest continuously running test garden in the United States. This rose garden is situated in a city famous for its residential plantings and love of roses. The annual Portand Rose Festival is one of the finest community festivals throughout the country.

Important Rose Pests

- Rose Aphid

Aphids are probably the least valued creatures in the garden. They are so used and pursued by bacteria, viruses, predators, predator parasites, and humans that they can survive only by means of an extremely high rate of reproduction. In so doing, they make use of a timesaving trick with so-called telescoping of the generation sequence. This term describes the fact that the embryo of the next aphid generation is already inside the embryos of the current generation.

There are over 3,000 different aphid species, most of which are highly specialized. Generally, aphids pierce the filtering tubes (sap conductors) of the plant. Before an aphid opens the "juice barrel," it considers the tap very precisely. Only after several test stabs does it make the main jab, which costs the aphid a great deal of energy and may take up to a day. If the tap is successful, the plant juice gushes out in abundance because of the high pressure in the filtering tubes.

The nitrogen compounds contained in the juice are vital to the survival of the aphid. It can only fulfill its great need with appropriate quantities of juice, but the sugar the juice also contains is not necessary to the aphid, which excretes it. So as not to cover its drinking companions with a sugar glaze, it kicks the sugar juice away with its rear feet. If the glaze lands on any rose foliage that happens to be in the way, black spot spores will colonize on it immediately.

The common spider mite can be recognized by its webs.

Signs: Green aphids attack young, still-soft canes, which they can tap easily with less expenditure of energy.
Occurrence: In warm, dry weather starting in April, the aphids reproduce massively.
Prevention: No rose is immune to attack, but too much nitrogen fertilizer and too little water make them more susceptible. Both cause soft, juicy tissue, which can easily be tapped. In addition, the high nitrogren values in the sap of the rose promotes the fertility of the aphid. Especially high concentrations of nitrogen are also found in the sap of roses suffering from the stress of dryness. Therefore, avoid hot sites with dry air.
Treatment with Commercial Preparations: You can use insecticidal soap, sabadilla, neem extract, rotenone, pyrethrin, and applications of beneficial insects.
Home Remedies: Small quantities of aphids can be washed off with a hard spray from the hose or snapped off with the fingers. Attempts with onion and garlic juices have proven acceptably effective. Stinging nettle broth is very often recommended, but there is argument about its effectiveness. In comparative attempts with plain water, it has not clearly shown to be any more effective.

Rolled leaves: Rose leaf roller wasp.

● Japanese Beetles

Signs: These beetles are $^1/_2$ in. (1 cm) long and have a brown metallic color. They have green heads and can do major damage to rose foliage and flowers. Insects can be easily seen on the plant. They skeletonize foliage and chew holes in buds and flowers. Larvae are 1 in. (3 cm) long, gray-white grubs with brown heads.
Occurance: Japanese beetles can be a problem in the eastern and especially northeastern areas of the United States. They start in midsummer and continue through late summer to early fall.
Prevention: Apply milky spore disease, which will multiply on its own to help kill grubs.
Treatment with Commerical Preparations: Sevin, neem, insecticidal soap, and pyrethrum can be used. Milky spore disease can also be applied in the spring on the lawn and ground according to directions. Several commercial names for milky spore disease are Doom and Japademic.
Home Remedies: Handpick beetles off plants. Collect into containers of soapy water to kill them. Turn over the soil to expose grubs for birds to feed on.

● Red Spider

Signs: Tiny little orange-red creatures, which are situated on the undersides of leaves, suck on the plant there, and are visible only with a magnifying glass. The upper sides of the leaves turn irregularly brownish yellow until the leaves drop off. In contrast with the common spider mite, red spiders do not make any webs.
Occurrence: In dry, hot weather, massive infestations start in May.

Rose leafhoppers are often found on climbing roses.

Prevention: Avoid hot, extremely dry sites. Especially at risk are (climbing) roses in front of hot south walls by means of which reflected heat strikes the foliage from all sides.
Treating with Commercial Preparations: Use a miticide, Plictran, application of predatory mites, insecticidal soap, or sulfur.
Home Remedies: Use a horsetail infusion or persistantly wash the top and bottom foliage with water.

● Rose Leaf Roller Wasp

Signs: Typical of leaf roller wasp infestation on roses are the rolled leaves. Depositing eggs on the edges of the leaves causes them to roll up, and the larvae then develop inside the little rolls.
Occurrence: Infestation can begin in May. The larvae leave the little rolls in July to pupate in the soil.
Prevention: Prevention is difficult. Observe the leaves from the beginning of May. Ultimately, the only remedy is spraying with systemic preparations, but these also endanger the beneficial insects.
Treatment: Remove affected leaves immediately and destroy them, collect the larvae by hand. Spray with Sevin only in cases of rampant infestation.

● Rose Leafhoppers

Signs: The leaves are dotted with white on the upper sides. On the undersides, one finds greenish white, aphid-like insects, which spring away immediately.
Occurrence: Since the leafhoppers prefer to winter over in the cracks of walls, the attack often appears on climbing roses growing there.
Prevention: Avoid hot, dry locations.
Treatment with Commercial Preparations: Winter spraying with oil emulsions, Sevin, Diazinon, or pyrethrum in cases of infestation.
Home Remedy: Stinging nettle infusion.

● Thrips

Signs: Thrips primarily attack buds about to bloom. The petals curl and show brown spots at the edges. If the flower is tapped, some of the tiny insects will jump out of the flower onto the surface of the hand and are recognizable under a magnifying glass.
Occurrence: From June in warm weather.
Prevention: Prevention is difficult. After an attack the previous year, be very alert during phases of hot, dry weather.
Treatment with Commercial Preparations: Use a systematic insecticide. Spray flower buds in time, remove infested flowers immediately, and destroy them. Application of beneficial insects only promises success in greenhouse culture.

Other Pests

● Rose Gall Wasp: The rose gall wasp often, but not only, affects the wild roses. Mosslike, clearly visible swellings appear on the canes, which are noticeable because of their hairlike outgrowth, the so-called **sleep apple.** Remove affected canes.

The sleep apple, which can be up to 2 in. (5 cm) in diameter, is considered a sleeping medication in folk medicine. When placed under the pillow, it is supposed to promote sleep. See picture on page 138.

● Snout Beetle: While about only $^4/_{10}$ in. (10 mm) long, the black snout beetle gnaws on the buds, leaves, and canes and leaves indented eaten places. Collect them at night; it is possible to combat the larvae with beneficial nematodes.

● Cane Borers: The canes begin to wither at the upper end. Splitting lengthwise reveals a bored tunnel in which a small caterpillar can be found. If it is eating the cane from top to bottom, it is a downward rose cane borer. When the direction of destruction is the other way around, it is the upward rose cane borer at work. Cut the cane back to healthy wood and remove the caterpillar.

● Rose Scale: Brown scale insects, which are protected by a hard shell and which excrete a sticky juice, occupy woody canes and branches. A massive infestation is possible in dry sites with still air; climbing roses on house walls are particularly at risk. Use mineral oil sprays. Protect lacewings, the natural enemies of the scale insect.

- **Root Nematodes:** There are countless species of these tiny threadworms, among them some beneficial ones. However, most are injurious since they suck on the roots of rose plants. They are difficult to combat. Green manuring with *Tagetes erecta* at least works against their further spread (see also page 148).
- **Garden Chafer:** Although occurring fairly rarely, the garden chafer beetle (4/10-inch [10 cm] long) preferentially gnaws leaves and flower buds. If it does not actually appear in masses, it does no damage. The garden chafer is not to be confused with the twice-as-large **rose chafer.** The rose chafer loves garden roses, in whose flowers it literally burrows into the pollen and eats the stamens.

General Notes About Controlling Diseases and Pests

Among responsible rose lovers and gardeners the basic motto should be, **"Prevention is better than spraying."** Anyone who pursues every aphid cannot expect to build up any beneficial fauna worth mentioning. Only when a threshhold of damage is crossed—i.e., the damage to the plants surpasses the costs to be expected from fighting—is the use of sprays worth considering. You must absolutely follow expert advice when buying and using commercial formulas.

In using the preparations, bear in mind the following. Never spray on open flowers or when there is wind. Use only materials that are not toxic to bees and other beneficial insects. Pay strict attention to the directions on the package labels. Dispose of any leftover chemicals in the hazardous waste collection. Carefully clean any tools and equipment after use.

A Request from the Author: Please do not shoot at sparrows with a cannon! The sprays available for hobby gardeners in garden retail stores are completely effective in their range. For large areas, commercial gardeners sometimes apply other preparations that often promise true miracles to laypeople and that are therefore sometimes sought after by them. As a rule these preparations are highly effective and are subject to very strict rules for application, the most recent among them demanding that equipment allowing no drift be used and expensive protective clothing be worn. Anyone who uses this material without appropriate protection and expert knowledge in the home garden unecessarily endangers himself or herself and others. No plant friend and nature lover can desire this. Even rose-producing firms are increasingly limiting their use; some companies even do without them entirely and produce their rosebushes within the framework of biological growing guidelines. Note: all the details given about substances are as of 1997. Always check with your local county extension agency for the proper chemical pesticide to use for disease or insect problems.

Home Remedies

Many times, the old home remedies have proven to be effective in controlling pests and diseases. Currently, they are increasingly being rediscovered and can also provide good service to the rose lover with their regular application.

> **Note** It is not advisable to use home-made soft-soap solutions. They often contain perfumes and other chemicals that can lead to burning of the rose foliage.

However, you should be warned against expecting too much. Home remedies work no miracles—especially when they are used only once. Their effect is primarily based on the strengthening of a plant's inner powers to fight off disease. The pests should be reduced to a tolerable number so that the food that is important to maintain the beneficial insects is always available in sufficient quantities.

Here are some general tips to make home remedies:

- **Infusions:** The plant material (fresh or dried) is soaked for one whole day and then boiled for about half an hour. It is allowed to cool in the covered pot. After straining, the infusion is thinned in water in the proportion of 1:5 and the affected part of the rose plant sprayed with it.
- **Extracts:** In contrast with an infusion, an extract is made exclusively with cold water and is also more time-consuming to prepare. The plant material (fresh or dried) is chopped and placed into a large vat of ceramic, wood, or plastic (no metal), has water poured over it, and is left in the sun. After several days, during which it is regularly stirred, small air bubbles rise to the surface as it begins to ferment. The addition of a small amount of lime may diminish the unpleasant smell. After fermentation, one has either fresh (stinging) or well-fermented extract. As a rule, the fermented extract is thinned with water in the ratio 1:20, the stinging extract in the ratio of 1:50. Then the extract is sprayed. Caution: too much can injure the rose more than it helps; therefore gather your own experiences step-by-step.

For more information on growing roses organically, read *The Rose Book* by Maggie Oster (Rodale Press, 1994). It lists a number of nonchemical remedies for insect and disease problems.

- **Neem Oil for Insects:** The seeds of the neem tree (*Antelaea azadirachta,* syn. *Azadirachta indica*) contain substances that have been used in Asia, South America, Africa, and India for many generations against parasites like ticks and lice. In agriculture and horticulture, people have known of the insect-repellent and insect-combatting effects of neem oil extracts. Neem oil is nontoxic and is classified as safe for beneficial insects. The pests, of course, do eat the plants treated with neem oil and take up its active substances. However, the concentration remains so small in the animals that it does no harm to the beneficial insects when it eats these pests.

Neem oil affects the hormone systems of the insects and prevents their molting. It is harmless to humans. The extract is produced as follows: Ground neem seeds are mixed with water, 1¾ ounces (50 g) of

Beneficial insects at work: Seven-spotted ladybugs decimate aphids.

Feathered benefits: An English robin feeds a caterpillar to its young.

neem seeds are needed per quart (liter) of water. Allow to steep for 5 hours and stir now and then. At that point, the active substances have dissolved. Pour the extract through a coffee filter. It can be applied either by brushing it onto the rose leaves or spraying with a flower sprayer.

● "Baking Powder" for Powdery Mildew: A baking-powder-like chemical compound, sodium hydrogen carbonate, is said to have a strengthening and preventive effect against powdery mildew in roses. It is dissolved in concentrations of 0.18 to 0.4 ounces (5 to 10 g) per quart (liter) of water and applied to the plant. When mixed with neem oil, the strength of both preparations is increased even more.

● Goethe's Advice: Anyone who has difficulty identifying a particular pest or disease can get advice from his or her state agricultural extension or farm bureau. Also, local nurseries are glad to help the gardener along with expert advice. Otherwise he or she has to rely on Mephistopheles' satiric comment, "You study the large world and the small, at the end to let it go as God pleases."

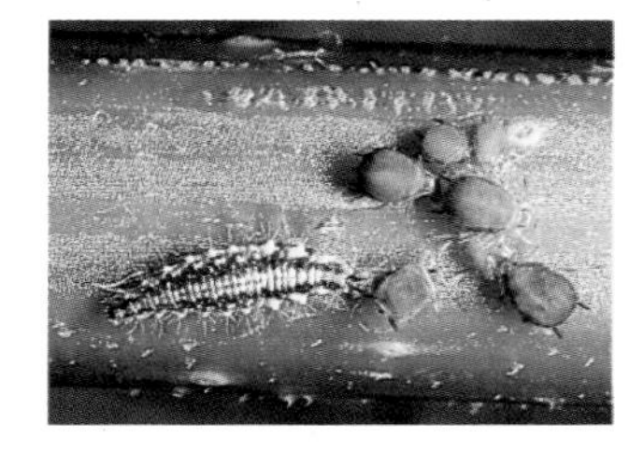

The larvae of the lacewing dining on aphids.

Insects Beneficial to Roses

The natural way of combatting pests with beneficial insects can function only if it is employed over the long term and is subordinated to the cycles of nature. Some time is needed for a population of beneficials of sufficient size to develop. Also, they can survive only if they find pests available in sufficient quantities.

● Ladybugs: The larvae of the ladybug are tremendously voracious. During its approximately three-week-long period of development, each one eats a total of up to 600 aphids and scale insects. Ladybug larvae are brownish black, up to almost ½ in. (1 cm) long, and have six legs. After they pupate, ladybugs appear in the garden starting in May. The adult insects also prefer to eat aphids and other soft-skinned insects. Their daily requirement is about 100 aphids. All the life stages of this insect are worth protecting. Protection is guaranteed if, for instance, they can overwinter under leaves that have been left lying on the ground.

Voracious aphid hunters: a syrphid fly larva.

● Lacewings: Lacewings are also called aphid lions. Their larvae are similar to those of the ladybug. With their sucking maxillae they grasp the aphid and suck it dry. Besides, they combat thrips and mites. The adult insect looks like a dragonfly; its flying style has a somewhat clumsy look. Every year several generations of lacewings are produced.

● Syrphid Flies: The larvae of the syrphid flies are greenish yellow to gray worms of about ¾ in. (2 cm) in length. They are not very active. Therefore, the mother fly places them directly into the aphid colony. The larvae spear the aphids with their mouthparts and suck them dry. A larva destroys a total of about 400 to 800 aphids during its brief, two-week development period. The adult fly is about ½ in. (2 cm) long and attracts attention with its black-and-yellow-marked lower body and its ability to tread air while dangling in one place, often for minutes at a time. Sometimes the creatures are confused with wasps and are thus persecuted. Several generations are produced every year. Syrphid flies can recognize colors, and they are especially attracted to yellow. *Rugosa* hybrids are popular landing spots, but parsley, caraway, and coriander also attract these beneficial insects into the garden.

Earwig stalking aphids.

● Predatory Mites: Predatory mites suck the eggs of the red spider (spider mite) or the animal itself. The predatory mite, a close relative, is larger and stronger than the red spider and does not create webs. Above all, it is very active and finds its prey unerringly. At a rough count, it is thought that a predatory mite can suck out about five adult red spiders or twenty eggs per day. That is quite a lot and produces a marked reduction in red spiders since the predatory mites also develop very quickly. For this reason, predatory mites are purposely and systematically introduced into commercial gardens.

● Earwigs: These brown insects, about ¾ in. (2 cm) long, are easy to identify because

Decorative and useful: Rose balls provide nesting places for earwigs.

of the pincers situated on their rear ends. The contention that they crawl into the human ear is a fairy tale and belongs to the realm of fable. Earwigs hunt aphids, primarily at night. Anyone who wants to create a useful as well as aesthetically appropriate house for them fills **rose balls** with excelsior. The mouth-blown balls in green, gold, red, blue, or purple are warm inside and so create the ideal nesting place and exit point for the hunting expeditions of these aphid destroyers. However, the earwigs can also feed on the flowers of the roses.

● Apropos Rose Balls: Their reputation for being effective scarecrows can also probably be relegated to the realm of legend. Supposedly, the birds take to their heels at the sight of their own distorted reflections.

● Aphid-Parasitic Hymenopterans: These insects lay their eggs in the egg, the larva, or the pupa of another insect, for example, of aphids. Even ants are hardly able to protect "their" aphids and prevent the egg deposits of the hymenopterans. A female parasitic hymenopteran can pierce up to 1,000 aphids and lay eggs into them.

● Birds: Insect-eating bird species like warblers, titmice, and house sparrows feed their young in large part with insects. Titmice give preference to voracious caterpillars when hunting to feed their brood. Shrubs and perennials offer birds protected living spaces. Hanging houses for titmice has proven especially helpful. Also, a bath in summer and a feeder in winter invites the useful visitor. However, if many cats stroll the garden, all the attractive houses and baths will not lure the insect-eating birds.

Winter Protection

Inner Winter Protection: The rose varieties offered commercially are sufficiently winter hardy in the local climate zone when their plant physiology is properly respected. It sounds like a paradox, but **winter protection begins in high summer.** Anyone who gives his or her roses growth-stimulating applications of nitrogen after the first of July decreases the maturity of their wood. Doing so promotes the growth of the plant into the fall and thus the development of a bloated, severely water-retentive, very frost-sensitive tissue.

Utterly in contrast with nitrogen, potassium—applied from the end of August—promotes hardening of the canes. This increases the maturity of the wood and winter hardiness.

Special attention should be paid to rose varieties that have severe infections with spot anthracnose. If as a result of the fungal infection the bush drops all its leaves in August, the rose responds with an emergency program, puts out new growth, and is certainly not dormant in time for the cold season of the year. A late-active plant of this sort is at risk of winterkill to the highest degree.

External Winter Protection: The best natural winter protection is snow. No, one should not artificially pack the rosebushes firmly with wet snow. Doing so is not advisable. Loose powdered snow, however, insulates best and protects ideally from frost. Unfortunately, one can not rely on it, so the roses must be protected from dangerous bare frosts—that is, frosts without snow—with suitable additional measures.

Good frost protection: Surround shrub roses with straw matting; fill the space between bush and mat with fine leaves (birch, beech, and so on).

External winter protection begins with the planting of the rose. If the bud union is 2 in. (5 cm) below the soil surface, this important junction between the grafted variety and the wilding understock is sufficiently protected. Danger for the rose principally occurs in extremely cold locations and with prolonged low temperatures. This is especially true for the period of late winter, when the juices are beginning to flow in the rose again and it is warmed by the sun after a cold night.

Hilling: In particularly frost-prone areas, with sensitive varieties, as well as in severe climates over 1,640 feet (500 m) above sea level, the roses should be hilled in December so that mature canes under the soil are optimally protected against frost. Hills about 6 to 8 in. (15 to 20 cm) high of loose leaf mold, garden compost, or the like are heaped

Hilling: Sufficient winter protection for bedding roses.

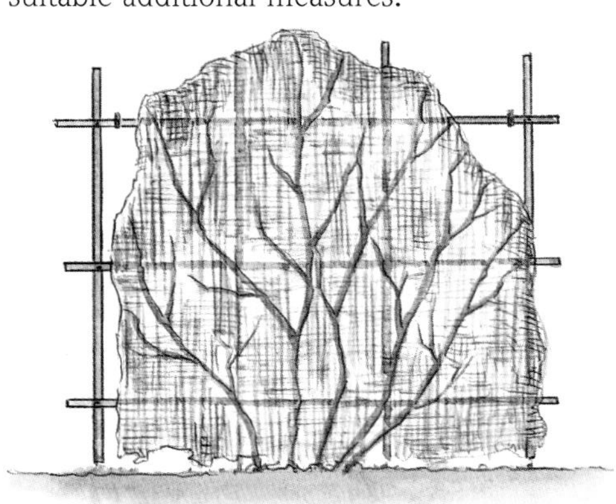

Climbing roses can be protected with burlap or . . .

. . . by tying evergreen boughs to them.

Rose arches are wrapped with evergreen branches . . .

. . . as are crowns of standard roses.

up around the base of the bush. Additional evergreen brush is piled over the now protruding canes. This looks pretty and protects the canes from drying winds. It is not advisable to use the soil in which the rose is growing to pile up for hilling, for it exposes too many roots and injures them. Never use peat for hilling due to environmental considerations and because frozen, wet peat becomes an icy vise that strangles the rose.

Shrub and Climbing Roses: Shrub and climbing roses, sometimes the entire rose arbor, are packed with evergreen branches for protection from strong sunlight. Heavy frost damage occurs primarily when roses are warmed intensely and directly by the winter sun but the water supply is lacking because the ground is frozen. Thus, the canes actually are **parched.**

Where the natural brush is not at hand or too expensive to use, one falls back on sacking or loosely woven jute. In nurseries, it is possible to buy so-called balling cloth, an ideal draping or wrapping material for trellises and rose arbors.

The materials named have the advantage of keeping off the drying winds but allowing cooling air to pass through and thus preventing trapped heat, especially in late winter. Such trapped warmth will induce premature new growth of the rose within the protection of its covering. This can cause the rose injury, especially in periods of weather with sunny days turning to crisply cold nights. Therefore, do not wrap the canes too thickly either, and remove the wrapping on a dull day toward the end of March.

One should never wrap roses, especially the crowns of standard roses, in plastic, not even if it is perforated. Warmth collects, and rotting inside this winter greenhouse is certain to result. Weakening of the plant and damage from night frost are preprogrammed.

Standard Roses: Like shrub and climbing roses, the crowns of old standard roses, regardless of their trunk height, are packed in sacking or evergreen boughs. This is the simplest method, but it is not the most secure.

With standard roses, the bud union is well above the ground. Young standard roses, especially, are thus particularly at risk of freezing in winter. A time-tested protection—when there is enough space—is **laying down the standard.** This involves bending the standard down over the snag.

The bare crown and the bud union are now placed flat on the soil surface and loosely covered with leaf mold or garden compost. In addition, the trunk is wrapped with brush or sacking. The laying down of tree roses is generally possible without any problems up to a plant age of ten years and is a sure frost protection even in severe climates. Do not jerk the trunk up again in the spring but carefully raise it.

Older standards, which can no longer bend, can also be protected with a wire cage—somewhat more time-consuming. The cage is erected around the trunk and filled with straw or leaves to over the top of the crown.

Harsh Altitudes: Finally, here are some more general tips for rose growers who live in altitudes above 1,640 feet (500 m) above sea level:

—Do not plant any varieties particularly prone to spot anthracnose, which diminishes the maturing of the wood through premature loss of leaves.

—Give preference to once-flowering shrub and climbing roses.

—Yellow-flowered bedding and hybrid tea roses have proven to be particularly winter hardy.

—Never go without hilling and covering with brush or sacking, even if several mild winters in a row tempt neglect.

—In snowy areas, protect shrub roses by staking against too heavy a burden of snow.

Transplanting Old and Young Roses

When an unavoidable construction project or a move is looming, many times the question arises for the rose lover: Can old, settled, sometimes fondly regarded roses be transplanted?

Yes, they can, even if a rose is over ten or more years old. The ideal time for transplanting is from November until the onset of frost. If necessary, early spring is still possible. The basic principle of transplanting should be to get as many fine roots out of the ground undamaged as possible.

With bedding and hybrid tea roses, the plant is carefully dug free, the spade set as deep as possible under the rose roots. Then the rose plant is lifted out of the hole without

Usually, standard roses are laid down in particularly cold regions. It is essential that the trunk be protected from the direct rays of the sun—which it is unused to—either by evergreen branches or by soil.

any soil. Broken roots are cut back before replanting and also the above-ground canes are taken back to a handbreadth. The old bushes are planted and hilled like new roses. Hilling is especially important for old rosebushes in spring, for it protects the rose from drying out. Hilling is removed after about eight weeks, that is in May/June.

With shrub, wild, and climbing roses, the above-ground canes are also pruned back very severely, to about 6 in. (15 cm). Old, dead wood is removed back to the base. Often, in digging out these rose species, it is possible to retain a root ball, which is marked out in a circle of vertical jabs with the spade and lifted out of the soil with rocking motions. Such a natural ball is ideal, but it is not urgently necessary to transplant roses successfully. As with the bedding and hybrid teas, retaining as much root mass as possible and not earth mass, is of first importance. Thus, if the soil drops off as the plant is moved, it is not a tragedy. Old shrub, wild, and climbing roses are planted like new roses.

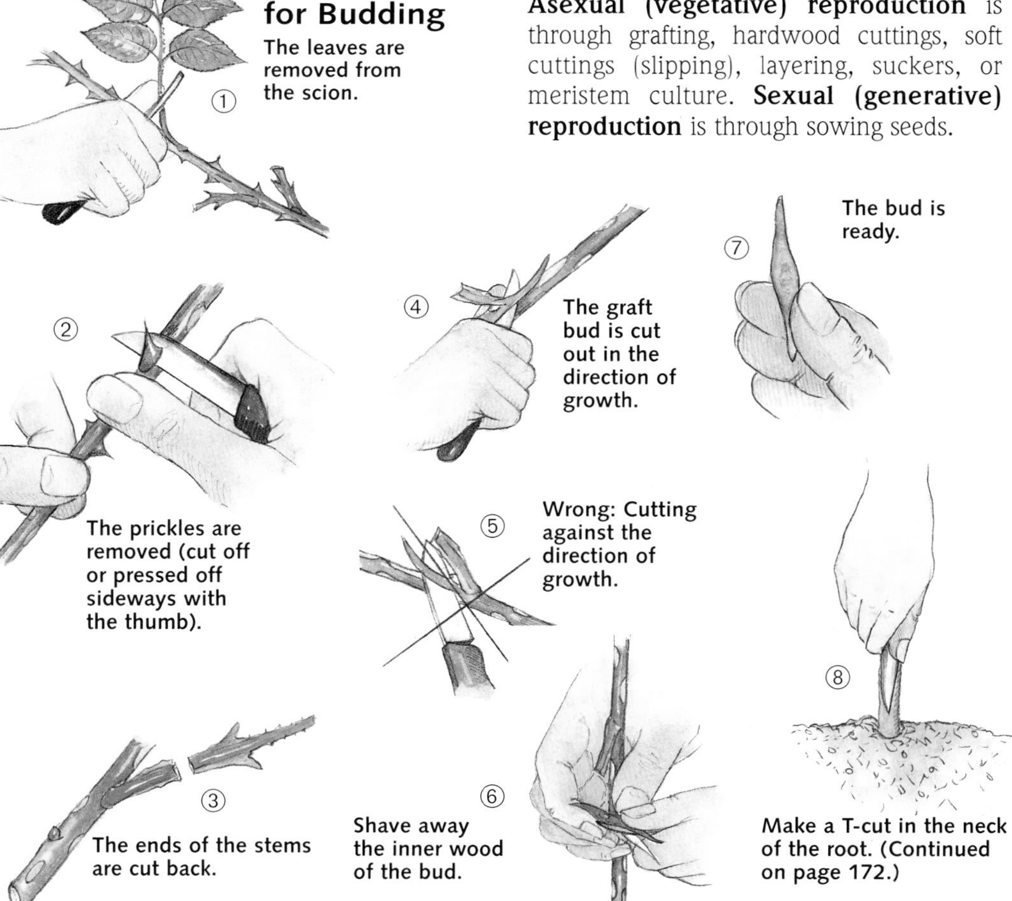

Propagation

We should be honest. Although getting plants from the professional grower is usually much cheaper, it is exciting for a good hobby gardener and rose lover to propagate a variety for himself or herself once in a while. Even if the desired result is achieved after many a disappointment, one has a very personal relationship with a plant one has propagated oneself.

Propagation of very rare rose varieties can, in addition, have a quite ordinary reason, since obtaining them commercially, especially without a precise name, is often difficult if not entirely impossible.

Anyone who has access to material for propagation from a friend can then fulfill his or her own particular rose dream.

Thus, the attempt is worth it. So that the rose desire does not turn into rose frustration, the methods that promise the most success are gathered here.

Two propagation principles are involved. **Asexual (vegetative) reproduction** is through grafting, hardwood cuttings, soft cuttings (slipping), layering, suckers, or meristem culture. **Sexual (generative) reproduction** is through sowing seeds.

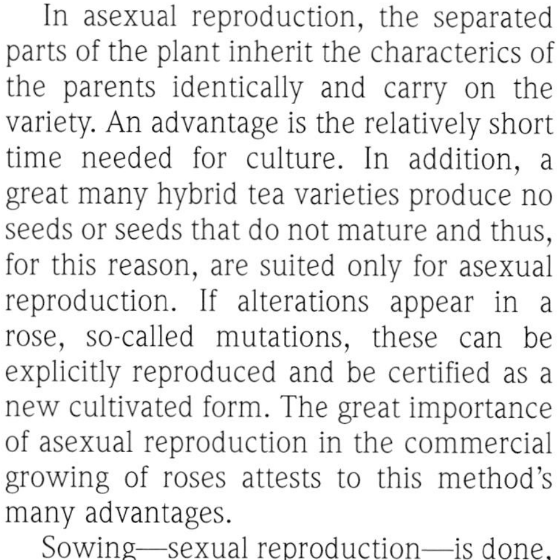

In asexual reproduction, the separated parts of the plant inherit the characterics of the parents identically and carry on the variety. An advantage is the relatively short time needed for culture. In addition, a great many hybrid tea varieties produce no seeds or seeds that do not mature and thus, for this reason, are suited only for asexual reproduction. If alterations appear in a rose, so-called mutations, these can be explicitly reproduced and be certified as a new cultivated form. The great importance of asexual reproduction in the commercial growing of roses attests to this method's many advantages.

Sowing—sexual reproduction—is done, as a rule, only with the pure wild species. These are used for plantings in natural settings—for which the seedlings supply the necessary genetic variability and vigor—and for the wilding understocks onto which the more refined rose varieties are grafted so they can profit from the vigor of the wild rose. In addition, rose breeders use sowing to gain new seed material from the results of their crossings to obtain new rose varieties.

Budding

Grafting, or budding, is the most commonly used method of propagatimg roses. Budding means the insertion of a cane bud, or an eye, of a refined rose variety onto an understock.

By means of budding, a wild rose can be turned into a bedding rose, climbing rose, shrub rose, or any other more refined rose plant.

How does budding work? Anyone who wants to try budding must first get a wild understock from a nursery. These are generally bundled in groups of fifty and are available only bundled in this way. With some luck, the nurseries will sell the rose lover these understocks for use in their own rose propagation. The best known and most commonly used wild species for stock is *Rosa laxa.*

Once one has the wildings, as they are called, the canes and roots are first pruned back lightly in the fall and are healed into sand, still bundled. They will be planted in the spring. If there are no seedling understocks available, the rooted hardwood

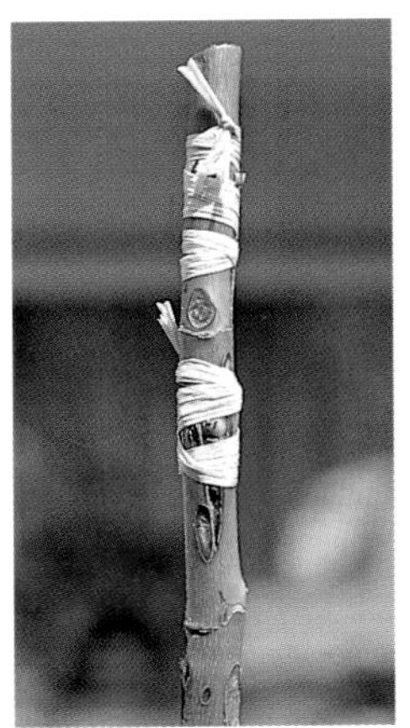

In the budding of standard roses, two to three grafts are made (left). Right: New shoots.

cuttings of wild roses, e.g., of *Rosa multiflora,* can be budded.

The actual budding takes place in the hot summer days in July. Professionals use a special grafting knife. It costs about $12.00 and is worth buying only if a number of plants are going to be budded. Otherwise, one can use a good sharp, clean knife.

Now the budding can begin:

- First, a so-called T-cut is made in the cleaned, polished smooth, and finger-thick root neck of the wilding.
- Then the scion—that is, a piece of a mature cane that has finished flowering and of the variety to be budded—is cut. The scion has its leaves and prickles removed. Only the leaf stem, about ½ in. (1 cm) long, is left at the eye.
- With a drawing cut, which is made about ¾ in. (2 cm) under the eye, a piece of bark including the eye is removed. The eye is held so that it faces the budder. Behind the eye is a shaving of wood, which is slowly and carefully removed.
- The bark around the T-cut is carefully loosened, and the scion bud is inserted into the cut. The bark will loosen well only in the hot weather since the understock, because of its great physiological activity, is full of juice.
- After inserting the scion, one cuts the part of the bark of the eye that is sticking up back to the height of the T-crossbar.
- Finally, the bud union is wrapped with raffia, rubber tape, or a rubber band and kept free of dirt.

Budding (continued from page 171)

- In the fall, the bud union is well hilled. Thus it goes through the winter.
- In spring it is uncovered again and the wild crown growing over the T-shoulders on top of the growing eye is cut away. Nursery owners call this process *topping.* Now the entire sap stream of the understock concentrates on the high-grade, still-dormant eye and compels it to burst forth. A new, upgraded rose begins to grow.
- Suckers around the bud union must be continally removed during summer and this means removed cleanly at their base.
- Some rose varieties do not develop branches very readily, e.g., the group of hybrid teas. To promote branching, they are pinched back more often in May. To do this, the overlong individual canes are shortened, and their dormant accessory eyes are forced to make new growth (see page 115).

Standard Roses: Standard roses are budded at the desired crown height. So that the crown can develop round and full, two or even three buddings are done around a stock.

Prerequisites for doing one's own propagation of tree roses are understocks in the form of crowned little standards. These understocks should be straight and as long as possible.

As a rule, nurseries use special selections of the species *Rosa canina.* Naturally, one can also prune a wild bush so that a cane only about 59 in. (1.5 m) long is left for a future trunk. Anyone who has an old wild rose in the garden can even, with a little luck, get a rooted wild understock from this bush. In earlier times, the standard roses were always grafted onto wild standards from the forest anyway.

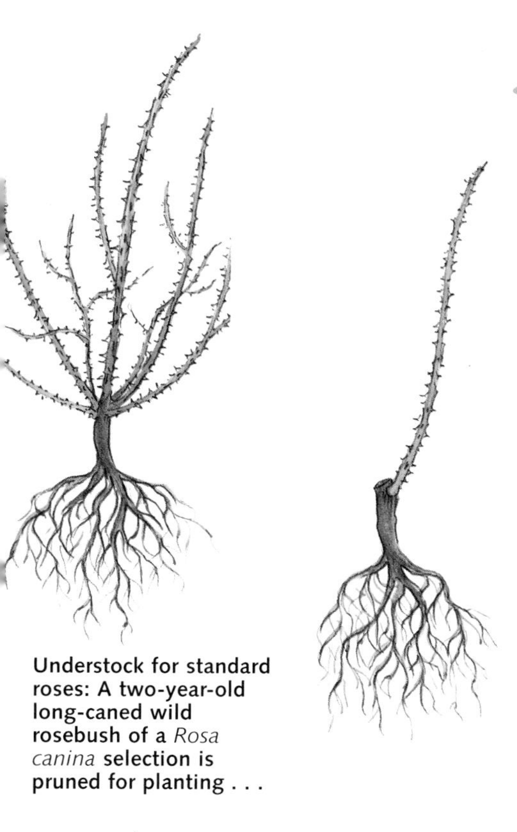

Understock for standard roses: A two-year-old long-caned wild rosebush of a *Rosa canina* selection is pruned for planting . . .

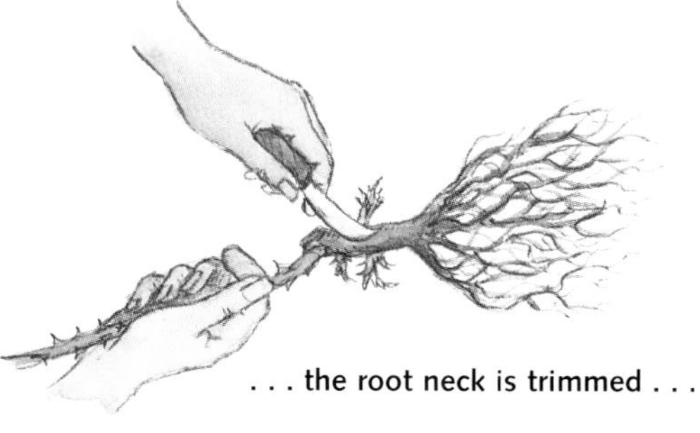

. . . the root neck is trimmed . . .

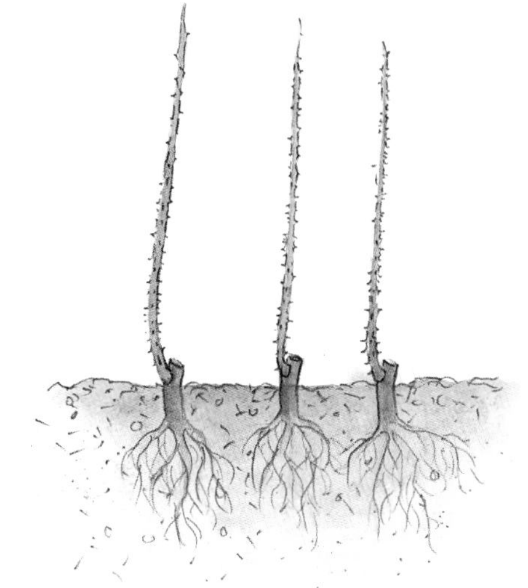

. . . and the stock is planted for summer budding.

TIP Older standard roses whose crowns have been damaged by frost or breakage can be repaired with a new graft and thus saved.

Winter Grafting

In the interest of completeness, the possibility of winter grafting should be mentioned at this point. It is only interesting for gardeners who have a small, heatable greenhouse. Professional gardeners use this method of grafting mostly for the propagation of greenhouse cutting roses. Grafting is done with diagonally cut scions on sturdy, finger-thick understocks of the variety *Rosa canina* 'Inermis' and, obviously, in the period from December to April.

Hardwood Cuttings

A very successful propagation method, especially for shrub, area, and climbing roses, is taking hardwood cuttings.

Especially with climbing roses, rooting hardwood cuttings is a lot of fun. They offer sufficient material for the approximately 8-in.- (20-cm-) long, pencil-thick woody cuttings, which are cut from now leafless woody branches and canes. These cuttings are stuck into loose soil either in late fall or early spring directly after cutting. Preferably, they are rooted just where the desired rosebush is to grow, where they will not have to undergo developmentally disruptive transplanting of the rooted cuttings.

After they are planted, only the topmost, last eye on the cutting should still be showing above ground. If the cuttings stick up too far, they dry out easily. The hardwood cuttings can be cut with a knife, but cutting is easier with a pair of sharp scissors.

As in budding, the cuttings must be regularly **pinched back** (see page 115).

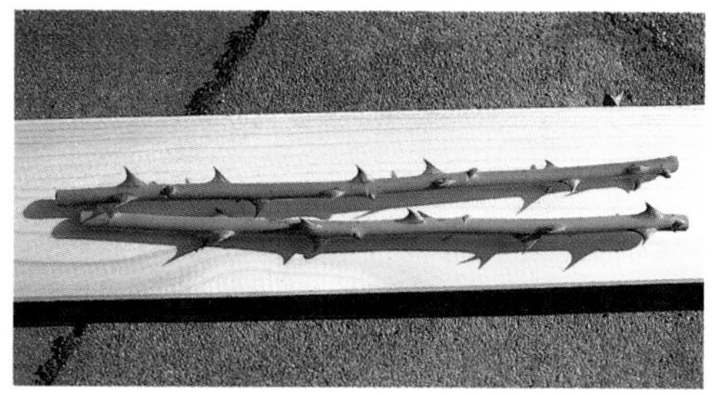

Cuttings of a climbing rose ready to plant.

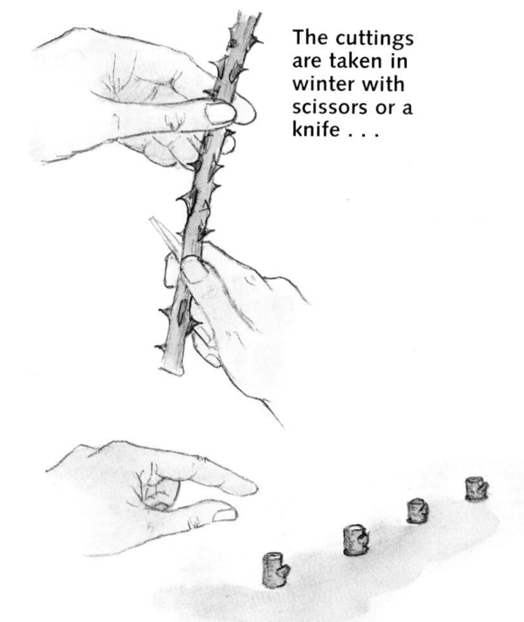

The cuttings are taken in winter with scissors or a knife . . .

. . . and stuck into the soil up to the topmost bud.

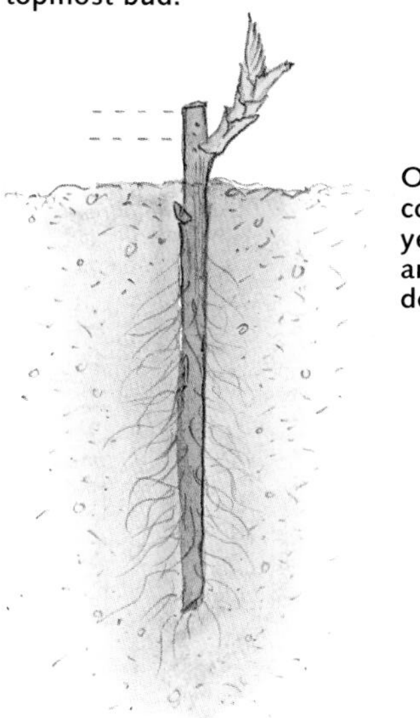

Over the course of the year, shoots and roots will develop.

Otherwise, the new rose will have only one cane and will not develop into a bush. **Fertilizing** is not necessary the first year; it will only negatively influence the hardening of the wood. Also, so that they will develop plenty of roots, the young plants should seek nutrients for themselves. From the second year on, the roses are then fed normally.

In principle, hybrid tea roses can also be reproduced through hardwood cuttings. However, they lack the genetic constitution of sufficient vigor to be able to thrive as own-rooted plants over the long term. Therefore, grafting them onto vigorous understocks is better.

Soft Cuttings

The propagation of roses by soft cuttings (slipping) is one of the oldest methods of reproducing roses. The gardeners who were working in the youth of what are today termed Old Roses knew no other way at all of reproducing roses true to variety. So in about 1845, C. Nickels, in his book published in Pressburg, *Cultur, Benennung, und Beschreibung der Rosen* (*Culture, Naming, and Description of Roses*), described very precisely which rose varieties were best suited for propagation by cuttings.

Various reasons explain why reproduction of roses from cuttings dozed for almost the entire twentieth century and began to enjoy a true renaissance only just a few years ago. For a long time, the image of being plants of weak vigor, early aging, and frost sensitivity has clung to roses reproduced by cuttings. Experiences with hybrid teas reproduced this way may have contributed to this image. In addition, since the beginning of this century, budding became established as the leading propagation method. It can take place entirely outside without greenhouses and produces sturdy, thick-caned plants for sale. These arguments in favor of budding are just as apt today as they were years ago. For most rose varieties, therefore, budding remains propagation method number one.

A rose cutting before and after planting.

On the other hand, the earlier judgments about roses from cuttings can today be considered outdated. Reproduction by cuttings appears interesting and sensible especially for many miniature, area, and wild roses. Why? Own-rooted miniature rose varieties without strong wilding roots fit much more easily into small containers like balcony boxes and troughs. Own-rooted area rose varieties do not become troublesome by persistently sending up suckers. The removal of suckers, especially in tightly enclosed public plantings, is not only tiresome but also involves expensive hand work that is hardly affordable anymore. By continuing to grow endangered native wild roses, propagation by cuttings helps to bridge the gap caused by missing seeds and contributes to the saving of these rose species.

Naturally, the varieties from other rose groups can also be reproduced from cuttings. As mentioned in the description of hardwood cuttings, the own-root reproduction of hybrid tea roses, for example, is not very practical since they lack the vigor powered by an understock.

When to Take cuttings: Basically, rose cuttings are taken just like cuttings from any other plant. An early planting date is important, preferably in June or even July at the latest. Then the young, graceful little roses have a real chance to develop suifficient wood until winter, which represents an important precondition for the tricky wintering over of the cutting.

Execution: Make cuttings from mature but not hard canes whose flower buds show color. These herbaceous cuttings have two to three leaf bases and a maximum length of 4 in. (10 cm). The lowest leaf is removed, and the cutting is stuck ¾ to 1¼ in. (2 to 3 cm) deep—up to the next leaf base—in a shallow pot or dish of potting soil mixed with sand (substrate). Garden compost can be added to the substrate if the material has been well aged and strained. Cuttings and substrate are now moistened with the watering can with spray head on. Then the pot or dish is covered with plastic wrap. (Also suitable is a hard plastic flat with a transparent, tightly closing cover, such as those attainable from the supermarket salad bar.) Thus closed, the box turns into a minigreenhouse in hot weather. The high humidity inside provides the cuttings with water for developing their own roots.

One need not use rooting hormone to help along the cuttings of suitable varieties. With enough moisture, but not collected water, over the next three to four weeks the new roots will develop swiftly. Then the little new roses are transplanted to small pots; this is called *pricking out.* They are further cultivated outdoors, protected at first from the blazing sun, however. Anyone who wants bushy plants the first year—perhaps as a personal gift for a friend—puts three or four cuttings into one pot.

The cuttings must be wintered over without frost the first year. Then in the spring, they can be moved to their final site and tagged as homegrown.

Layering

All rose varieties with long, flexible canes—thus shrub and climbing roses, especially ramblers— can be propagated by layering. (Many grape and rhododendron varieties can be propagated this way, too.)

In fall or early spring one selects a long, well-matured, and woody cane. The gardener then carefully removes the foliage around the the middle of the cane. At least three eyes must be on this leafless part of the cane. Now this cane is carefully bent down to the ground. A slight cut must be made at the leafless bend, which is laid into the loosened soil and fixed with a hook. Now some additional loose soil mixed with compost or bark humus is placed over it.

Over the course of the summer new roots develop at the injured place on the bent cane under the hill of soil, which is kept constantly moist but not wet. After sufficient growing time, a new rose arises. The next spring, the rooted piece of cane is separated from the mother plant with the pruning shears and planted into a new spot.

One more piece of information must be discussed about the subject of rooting. Some ground-covering area roses like 'Pheasant' or 'Immensee' form new roots when their canes touch the ground. They do not

Plastic containers like this make terrific mini-greenhouses. In front of the container is a cutting ready to plant.

require any human help to accomplish this. This ability allows these varieties to propagate themselves.

Runners, Division

Other rose species—like blackberries and raspberries—develop so-called **runners** (also called suckers). In early spring, the gardener can easily cut off these runners by using a sharp spade and can plant them into a new spot. The young shoots of the new plants should be cut back to about 8 in. (20 cm) so that they will branch into a bush. Then they are planted immediately and watered well. Such running roses can—like perennials—be regularly divided.

Primarily, wild roses are best suited for this special propagation method. These varieties include *Rosa nitida, Rosa rugosa* and its own-rooted hybrids, *Rosa rubiginosa, Rosa moyesii, Rosa spinosissima,* and also *Rosa gallica* varieties on their own roots.

Sowing

Roses are only rarely grown from seed. The few reasons they are grown from seed include growing understocks for grafting, obtaining wild roses, and producing new varieties in rose breeding.

In principle, all roses that produce hips can also reproduce from seed. This statement has only one hitch. In the hybrid varieties, the seed does not develop true, and the offspring are more than nonuniform in appearance and resemble the parent varieties only in part, if at all. The reason for this is, among other things, that roses exert a powerful attraction for bees, bumblebees, and other insects when they flower. As a result, the roses are pollinated with all kinds of pollens of numerous other varieties or species. The resulting seed is thus extremely mixed up genetically.

Fundamentally, the development of a multitude of new rose varieties through breeding is based on this mixing phenomenon. Namely, someone who knows with what probability certain characteristics of which parent species are passed on has a good chance of achieving new, improved varieties. Read more about this in the section "Rose Breeding" (see pages 178 through 182).

Note

Species and varieties that send out runners can also become a nuisance, e.g., if they travel several yards from the mother plant on their own and turn up uncontrolled in the perennial bed. Someone who places little value on runners should therefore either resort to grafted plants or place a runner barrier around the plant. A runner barrier is a very thick plastic pond liner (or a cement ring), which is buried in a circle around the plant and cannot be overcome by the surface-traveling runners. This method has produced the best results with runner-forming bamboo species.

Stratification: The seeds are harvested in the fall and separated from the meat of the ripe hips. Unfortunately they are not yet ready for takeoff. The seeds of the wild roses lie over. This means that before the seeds can germinate, they must pass through certain natural fluctuations of temperature that, as a rule, correspond to two winters. Only after 18 months in the soil has the largest portion of the seed achieved sufficient germination capacity. Nursery owners simulate these temperature conditions for the rose seed with a trick. They layer the seeds in moist, cool sand. This procedure is called **stratification.** With this method of accelerated decay, the ordinary germinating capacity of the seed is considerably increased. Stratification is like a kind of seed composting. Nurseries stratify in large, deeply-sunk concrete boxes out-of-doors.

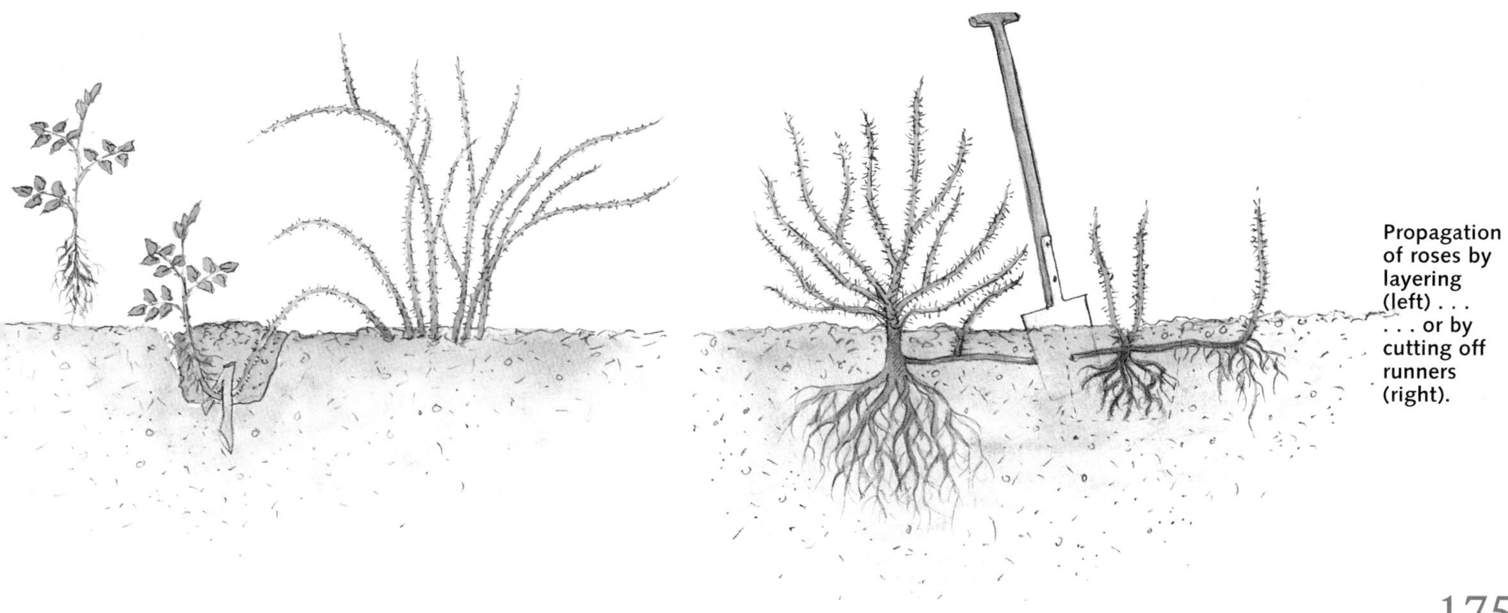

Propagation of roses by layering (left) . . . . . . or by cutting off runners (right).

Of course, the home gardener can get rose seeds to germinate in his or her own garden without stratification. However, some patience and pleasure in a natural experiment is needed. For the rose hobby breeder, the results of seeding should not be the main goal but rather the fun in working with the secrets of nature.

Practical Sowing: First, the red, ripe hip is opened with a clean, sharp knife and the seed carefully removed. It is placed into a bag with damp garden compost, which is again stored for several days at room temperature. Then it disappears, tightly closed, into the refrigerator for six weeks. During this period, one should watch for the development of mold and remove it if that occurs.

The seed thus prepared is spread in a dish with sandy soil and covered with a very thin layer of sand or very fine gravel. Under no circumstances should the seed be fertilized. The dish is placed somewhere cool and, above all, safe from mice. If it is the seed of a hybrid tea rose and the result of an attempt at crossing, put the container into a frost-free area, perhaps a garage. Wild roses, on the other hand, need a freezing phase.

Beginning of Germination: Now everything is in nature's hands. Germination can begin after several months with seeds from one's own crosses. However, in wild roses, it may not happen until after a year in some cases. Often, the germination success with unstratified seed is less than 15 percent. After successful germination, the oval seed leaves develop first. Necessary for further development of the seedlings are daytime temperatures of around 68 °F (20 °C) and adequate daylight. In this phase, the rose gardener must not fertilize with compost or similar organic material. The addition of such materials increases the humidity of the microclimate around the rose seedlings and quickly leads to fungal disease. Only when the typical rose leaves are recognizable are the seedlings transplanted to their own little individual pots. When the young plants are established and acclimated after several weeks, they are placed outdoors in spring temperatures.

Pot Roses from Seed: In the stores, seeds of pot roses are offered along with other seeds. These seeds are from offspring of the species *Rosa multiflora,* which germinate after several weeks and appear relatively true to variety. These seed packets are worth a try in any case. During culture on the windowsill, natural processes can be observed up close and live, especially by children.

'Scarlet Meidiland' from meristem propagation.

Meristem Propagation

Meristem propagation is still a young method of propagating roses. The principle behind it is to achieve a viable rose identical to the mother variety by using an isolated plant cell.

To this end, certain tissue sections (meristems) are removed from the axil buds of young shoots of the rose. Meristem is a formative plant tissue. It has the ability to divide indefinitely. As a result, meristematic tissue can give rise either to similar cells or to cells that eventually differentiate into definitive tissues and organs. During meristem propagation, meristems—disinfected and prepared—grow on in the laboratory in a nutrient solution and become a rose that is true to variety. Within twelve weeks after the tissue is harvested, the tiny rose is ready to prick out, as a rule. After a period of hardening off, it leaves the special laboratory and is further cultivated in the nursery like a normal plant rooted from a cutting. This reproduction in a petri dish (in vitro culture) is primarily possible for area and pot roses. This method is used, of course, primarily for commercial horticulture, for which mass propagation makes sense and the high labor costs can be justified.

The great advantage of in vitro plants lies in their absolute freedom from disease and their quick development. Above all, new varieties can thus be offered to the rose market very fast and in much larger quantities.

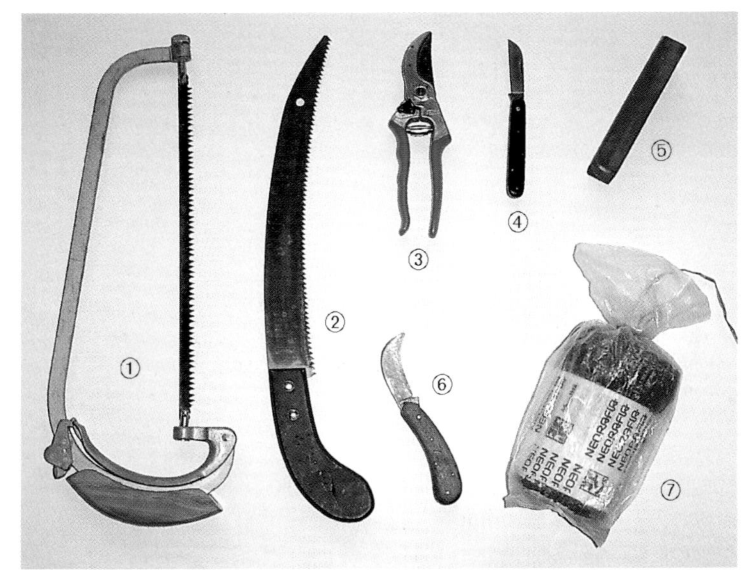

Useful Tools:
① **Tree saw**
② **Hand saw**
③ **Pruning shears**
④ **Grafting knife**
⑤ **Sharpening stone (Belgian stone)**
⑥ **Pruning knife**
⑦ **Tying wire**

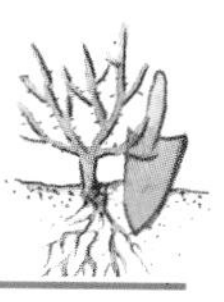

Times Table of Tools

Dealing with the rose, its care, and its propagation is made considerably easier with a battery of garden equipment and tools. These need not always be the most expensive and varied equipment. The most important qualification is that they fulfill the desired function perfectly.

A portion of the items introduced in this section belong to the basic equipment of the hobby gardener anyway. This equipment should be handy and in a usable, well-maintained condition so that working with it is not a torment. Cutting tools should be sharp. Hoes and spades are helped by being rubbed with a few drops of oil to get through the long winter pause without harm.

Anyone who gets special rose tools should check beforehand to see if the financial outlay for it will pay. This goes for all special knives and shears for grafting or pruning roses.

Claw: This tool is outstandingly suited for unhilling the roses in the spring. After that, it can be used to loosen soil.

Dethorner: The dethorner is a very useful implement for the lovers of cut roses. It is used to strip the prickles from the lower ends of roses easily, with a flick of the wrist. Also, with its attached sharp hook, the ends of the rose can be cut into a slant in the next move. The rose stem is then ready for vase or arrangement.

Flower Gatherer or Cut-and-Hold Shears: These shears are a blessed invention for all lovers of cut roses and other flowers. With them, a stem can be cut and held in place by the shears so the stem does not fall to the ground. Because they have a unique blade design mechanism, these shears hold a stem strongly in place after cutting, thus allowing a gardener to cut a stem and then place it directly into a container. This is especially handy with roses since the gardener does not have to grasp the prickle-filled stem of a rose after cutting it. Thus, a whole bunch of cut roses can be gathered very comfortably.

Gloves: Roses have prickles that can be quite sharp. The fine bristles of *rugosa* hybrids are really mean. Good garden gloves protect the hands while you work.

Practical tool for flower arranging: the thorn stripper or dethorner.

Grafting Knife: For those who would like to bud lots of roses themselves, getting a grafting knife will pay. Like other specialized tools, these knives are designed for budding work. Their blades can be brought to razor sharpness.

A distinction is made between grafting knives with the loosener on the blade and those with a separate loosener made of plastic at the end of the knife. The loosener facilitates the loosening of the bark at the root neck of the wilding before the placement of the eye in the T-cut. Since a grafting knife is used only a few weeks out of a year, it must be cleaned after being used and before it is stored again, rubbed with oil, and wrapped in cloth. Dealers of horticultural supplies offer outstanding knives labeled with the trademark Tina.

Leaf Vacuum: All the rose books advise the gardener to remove fallen rose foliage extremely carefully from the beds in the fall. This is good advice, for the resting spores of spot anthracnose and other diseases are harbored on the leaves. If they are left lying, the new infection of the roses is preprogrammed. However, whether it is absolutely necessary to remove the fallen leaves with a vacuum or whether the good old rake does it just as well must be decided by every gardener for himself or herself. In any case, a leaf vacuum does not suck up selectively, so that all kinds of useful creeping and scratching animals also land in the catching bag. Therefore, after using a leaf vacuum, one should go through the bag and remove and destroy the rose leaves. The useful remainder can be returned to the garden again.

Acquiring a grafting knife is only practical for a lot of grafting.

Loppers: Anyone who has many older shrub and climbing roses probably is glad to call on the services of a pair of loppers, which are heavy-duty, long-handled pruners. With them, canes up to 1 1/2 in. (4 cm) thick can be easily, cleanly, and smoothly removed.

Pruning Knife: The pruning knife is a curving garden knife that is outstandingly suitable for removing suckers from roses. With one drawing cut, the sharp knife cleanly removes the sucker right at the base of the cane.

Sharpening Stone: The best pruning tool is no good if it is not sharp. For sharpening blades, a whetstone is appropriate. For final fine finishing, use a so-called Belgian stone. Anyone who is unpracticed at sharpening knives and shears should have this done by an expert before the beginning of the season.

Tree Saw: This is useful for rejuvenating old, overaged shrub roses, for example. In addition, hard branches are not easily cut with pruning shears, so a gardener can use a tree saw instead. The teeth of the saw blade should saw finely enough so that no fringed sawed places are left behind. Especially handy are foxtail-like hand saws with a narrow, curving blade that sometimes folds up for storage.

Watering Can: Sometimes, such as after planting, roses can be watered better with the watering can than with the hose. Most effective is the oval garden can with an easily graspable, long, curved handle. Round cans or those with rectangular handles are hard to

For anyone who gathers a lot of cut roses, cut-and-hold shears (top and right) make the work considerably easier.

handle for continued use. If the can is metal, one should make sure it is galvanized. Plastic cans are not bad so long as their shape allows for working with them comfortably.

Rose Spading Fork: A genuinely special tool for the rose gardener, and his or her most often used piece of equipment, is the two-tined rose spading fork. With it, the soil can be loosened without stressing the rose roots unduly.

Shears: The price of the shears is not important. However, the shears' sharpness and their ability to cut is most decidedly of importance. Shears should cut and not crush. Roses have a very pressure-sensitive wood since it is relatively soft. There is great danger that by using bad shears, larger wounds will be made, which are only more difficult to heal. With the rose, this is especially fatal, for the wound-healing of the bark proceeds very slowly in roses. Even years later, one can still easily identify large wounds in them. The Swiss firm of Felco is considered a certified maker of good pruners.

Spade: Without this classic tool, the planting of roses and the thorough working of the soil is unthinkable. It is a part of the basic equipment of every gardener. Thus, he or she should pay special attention to the sturdiness of spades available when shopping for one. For example, Ideal spades have shown themselves to be good.

Sprayer: Good spraying equipment is needed for effective application of biological and bee-preserving preparations as well as home remedies. For anyone who has many roses in the garden, it pays to get one. When buying, one should pay attention to how the sprayer handles and, above all, the weight.

An outstanding special tool for rose fans: The two-tined rose spading fork.

Rose Breeding

The magnificence of roses as we know it today is still not 200 years old. Until the beginning of systematic rose breeding in the middle of the last century, horticulturalists and hobby gardeners knew of just about no everblooming roses. Garden roses in orange, copper, yellow, or fire red were strange to them. Shining, robust foliage was unknown. A single open flower that held for ten days and longer would have seemed a suspect record to them.

For the rose lover of our time, these many special characteristics are a matter of fact and often not worth mentioning anymore. For a comparatively very small price, he or she can buy roses and enjoy them. Anyone who gets more involved with roses, however, will quickly learn that the rose, as we know it today, has developed from the industry of many breeders and their work.

The Genetic Foundations

The special characteristics of each rose variety are fixed in thousands of genes. These genes are the working materials of the breeder, who tries to find new rose varieties by putting together new gene combinations. Each characteristic of a rose is rooted in one or in several genes. An analogy comparing the rose with an all-embracing library gives some idea of the tremendous dimensions of the gene world. The thousands of genes of a rose correspond to the number of books of a library. Every book (every gene), exists in two copies. One volume is always donated by the father, and one volume is always donated by the mother. These volumes may be different editions of the same book. This corresponds to different varieties (alleles) of the same gene. The books are spread over several dozen departments, the chromosomes, and their pages are closely printed with ever-new combinations of the same few letters, the nucleotides. When different editions of books are placed next to each other (different alleles combined in successive generations), new combinations result. Therein lies the difficult task of the breeder. As might be expected, the combinations do not always result in the idea—the new hybrid does not fulfill the stated expectations. Now and again, however, out of the old books grows a new literary pearl, a new rose variety. To find it fairly unerringly is the secret of rose breeding.

The Augustine monk Gregor Johann Mendel was the first on the trail of genes barely 150 years ago. The Mendelian laws of inheritance he stated in 1866 are common knowledge today. Mendel established in his experiments the existence of dominant and recessive genetic characteristics. A dominant characteristic can suppress a recessive, thus weaker, one completely for a generation. The suppressed characteristic is not lost, however, but reappears with a precisely determined probability in the next generation.

Many geneticists have since restated the correctness of the Mendelian laws over and over again. These also apply to the rose, but the genetic circumstances with the queen of the flowers are extremely complex. The dominance of a characteristic is present in numerous gradations; a single characteristic is often governed by many genes. The rose breeder is the librarian in our analogy who is looking for a certain title in the gigantic stacks. No alphabetically arranged catalog or databank helps the breeder. Only through thousands upon thousands of visits to the library, the so-called crossings, can he try to reach his goal.

Crossing—Ladies First

Rose breeding with all its facets is enormously time-consuming and involves high expenditures of money, time, and patience. Nevertheless, all over the world professionals as well as private breeders are always trying to find new varieties—in the hope of finding the perfect rose. Anyone who has ever tried rose breeding and has quickly been seized by this fever and the rose passion will have a great deal of fun but will also suffer bitter disappointments.

Wilhelm Kordes, father of countless world-famous varieties, describes the actual crossing process with a lapidary sentence, "Thus we need only go into the garden, even better into a greenhouse, unman the roses from which we wish to have seeds, then convey pollen from another variety to the pistil, and leave the rest for the gods to do."

Every breeder first selects crossing partners that he or she hopes can provide certain desirable characteristics. Professional breeders use greenhouses for the crossing work. For all who cross out-of-doors, selecting crossing partners that produce red, fully ripened hips in September is recommended. If the gardener does not use ripened seed, the breeding work will not be successful.

A result of high breeding skill: The bicolored Tantau variety 'Nostalgie'®.

The crossing countdown starts in summer. Two days before the actual crossing, the yellow stamens of the father variety are cut off with shears, placed into a saucer, and dried with the light excluded. A fine yellow pollen dust remains. One day before the complete opening of the flower of the mother variety, all the flower petals and the stamens are removed by hand in the early hours of the morning. This prevents self-pollination. The castrated flower is left.

On the day of crossing, the breeder dips the castrated flower of the mother variety into the pollen of the father variety. Immediately after the crossing, a bag of paper or aluminum foil or plastic is placed over the flower, which protects it from insects and other pollination, and is removed after a week. The parent varieties of the crossing are noted in a notebook and also on a tag that is attached to the pollinated flower. The manner of writing this crossing notation is internationally established. It requires that the mother variety be listed first and the father variety be listed second—ladies first. Such an entry looks like this example: 'Birgits Beste' x 'Bonica'.

Now hip development begins. It really does lie in the hands of the gods, who have, as everyone knows, prescribed sweat before success. In the practice of rose breeding this means that the more crossings undertaken, the higher the quota of hits among the seedlings later. Europe's largest rose nursery, W. Kordes Söhne in Schleswig-Holstein, for example, carries out more than 80,000 crossings annually.

In fall, the ripe hips with their seeds are collected. As described in the section "Sowing" in the section "Propagation" (see pages 175 and 176), they are prepared and sown.

One year after the crossing, the first seedlings bloom. The selection continues, in the course of which one should pitilessly—if also often with heavy heart—part with the unpromising seedlings. Of thousands of little plants sown, usually only a few are left, often only one, sometimes even none at all.

In the big nurseries, which have their own breeding departments, experienced specialists meet over the decision as to which seedlings should be further propagated. Having been schooled through many years of selection work, they separate the chaff from the grain. Among the professionals, all very promising seedlings are budded onto wilding understocks in August. The rest are thrown away. For the hobby breeder, the further culture of interesting seedlings by cutting is fast, without problems, and less time-consuming than budding.

The developmental history of a new variety by a large rose breeder can look like this:

Winter 1997/98: Planning of 50,000 crossings.

Summer 1998: Execution of pollination, harvest of 30,000 hips.

Spring 1999: Germination of seed.

Summer 1999: First selection of seedlings for flower form and color; 15,000 seedlings left.

Summer 2000 to 2004: Further selections over several years for robustness, vigor, winter hardiness, and so forth. Three or four varieties left.

Summer 2004: Commercial reproduction of these varieties in large numbers.

Fall 2005: Introduction and sale of the new roses more than seven years after crossing, hoping for a bestseller, which would recover the breeding costs.

Thus, at least eight years pass from the planning to the sale of a new rose variety. This is an eternity in the fast-living modern world.

However, this scenario should not scare away the hobby breeder, for the many years of selection are not a necessity for him or her. In the Verein deutscher Rosenfreunde there exists a committee of amateur rose breeders that one can join to exchange information. This can help avoid many mistakes. In Great Britain, there is even a test garden in St. Albans for new varieties by amateur breeders. After three years of trial, the best variety receives the Trial Ground Certificate, a high distinction, which is awarded during the annual convention of the British Rose Society.

Selection field of thousands of new roses that are awaiting discovery by the market.

Variety Protection, Trademarking

Variety Protection

Anyone who breeds new roses with much effort and tremendous investment of money would understandably like to enjoy the financial fruits of his or her labor. Variety and/or cultivar protection helps the breeder to do this. It can be compared to a patent for a new invention. If a rose is bred to be different, homogeneous, reliably propagated, and new, it can be awared variety protection. The new rose variety is the invention of the breeder. He or she can protect the variety and have exclusive rights to propagate and sell it. If a patent is granted, it remains in effect for 17 years. During this period, neither commerical nor backyard gardeners are allowed to propagate the plant unless given legal permission and/or pay a royalty fee to the patent holder. To apply for a plant patent, papers must be filed with the United States Patent and Trademark Office in Washington, D.C.

Some rose variety names have an ® after their name. This indicates it is officially registered and protected from unauthorized propagation. Protected or registered rose names may also have the first three letters of the name begin with the breeder's name or firm, for example, 'Noatraum' for the breeder Noack or 'Meidakir' for the firm of Meilland.

The first rose patent ever was awarded in the United States for 'New Dawn' in 1930.

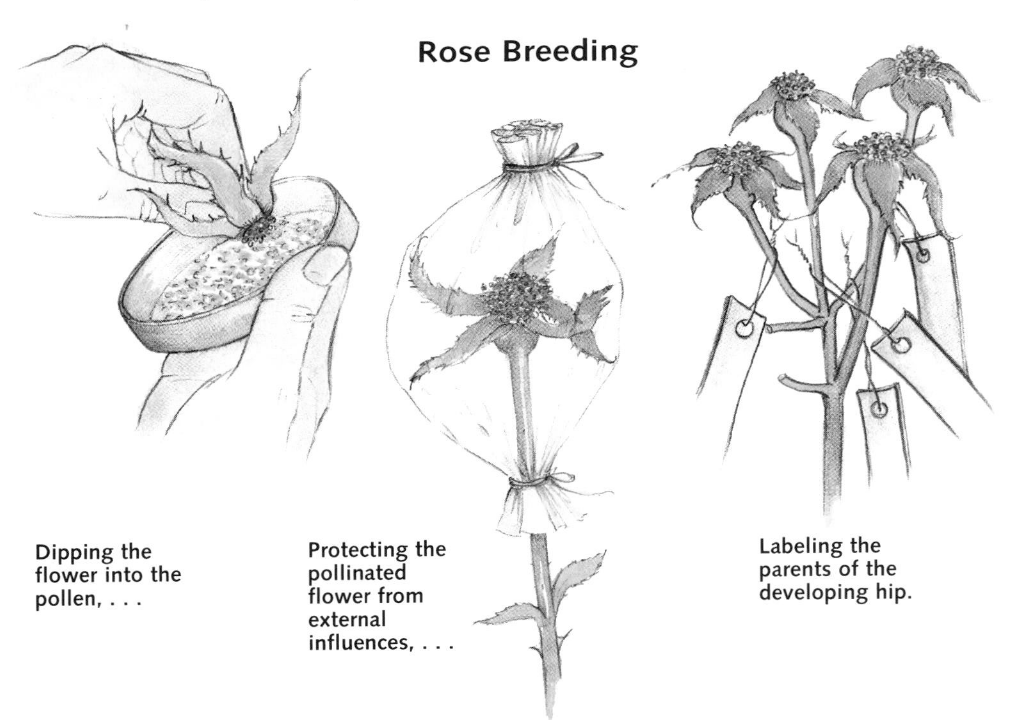

Rose Breeding

Dipping the flower into the pollen, . . .

Protecting the pollinated flower from external influences, . . .

Labeling the parents of the developing hip.

Trademark Protection

Another way to protect a cultivar or variety is to trademark the new name. A cultivar name identifies certain characteristics of a specific species of plant. Cultivar or variety names can be used worldwide to distinguish those characteristics without illegal infringement if the name is not a registered trademark. However, if the new name is trademarked, it cannot be used in commerce without the owner's explicit permission. Trademark rights are a valuable commercial item to protect the initial breeding efforts and subsequent sales of the plant.

A plant trademark name also often indicates to consumers the origin of the plant (a specific nursery company that bred, propagates, and sells the variety). The trademark name also indicates the quality associated with the product.

A trademarked name can be recognized by a ™ just after the name. For example, the following rose names have been trademarked: 'All That Jazz'™, 'Breathless'™, and 'Diana, Princess of Wales Rose'™.

Some trademarked names are also distinguished by being spelled out in capital letters as opposed to nontrademarked cultivar names, which are in single quotes only. Trademarks are also granted by the U.S. Patent and Trademark Office in Washington, D.C.

Thus, there are two levels of protection with rose varieties. In the first type, variety protection, protects the product rose—one could even say the product content. The second type, the trademark, protects the name, which differentiates the plant and thus represents the rose as a marketable "product package." Trademarking and variety or content protection are completely independent of one another and are interchangeable. An example is Tide™, a well-known laundry detergent for many decades that has a registered trademark name. Although the exact chemical composition may change as the company comes out with new and improved versions, the product name always remains the same for promotion: Tide™.

While not carrying any legal weight, new rose names are also registered with the American Rose Society, which maintains a comprehensive listing of rose names. This listing helps introducers know what names have already been chosen for roses when they are searching for new names.

In the same way, trademarks of an old, superseded rose variety are sometimes carried over to a new variety. For example, in the 1950s there was a rose named 'Eden Rose'. When this variety was taken out of trade, the protected name was transferred to a new variety. The old variety had of course vanished from the market, but its valuable trademark remained. So it happens that the modern variety 'Eden Rose '85' has nothing to do with the old 'Eden Rose' although both share the same trademark. One can recognize this trademark recycling, as a rule, by the year numbers that have been added. In the name 'Eden Rose '85', the '85 means the introduction year of the variety, 1985. Anyone who thumbs through an old rose catalog will sometimes stumble on varieties that he or she can hardly know of but whose names are familiar. This occurs because these names once belonged to old varieties and now adorn modern roses.

Trademark and variety protection law is a complicated matter and a science in itself. It is upheld and untangled by specialized patent attorneys the world over. Nevertheless, rose lovers should at least hear about it once. It is then easier to understand that innovation in rose selection has its price and is tied to high advertising, protection, and breeding costs. New varieties do not just fall from heaven.

When nurseries want to propagate a variety-protected rose commercially or offer a trademarked variety, they must pay the owner a licensing fee for it. This is reflected in the selling price and explains the higher price for protected varieties in trade. Commercial propagation of the variety and use of the trademark without permission of the particular owner is forbidden on good grounds and is subject to penalty. The consumer can recognize a protected rose by the apprpriate protection ticket. As a result, only if innovation pays will there be innovations. Anyone who wants to have the best rose varieties of the moment in his or her garden should also be ready to make a contribution for this advance.

'Helmut Kohl-Rose' (a cross from Rosen Tantau): A well-known name is a marketing force for a new rose variety.

A red that is so dark it is almost black, the fragrant rose, 'Deep Secret'.

Blue and Black Roses

Although rose breeding over the last 200 years has produced very many new growth and color combinations, really blue or black roses have not yet been achieved at this time. Nevertheless, or more correctly, primarily because of this, roses with these flower colors continue to exert a particularly strong fascination on rose lovers. Pictures of these flowers turn up regularly as supposed sensations in color catalogs.

Blue Roses: The Arabs solved the problem of the demand for blue roses in a special way. Old, reliable sources report of a process used by Arab gardeners to achieve the blue flower color artificially. One loosened the bark on the roots of the rose, let indigo run in, and reburied the roots again. The rose flowers did actually turn blue but not permanently and not for long.

Evidently, this method is not very practical economically. Scientific experiments therefore sought to get to the bottom of the phenomenon of blue flower color in order to find a start in breeding lasting blue roses. The color blue is produced by the inner pigments of the rose, the so-called anthocyanins. These are under strong genetic control. The anthocyanin delphinidin is responsible for the pure blue displayed in other plant species. Until now, breeding has not succeeded in transferring delphinidin to roses. Until this happens, truly blue roses are real only in promisingly colored prospectus pictures.

Black Roses: Truly black roses also still belong to the realm of fantasy. Again and again, catalogs offer very dark red roses, which with considerable goodwill may be termed black. As a rule, these varieties are weak growers and are prone to disease.

The Rose Market

The Romans dedicated the rose to the goddess Venus, the Greeks chose Aphrodite as the rose symbol, Roman banquets without crowns of roses for the heads of the guests were considered mediocre. These are only some examples of an extended ancient rose cult, which was to unleash an immense demand for rose flowers and plants (see page 11). The growing of roses in the areas around large cities became a profitable business. The commercial rose market thus had its beginnings in ancient times.

Evidence for an active import trade in cut roses, which probably reached Rome as rosebushes growing in pots from Egypt, is found in some sources. This importation did not have an entirely positive effect on the Roman trade balance as is evidenced by a citation from Martial, "O Egypt, send us grain instead of roses. We will repay you with our winter roses." By winter roses he meant rosebushes that by forcing, with winter protection and watering with lukewarm water, were induced to flower outside the regular flowering season. Also at that time, critics of this "unnatural" method of culture were already establishing their own camp. In a diatribe, Seneca lashed out *contra eos qui naturam invertunt*—against those who turn nature upside down.

Today, the rose is not only an important cultural item but also a commercial one all over the world. Many thousands of people work in the production of cutting, potted, and garden roses.

■ Where Do Garden Roses Come From?

In the section "Propagation," the budding of roses is described as the most widely used method of propagating rose varieties. In America, tens of millions of roses are budded annually. The majority are garden roses, that is, roses for the private garden market.

Some of these roses are produced in large nurseries in the West. For example, Jackson & Perkins (the world's largest producer of roses) is headquartered in Oregon but grows much of its roses in California. Indeed, California and the Northwest are important growing areas for a great deal of the garden rose production in the United States. However, many middle- and smaller-sized nurseries are important in producing the total number and diversity of roses that are grown for sale in the United States. A number of these firms specialize in growing a particular kind of rose. For example, firms like Nor'East Miniature Roses, Inc. in Rowley, Massachussetts specialize in miniatures, while Hardy Roses for the North, a company in Grand Forks, Washington, is known for cold-hardy roses. A number of nurseries, like Antique Rose Emporium in Brenham, Texas and Heirloom Old Garden Roses in Saint Paul, Oregon are specialists in supplying old garden roses.

The culture and care of roses involve hard hand labor. The wildings are planted in all kinds of weather and must be kept free of weeds. Budding in high summer in glaring heat is ideal for the growth of scion eyes. However, for the professional rose grower, this work, which must be performed with back bent, saps the strength.

While the endless budding process can be tedious, it needs to be done by motivated, careful, and knowledgeable workers. In smaller firms, much of the work is done by family members who work long hours to get the job done in the relatively short seasonal window of opportunity.

■ Where Do Cut Roses Come From?

Cut roses are available throughout the year and are extremely popular. They are a must-have flower for many holidays and celebrations such as Valentine's Day, when over 88 million rose stems alone are sold. Giving a dozen roses is still one of the most romantic or celebratory gestures to make toward someone—even in our high-tech world. According to United States Department of Agriculture (USDA) figures, 1.2 billion cut roses were consumed in the United States in 1997. The total U.S. retail value of cut roses amounts to nearly $2.5 billion, making the rose, by far, America's most popular cut flower.

It may be interesting to the rose lover that only about 20 percent (350 million stems) of the cut roses sold in the United States are grown in the United States. Much of the other 80 percent (850 million stems) of cut roses that are consumed in this country are imported. Most of the imported cut roses now come from South America, notably Colombia and Ecuador. Other countries that produce significant numbers of cut roses for the world market are Italy, Holland, Kenya, Japan, and Spain.

The above numbers make clear how important the rose as a cut flower is to the rose grower. Anyone who lands a best-seller, which is then grown under many acres of glass, can rejoice in receiving significant royalties. This is also necessary in order to cover the constantly occurring, immense costs of breeding. Breeding new types of cut roses runs a completely independent course than that of the search for new garden roses. Above all, the multiyear testing and selection of a new cut rose under hothouse conditions require huge greenhouse areas in which nothing but new varieties are produced but that, nevertheless, must be maintained.

The goals of modern cut rose breeding are primarily high productivity (many stems per square foot [m^2]), great mildew resistance, good durability, good growing behavior, and good transportability in a rose. Recently, fragrance was added to this list.

The actual supply of varieties for commercial cut rose growing has become, even for those who know the scene,

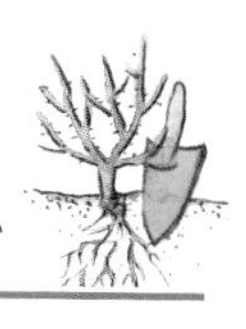

Cutting rose trials at W. Kordes' Söhne in northern Germany.

beyond grasp. The days of the great varieties like 'Baccara', 'Mercedes', or 'Sonia' are past. The flower trade wants a large assortment in many color shades and flower sizes. Today, varieties such as 'Europa', 'First Red', 'Frisco', 'Grand Gala', 'Ilseta', 'Jacaranda', and 'Madelon' are considered important and particularly successful cut roses.

The summer sale of cut roses is playing an increasingly important role, with a growing trend toward the so-called field-grown roses. In contrast with hothouse roses, they come in large part from domestic production. Above all, spray roses, that is, very bunch-flowered varieties with many flowers per stem, are runaway sellers. They provide the customer with the image of an especially rustic cut rose.

Where Do Potted Roses for Houseplants Come From?

Potted roses for houseplants are a popular crop especially in spring for Mother's Day, when they are for sale in florists' and grocery shops in pots of 4 to 6 in. (10 to 15 cm) in diamter. The growing business for these roses is becoming more and more industrialized, like other manufactured products.

An interesting example of how rose production is becoming more high tech is in Denmark's production of potted rose houseplants. Denmark produces 35 million potted roses every year. A visit to a firm there feels like an excursion to a high-tech factory floor of an automobile manufacturer.

First, fully automated potting machines fill the empty pots with soil. From there the pots move—placed row on row onto mechanized moving tables—into a room for placement of cuttings. Planting of the cuttings is done (still) by hand. Usually, three to four cuttings go into one pot so that a bushy, many-caned rose develops. Finally, the moving tables of roses move into a hothouse with very high humidity. Under an extremely fine fog and temperatures of 81 °F (27 °C), the cuttings root within eight to fourteen days. Now the tables move again into the greenhouse where actual cultivation takes place, where the roses continue to grow. The pots are moved along, after a certain time, to another position where, as they run through the moving machine, compressed air blows away any wilted leaves. During the entire growth period, the new roses are repeatedly cropped, three times by a special machine and twice with growth inhibitors. This is necessary to produce compact, bushy plants in a period when light is poor.

More than 35 million potted roses are grown annually in Denmark.

The material produced by the pruning is used again for cuttings; the cycle begins anew. Mother plants thus are not necessary, the cuttings are obtained from the ongoing culture.

Watering and fertilizing of the roses take place through a flooding process. The rolling tables are briefly flooded with water until the pots have sucked up enough water, and the nutrients dissolved in it, to fill them. The remaining water is drained off and reused.

After twelve to fourteen weeks, the roses are ready for sale. The production goes on all year long.

Breeding for new potted rose varieties—like that for cutting roses—proceeds independently of garden rose breeding. This is obvious when one considers that conditions for the potted roses during their later sojourn in living rooms are not similar to those outdoors. Most important are the poor light conditions and the dry, warm house air with which the little roses must manage. The lover of potted roses should therefore not place his or her little roses near a heating unit and should choose a sunny windowsill. During the summer they belong outdoors, either on the balcony or the terrace. Note: Potted roses are scarcely suitable for planting later in the garden. They were selected in the first place for their characteristics as potted plants.

The Rose Atlas

Rosy maze—while creative gardeners explore the multiplicity of the rose world without concern, new rose gardeners are especially dismayed when faced with the tremendous number of varieties. Facts and practical guidance make choosing easier—reliable paths through the maze.

200 Rose Species and Varieties for Any Garden

This chapter is given the title "The Rose Atlas." There is a reason for this. An atlas is a guide and an aid to orientation in a strange landscape, in this case the broad landscape of rose varieties. Like a geographic atlas, it should give an overview, name the most important landmarks, and serve as a direction finder. Two hundred signposts in the gigantic land of roses—the number of varieties in the world is estimated to be more than 30,000—should point the way to finding the right variety and thus help not to lose sight of the most important goal: to have fun with roses.

The varieties described in the atlas have been collected primarily with an eye toward their sturdiness—a sturdiness that often corresponds to the vigor of a variety. Simply put, this means that vigorously growing roses can resist diseases and pests better than weakly growing varieties.

Understandably, pure Darwinism, with survival of only the fittest, cannot be the rule for growing a modern assortment of the oldest cultivated plant. We would have to do without most of the repeat-flowering miniature and scented roses. The assortment would consist primarily of pink-flowered shrub and area roses. The rose-filled garden would be dominated by death-defying fighters. The dainty, sensitive ones would fall to the ground: a rose assortment without variety. Thus the final decision is sturdiness, yes, but measured by that of the particular rose group.

Character and Uses

Each rose variety is an individual with its own character, which becomes more familiar the more one has observed the variety in various climate zones and planting situations. What a variety really has to offer becomes clear when the growing conditions have not been perfect. At the end of September, after a rainy summer, someone with some experience and a practiced eye can look deep into the soul of a rose variety. The ADR certificate awarded a rose only after years of trial under aggravated circumstances (see page 163) is a great help in orientation for this reason and also in this atlas. The AARS (All-American Rose Selections) winners are also noted (see page 163).

The Subjective Variety: Each description of a variety begins with this sort of characterization ("Character"), which tries to capture the individual features of the variety. It is a subjective personality description that is not measurable and depends on the credibility of the author.

The Objective Variety: This variety profile describes important, established characteristics. It names rose group, breeder or provenance, year of introduction, flowering pattern, height of growth, and growth habit. It states the number of plants per square foot (m^2) and ends with tips on use and the notation if and when this variety received the ADR certificate.

Plant Density

Finally, a word must be included about plant density—the number of plants of a variety recommended per square foot (m^2) in an all-rose planting. The numbers given are only a rough guideline since, obviously, the number of plants needed can vary with situation, purpose of the planting, and climate. Thus, certain hybrid tea roses grow like climbing roses in the South but in the harsh North, they grow hardly more than 39 in. (100 cm) tall.

Overview of Varieties (height in inches/centimeters)

	Yellow	Orange Apricot	Pink	Red	Violet	White/Cream
Bedding roses	'Goldener Sommer '83'® (16–24/40–60, double) 'Goldmarie '82'® (16–24/40–60, double) 'Mountbatten'® (32–39/80–100, double) 'Polygold'® (16–24/40–60, double) 'Rumba'® (24–32/60–80, double) 'Sunsprite' (24–32/60–80, double, fragrant)	'Amber Queen'® (16–24/40–60, double) 'Bernstein Rose'® (24–32/60–80, double)	'Ballade'® (24–32/60–80, double) 'Blühwunder'® (24–32/60–80, semidouble) 'Bonica '82'® (24–32/60–80, double) 'Diadem'® (32–39/80–100, double) 'Dolly'® (24–32/60–80, semidouble) 'Escapade'® (32–39/80–100, semidouble) 'Europas Rosengarten'® (24–32/60–80, double) 'Frau Astrid Späth' (16–24/40–60, double) 'Heidepark'® (24–32/60–80, semidouble) 'Leonardo da Vinci'® (24–32/60–80, double) 'Make Up'® (32–39/80–100, double) 'Play Rose'® (24–32/60–80, double) 'Ricarda'® (24–32/60–80, semidouble) 'Rosali '83'® (24–32/60–80, double) 'Royal Bonica'® (36–72/91–182, double) 'Schleswig '87'® (24–32/60–80, semidouble) 'Schöne Dortmunderin'® (24–32/60–80, double) 'Sommermorgen'® (24–32/60–80, double) 'The Queen Elizabeth Rose'® (32–39/80–100, double) 'Träumerei'® (24–32/60–80, double, fragrant)	'Chorus'® (24–32/60–80, double) 'Fragrant Cloud'® (24–32/60–80, double, fragrant) 'Gartenzauber '84'® (16–24/40–60, double) 'Happy Wanderer'® (16–24/40–60, double) 'La Sevillana'® (24–32/60–80, semidouble) 'Mariandel'® (16–24/40–60, double) 'Matthias Meilland'® (24–32/60–80, double) 'Montana'® (32–39/80–100, double) 'Nina Weibull' (16–24/40–60, double) 'Pussta'® (24–32/60–80, double) 'Rose de Rescht' (32–39/80–100, double, fragrant) 'Sarabande'® (16–24/40–60, semidouble) 'Stadt Eltville'® (24–32/60–80, double)		'Gruss on Aachen' (16–24/40–60, double) 'Edelweiss'® (16–24/40–60, double) 'La Paloma '85'® (24–32/60–80, double) 'Schneeflocke'® (16–24/40–60, semidouble)
Hybrid teas	'Banzai '83'® (32–39/80–100, double, fragrant) 'Duftgold'® (24–32/60–80, double, fragrant) 'Elina'® (32–39/80–100, double, fragrant) 'Peace'® (32–39/80–100, double)	'Christopher Columbus'® (24–32/60–80, double) 'Paul Ricard'® (24–32/60–80, double, fragrant)	'Aachener Dom'® (24–32/60–80, double) 'Carina'® (32–39/80–100, double) 'Pariser Charme' (24–32/60–80, double, fragrant) 'Silver Jubilee'® (24–32/60–80, double) 'The McCartney Rose'® (24–32/60–80, double, fragrant)	'Barkarole'® (32–39/80–100, double, fragrant) 'Deep Secret'® (32–39/80–100, double, fragrant) 'Eroica'® (32–39/80–100, double, fragrant) 'Hidalgo'® (32–39/80–100, double, fragrant) 'Ingrid Bergman'® (24–32/60–80, double) 'Loving Memory'® (24–32/60–80, double, fragrant) 'Papa Meilland'® (24–32/60–80, double, fragrant) 'Senator Burda'® (24–32/60–80, double)	'Duftrausch'® (32–39/80–100, double, fragrant)	'Karl Heiz Hanisch'® (24–32/60–80, double, fragrant) 'Polarstern'® (32–39/80–100, double, fragrant)
Native wild roses	'Limelight'® (32–39/80–100, double) 'Sunblest' (24–32/60–80, double)		*Rosa gallica* (32–39/80–100, single, fragrant) *Rosa majalis* (59–79/150–200, single) *Rosa eglanteria* (79–118/200–300, single) *Rosa scabriuscula* (59–79/150–200, single)			*Rosa arvensis* (32–39/80–100, single) *Rosa pimpinellifolia* (32–39/80–100, single, fragrant)
Area roses			'Ballerina' (24–32/60–80, single) 'Bingo Meidiland'® (16–24/40–60, single) 'Dortmunder Kaiserhain'® (32–39/80–100, double) 'Flower Carpet'™(24–32/60–80, semidouble) 'Immensee'® (12–16/30–40, single, fragrant) 'Lovely Fairy'® (24–32/60–80, double) 'Magic Meidiland'® (16–24/40–60, double) 'Marondo'® (24–32/60–80, semidouble) 'Max Graf' (24–32/60–80, single) 'Mirato'® (16–24/40–60, double) 'Mozart' (32–39/80–100, single) 'Nozomi' (16–24/40–60, single) 'Palmengarten Frankfurt'® (24–32/60–80, double) 'Pheasant'® (24–32/60–80, double) 'Pink Meidiland'® (24–32/60–80, single) 'Repandia'® (16–24/40–60, single) *Rosa nitida* (24–32/60–80, single) 'Satina'® (16–24/40–60, double) 'Sommermärchen'® (16–24/40–60, semidouble) 'Surrey'® (16–24/40–60, semidouble) 'The Fairy' (24–32/60–80, double) 'Wildfang'® (24–32/60–80, double)	'Fairy Dance'® (16–24/40–60, double) 'Marjorie Fair'® (24–32/60–80, single) 'Red Meidiland'® (24–32/60–80, single) 'Royal Bassino'® (16–24/40–60, semidouble) 'Scarlet Meidiland'® (24–32/60–80, double)	'Lavender Dream'® (24–32/60–80, semidouble, fragrant)	'Alba Meidiland'® (32–39/80–100, double) 'Apfelblüte'® (32–39/80–100, single) 'Snow Ballet'® (16–24/40–60, double) 'Swany'® (16–24/40–60, double)
Climbers	'Golden Showers'® (79–118/200–300, double)	'Salita'® (79–118/ 200–300, double)	'Compassion'® (79–118/200–300, fragrant) 'Lawinia'® (79–118/ 200–300, double, fragrant) 'Maria Lisa' (79–118/200–300, single) 'Morning Jewel'® (79–118/200–300, double, fragrant) 'New Dawn' (79–118/200–300, double, fragrant) 'Rosarium Uetersen'® (79–118/200–300, double)	'Dortmund'® (79–118/200–300, single) 'Santana'® (79–118/200–300, double) 'Sympathie' (79–118/200–300, double)		'Ilse Krohn Superior'® (79–118/200–300, double)
Ramblers			'Paul Noël' (118–197/300-500, double, fragrant) *Rosa rugosa* (118–197/300-500, semidouble, fragrant) 'Super Dorothy'® (118–197/300-500, double)	'Flammentanz'® (118–197/300-500, double) 'Super Excelsa'® (118–197/300-500, double)		'Albéric Barbier' (118–197/300-500, double) 'Bobbie James' (118–197/300-500, single, fragrant) 'Venusta Pendula' (118–197/300-500, semidouble)
Rugosa roses	'Yellow Dagmar Hastrup'® (24–32/60–80, semidouble, fragrant)		'Frau Dagmar Hartopp' (24–32/60–80, single, fragrant) 'Foxi'® (24–32/60–80, double, fragrant) 'Pierette'® (24–32/60–80, double, fragrant) 'Pink Grootendorst' (39–59/100–150, double) 'Polareis'® (24–32/60–80, double, fragrant)			'Repens Alba' (12–16/30–40, single) 'Schnee-Eule'® (16–24/40–60, double, fragrant)
Shrubs	'Frühlingsgold' (59–79/ 150–200, single, fragrant) 'Ghislaine de Féligonde' (59–79/150–200, double) 'Lichtkönigin Lucia'® (39–59/100–150, double) 'Maigold' (59–79/150–200, double, fragrant) *Rosa hugonis* (79–118/ 200–300, single, fragrant) 'Rugelda'® (59–79/150–200, double)	'Friesinger Morgenröte'® (39–59/100–150), double, fragrant) 'Kordes Brilliant'® (39–59/100–150, double) 'Polka '91'® (39–59/ 100–150, double) 'Westerland'® (59–79/ 150–200, semidouble, fragrant)	'Armada'® (39–59/100–150, double) 'Astrid Lindgren'® (39–59/100–150, double) 'Bourgogne'® (39–59/100–150, single) 'Centenaire de Lourdes' (59–79/150–200, semidouble, fragrant) 'Dornröschen Sababurg'® (39–59/100–150, double) 'Eden Rose '85'® (59–79/150–200, double) 'Ferdy'® (32–39/80–100, double) 'IGA '83 München'® (32–39/80–100, double) 'Ilse Haberland'® (39–59/100–150, double, fragrant) 'Louise Odier' (59–79/150–200, double, fragrant) 'Maiden's Blush' (39–59/100–150, double, fragrant) 'Marguerite Hilling' (59–79/150–200, semidouble) *Rosa centifolia* 'Muscosa' (32–39/80–100, double, fragrant) 'Raubritter' (79–118/200–300, double) 'Romanze'® (39–59/100–150, double) *Rosa sweginzowii* 'Macrocarpa' (79–118/200–300, single) 'Rosendorf Sparrieshoop'® (39–59/100–150, double) 'Rosenresli'® (59–79/150–200, double, fragrant) 'Rush'® (39–59/100–150, single) 'Souvenir de la Malmaison' (32–39/80–100, double, fragrant) 'Trigintipetala' (59–79/150–200, semidouble, fragrant) 'Versicolor' (39–59/100–150, semidouble, fragrant) 'Vogelpark Walsrode'® (39–59/100–150, double)	'Bischofsstadt Paderborn'® (39–59/100–150, single) 'Dirigent'® (59–79/150–200, semidouble) 'Elmshorn' (59–79/150–200, double) 'Grand Hotel'® (59–79/150–200, double) 'Gütersloh'® (39–59/100–150, double) 'Robusta'® (59–79/150–200, double, fragrant) 'Rödinghausen'® (39–59/100–150, double) *Rosa moyesii* (grafted) (79–118/ 200–300, single) 'Scharlachglut'® (59–79/150–200, single)		*Rosa sericea* f. *pteracantha* (79–118/200–300, single) 'Schneewittchen'® ('Iceberg') (39–59/100–150, double) 'Suaveolens' (79–118/200–300, double)
Miniatures	'Guletta'® (12–16/30–40, double) 'Sonnenkind'® (12–16/30–40, double)	'Peach Brandy'® (12–16/30–40, double)	'Pink Symphonie'® (12–16/30–40, double)	'Dwarfking'® (16–24/40–60, double) 'Orange Sunblaze'™ (12–16/30–40, double)		
English Roses	'Graham Thomas'® (39–59/ 100–150, double, fragrant)	'Abraham Darby'® (59–79/ 150–200, double, fragrant)	'Constance Spry' (59–79/150–200, double, fragrant) 'Heritage'® (39–59/100–150, double, fragrant) 'Warwick Castle' (24–32/60–80, double, fragrant) 'Wife of Bath' (32–39/80–100, double, fragrant)	'Othello'® (39–59/100–150, double, fragrant) 'The Squire' (39–59/100–150, double, fragrant)		

'Aachener Dom'®

Character: A hybrid tea rose with leathery, very robust foliage, also suitable for harsh conditions. A light, fine scent surrounds the round flower ball, up to 5 in. (12 cm) across. A problem-free and thus ideal rose-pink flower for beginners.
Group: hybrid tea
Breeder/Provenance: Meilland
Year of introduction: 1982
Flower color: salmon pink
Flowers: very double

'Aachener Dom'®

Flowering pattern: repeat flowering
Growth height: 24–32 in. (60–80 cm)
Growth habit: upright
Plants per 11 ft.^{2} (1 m^2): 6 to 7
Use: Gardens, in groups, tolerates frosty situations, tolerates semishade, very rain-fast, pots, cut flowers, full standards.
ADR rose: 1982

'Abraham Darby'®

Character: The strongly fragrant flowers charm with their unique play of colors from pink to apricot. The foliage is robust and glossy green. A sparkling solitaire for the appropriate garden context.

'Alba Meidiland'®

'Albéric Barbier'

Group: shrub rose with fragrance, English Rose
Breeder/Provenance: Austin
Year of introduction: 1985
Flower color: apricot yellow, overcast with pink
Flowers: very double
Flowering pattern: repeat flowering
Growth height: 59–79 in. (150–200 cm)
Growth habit: arching
Plants per 11 ft.^{2} (1 m^2): 1 to 2
Use: Gardens, parks, singly or in small groups, pots, cutting rose, also trellis rose.

'Alba Meidiland'®

Character: A dream in white, robust and especially effective on slopes with an attractive long-distance effect. The flowers are small, densely double, and numerous. A few stems are enough for enchanting bouquets and arrangements.
Group: area
Breeder/Provenance: Meilland
Year of introduction: 1987
Flower color: pure white
Flowers: very double
Flowering pattern: repeat flowering
Growth height: 32–39 in. (80–100 cm)
Growth habit: bushy
Plants per 11 ft.^{2} (1 m^2): 3 to 4
Use: Garden, individually or in groups, tolerates frosty situations, troughs, hanging baskets, very rain-fast, heat tolerant, cut flowers, yellow fall coloring of leaves, half and full standard tree roses.

'Albéric Barbier'

Character: Airy locations lessen the impact of mildew attacks. Then the luxuriant rambler is very ornamental on pergolas or growing loosely through thin-crowned trees with an opulent burst of summer flowers.
Group: rambler
Breeder/Provenance: Barbier
Year of introduction: 1900
Flower color: cream white
Flowers: double
Flowering pattern: once flowering
Growth height: 118–197 in. (300–500 cm)
Growth habit: flat growing without support
Plants per 11 ft.^{2} (1 m^2): 1 to 2
Use: Garden, specimens, tolerates semishade.

'Amber Queen'®

Character: The rose of the year in England in 1984 is striking because of its robust foliage and the roundish-opening, amber to orange flowers. A bedding rose with an eye-catching color.
Group: bedding
Breeder/Provenance: Harkness
Year of introduction: 1984
Flower color: amber yellow
Flowers: quite double

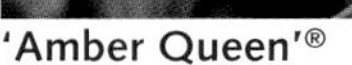
'Amber Queen'®

'Ballade'®

Flowering pattern: repeat flowering
Growth height: 16–24 in. (40–60 cm)
Growth habit: upright, bushy
Plants per 11 ft.^{2} (1 m^2): 6 to 7
Use: Garden, in groups in beds and borders, full standard rose.
AARS winner

'Apfelblüte'®

Character: The variety is not susceptible to mildew and spot anthracnose and rightly bears the ADR certificate. The low-lying canes with their imposing clusters of flowers cover the ground well. A robust variety with lush wild rose flair for natural garden designs.
Group: area
Breeder/Provenance: Noack
Year of introduction: 1991
Flower color: white
Flowers: single
Flowering pattern: repeat flowering
Growth height: 32–39 in. (80–100 cm)
Growth habit: low, bushy
Plants per 11 ft.^{2} (1 m^2): 3 to 4
Use: Garden, specimens, in groups, tolerates semishade, very rain-fast, heat tolerant, pollen source, hips.
ADR rose: 1991

'Armada'®

Character: A repeat-flowering shrub rose with a bushy growth habit. The cheery pink flowers are reminiscent in shape of elegant hybrid tea roses.
Group: shrub
Breeder/Provenance: Harkness
Year of introduction: 1989
Flower color: pink
Flowers: double
Flowering pattern: repeat flowering
Growth height: 39–59 in. (100–150 cm)
Growth habit: bushy, upright
Plants per 11 ft.^{2} (1 m^2): 1 to 2
Use: Garden, specimens or in groups.
ADR rose: 1993

'Astrid Lindgren'®

Character: A striking robust shrub rose to honor one of the best-known children's book authors of our time. A vigorous variety, which knows how to survive in the garden as well as in the higgledy-piggledy of a country home.
Group: shrub
Breeder/Provenance: Poulsen
Year of introduction: 1989
Flower color: pink
Flowers: double

Flowering pattern: repeat flowering
Growth height: 39–59 in. (100–150 cm)
Growth habit: upright
Plants per 11 ft.^{2} (1 m^2): 1 to 2
Use: Garden, specimens or in groups, ideal for free-form hedges.

'Ballade'®

Character: Pink counterpart of the world-famous rose 'Montana'. Robust, lushly flowering, with a pink color like a tender love letter.
Group: bedding
Breeder/Provenance: Tantau
Year of introduction: 1991
Flower color: pink
Flowers: double
Flowering pattern: repeat flowering
Growth height: 24–32 in. (60–80 cm)
Growth habit: upright
Plants per 11 ft.^{2} (1 m^2): 6 to 7
Use: Garden, singly or in groups, also as ground cover, tolerates heat well.

'Ballerina'

Character: The countless light, single flowers reminiscent of wild roses create an effect like a graceful ballerina and hardly suggest the sturdiness and vigor that lie concealed in this variety.

'Astrid Lindgren'®

Ballerina develops new growth uncommonly fast.
Group: area
Breeder/Provenance: Bentall
Year of introduction: 1937
Flower color: pink/white
Flowers: single
Flowering pattern: repeat flowering
Growth height: 24–32 in. (60–80 cm)
Growth habit: overhanging
Plants per 11 ft.^{2} (1 m^2): 3 to 4
Use: Garden, specimens or in groups, ideal for free-form hedges, pots, troughs, tolerates semishade, pollen source, sets hips, for half and full standards.

'Ballerina'

'Banzai '83'®

Character: The robustness of this variety has no rival among the yellow-red hybrid teas. A multicolored cut rose with sturdy growth and full flowers, which sparkle with the colors of a fire.
Group: hybrid tea with fragrance
Breeder/Provenance: Meilland
Year of introduction: 1983
Flower color: yellow with orange-red outer edges
Flowers: very double
Flowering pattern: repeat flowering
Growth height: 32–39 in. (80–100 cm)
Growth habit: upright
Plants per 11 ft.^{2} (1 m^2): 6 to 8
Use: Garden, in groups, very rain-fast, cut flower.
ADR rose: 1985

'Barkarole'®

Character: The dark-red flowers laden with dew shimmer in the morning sun like the finest velvet. A unique look, which is also a pleasure in a vase. Intoxicating scent.
Group: hybrid tea with fragrance
Breeder/Provenance: Tantau
Year of introduction: 1988
Flower color: red
Flowers: double
Flowering pattern: repeat flowering
Growth height: 32–39 in. (80–100 cm)
Growth habit: upright
Plants per 11 ft.^{2} (1 m^2): 6 to 7
Use: Garden, in groups, cut flower, full standard.

'Bella Rosa'®

Character: The canes bear luxuriant clusters of flowers, of which only a few are enough to fill a vase. A tried-and-true classic for the most varied garden areas, absolutely weatherproof.

'Bella Rosa'®

'Banzai '83'®

'Bernstein Rose'®

Group: bedding
Breeder/Provenance: Kordes
Year of introduction: 1982
Flower color: rose pink
Flowers: densely doubled
Flowering pattern: repeat flowering, late
Growth height: 24–32 in. (60–80 cm)
Growth habit: bushy
Plants per 11 ft.^{2} (1 m^2): 6 to 7
Use: Garden, alone or in groups, pots, troughs, heat tolerant, half and full standards.

'Bernstein Rose'®

Character: A color that is not your everyday one gives this robust variety its name, which means *amber* in German. Add to that an old-fashioned flower form, which with its rosette-like structure combines the flair of the English Roses with a compact bedding rose habit.
Group: bedding
Breeder/Provenance: Tantau
Year of introduction: 1987
Flower color: amber yellow
Flowers: well doubled
Flowering pattern: repeat flowering, early
Growth height: 24–32 in. (60–80 cm)
Growth habit: bushy
Plants per 11 ft.^{2} (1 m^2): 6 to 7
Use: Garden, in groups, full standard.

'Bingo Meidiland'®

Character: A civilized wild rose with controlled vigor and a great future. With an effect like a miniature version of 'Pink Meidiland', this desirable, ground-covering pollen source also lends a natural effect to smaller garden areas.
Group: area
Breeder/Provenance: Meilland
Year of introduction: 1991
Flower color: delicate pink
Flowers: single
Flowering pattern: repeat flowering
Growth height: 16–24 in. (40–60 cm)
Growth habit: bushy
Plants per 11 ft.² (1 m²): 3 to 4
Use: Garden, specimens or in groups, tolerates frosty areas, heat tolerant, pollen source, ideal for combinations with low perennials.
ADR rose: 1994

'Bischofsstadt Paderborn'®

Character: After more than 30 years, still a classic among the red shrub roses. Smashing in any garden, with unsurpassed illumination power and long-distance effect.
Group: shrub
Breeder/Provenance: Kordes
Year of introduction: 1964
Flower color: cinnabar red
Flowers: single
Flowering pattern: repeat flowering
Growth height: 39–59 in. (100–150 cm)
Growth habit: upright
Plants per 11 ft.² (1 m²): 1 to 2
Use: Garden, singly or in groups, free-form hedges, pollen source.
ADR rose: 1968

'Bischofsstadt Paderborn'®

'Blühwunder'®

Character: The name, blooming wonder, says it all—during its first flush of flowering the bush is overloaded with flowers. A prospective candidate for the book of records. The flower clusters sometimes contain up to 50 individual flowers.
Group: bedding
Breeder/Provenance: Kordes
Year of introduction: 1995
Flower color: pink
Flowers: semidouble
Flowering pattern: repeat flowering
Growth height: 24–32 in. (60–80 cm)
Growth habit: upright
Plants per 11 ft.² (1 m²): 5 to 6
Use: Garden, alone or in groups, pots.
ADR rose: 1994

'Bobbie James'

Character: A fragrant superclimber, which passionately loves climbing in thin-crowned trees. There the effect of the large, creamy-white clusters of white flowers is like little heaps of whipped cream. The right eye pleaser for fragrant summer days.
Group: rambler with fragrance
Breeder/Provenance: Sunningdale Nurseries
Year of introduction: 1961
Flower color: white
Flowers: single
Flowering pattern: once flowering
Growth height: 118–197 in. (300–500 cm)
Growth habit: grows flat without support
Plants per 11 ft.² (1 m²): 1 to 2
Use: Garden, alone or in groups, tolerates semishade, reddish fall leaf coloring.

'Bonica '82'®

Character: It flowers all summer long, a bedding rose with outstanding winter hardiness. Used multifunctionally as ground cover and field-grown cutting rose. Problem-free variety for novice rose growers.
Group: bedding
Breeder/Provenance: Meilland
Year of introduction: 1982
Flower color: rose, lighter in hot sites
Flowers: double
Flowering pattern: repeat flowering
Growth height: 24–32 in. (60–80 cm)
Growth habit: bushy
Plants per 11 ft.² (1 m²): 5 to 6
Use: Garden, specimen or in groups, tolerates frosty situations well, tubs, troughs, tolerates semishade, very rain-fast, pollen source, cut flowers, good hip development in hot summers, half and full standard rose.
ADR rose: 1982
AARS winner

'Bourgogne'®

Character: The hips of this variety are considered the most beautiful of the rose fruits. The brilliant red vitamin C sources ornament the picturesque overhanging canes of this Pendulina offspring and are a fall highlight for larger gardens.
Group: shrub
Breeder/Provenance: Interplant
Year of introduction: 1983
Flower color: delicate pink
Flowers: single
Flowering pattern: once flowering
Growth height: 59–79 in. (150–200 cm)
Growth habit: arching
Plants per 11 ft.² (1 m²): 1 to 2
Use: Garden, singly or in groups, pollen source, hip rose par excellence.

'Carina'®

Character: A velvety rose-pink cloak of flowers decorates this hybrid tea rose. For decades a proven field-grown cutting rose, which tirelessly produces long, sturdy stems all summer long—provided it gets appropriate care.
Group: hybrid tea
Breeder/Provenance: Meilland
Year of introduction: 1963
Flower color: rose pink
Flowers: double
Flowering pattern: repeat flowering
Growth height: 32–39 in. (80–100 cm)
Growth habit: upright
Plants per 11 ft.² (1 m²): 6 to 7
Use: Garden, in groups, outstanding for cut flowers, full standards.
ADR rose: 1966

'Centenaire de Lourdes'

Character: Why this fantastic, problem-free rose is one of the best-kept rose secrets may lie in its French name, which sticks in the ear only after repeated exposure. Overwhelming flower fireworks, also optimal for smaller situations.
Group: shrub rose with wild rose scent
Breeder/Provenance: Delbard-Chabert
Year of introduction: 1958
Flower color: pink
Flowers: semidouble
Flowering pattern: repeat flowering
Growth height: 59–79 in. (150–200 cm)
Growth habit: arching
Plants per 11 ft.² (1 m²): 1 to 2
Use: Garden, specimen or in groups, unpruned hedges, tolerates part shade, very rain-fast, heat tolerant, pollen source, in short: multifunctional garden specimen.

'Chorus'®

Character: Always a leading variety among the compact-growing, red-flowered bedding

'Centenaire de Lourdes'

roses, with great sturdiness. Needs much sun fuel for lush flowering, so plant in a sunny site.
Group: bedding
Breeder/Provenance: Meilland
Year of introduction: 1975
Flower color: scarlet
Flowers: double
Flowering pattern: repeat flowering
Growth height: 24–32 in. (60–80 cm)
Growth habit: upright
Plants per 11 ft.^{2} (1 m^2): 6 to 7
Use: Garden, single or (also in larger) groups, heat tolerant.
ADR rose: 1977

'Christopher Columbus'®

Character: In color and type, a copper-orange hybrid tea rose blessed with model sturdiness and a rosy "discovery" in the Columbus year of 1992. This rewarding cut rose has no scent—a perfect rose would have been boring, too, and the beginning of the end of all rose breeding.
Group: hybrid tea
Breeder/Provenance: Meilland
Year of introduction: 1992
Flower color: copper orange
Flowers: very double
Flowering pattern: repeat flowering
Growth height: 24–32 in. (60–80 cm)
Growth habit: upright
Plants per 11 ft.^{2} (1 m^2): 6 to 7
Use: Garden, in groups, pots, tolerates part shade, very rain-fast, cutting rose for the home garden.

'Compassion'®

Character: Robust, frost-hardy, and at the same time dreamily fragrant climber arrived under the lucky stars of rose breeding. With 'Compassion' it is a star that shines magnificently pink in the rose sky.
Group: climbing rose with fragrance
Breeder/Provenance: Harkness
Year of introduction: 1974
Flower color: salmon pink
Flowers: very double
Flowering pattern: repeat flowering
Growth height: 79–118 in. (200–300 cm)
Growth habit: trailing
Plants per 11 ft.^{2} (1 m^2): 1 to 2
Use: Garden, singly or in groups, pergolas, greening of walls, rose arches, flower pyramids.
ADR rose: 1976

'Constance Spry'

Character: The first English Rose to be sold commercially. Where it has sufficient space, it develops into a spreading shrub, overflowing with peony-like round flowers and many thorns. A prickery mountain of flowers with a special fragrance note. The name honors the famous English star florist who, in the 1930s, achieved fame for her striking fruit and vegetable arrangements.
Group: shrub rose with scent, English Rose
Breeder/Provenance: Austin
Year of introduction: 1960
Flower color: pink
Flowers: densely doubled
Flowering pattern: once flowering
Growth height: 59–79 in. (150–200 cm)
Growth habit: arching
Plants per 11 ft.^{2} (1 m^2): 1 to 2
Use: Garden, singly or in groups, cutting rose, extremely spiny, free-form hedges, also suitable for climbing rose.

'Deep Secret'

Character: After the almost black buds have opened, there unfold the lustrously velvety, elegantly shaped dreams of flowers. The flower perfume of this variety is unique and surprising. In sunny sites, it is a vital variety for lovers of fragrance who are looking for something special.
Group: hybrid tea with fragrance
Breeder/Provenance: Tantau
Year of introduction: 1976
Flower color: red
Flowers: very double
Flowering pattern: repeat flowering
Growth height: 32–39 in. (80–100 cm)
Growth habit: upright
Plants per 11 ft.^{2} (1 m^2): 6 to 7
Use: Garden, singly or in groups, full standards.
ADR rose: 1978

'Diadem'®

Character: Pure pink cutting flower wonder for the home garden as well as for professional growers. Every stem is a rose bouquet in itself, which is why the outstanding clusters of flowers are termed spray roses.
Group: bedding
Breeder/Provenance: Tantau
Year of introduction: 1986
Flower color: pink
Flowers: double
Flowering pattern: repeat flowering, late
Growth height: 32–39 in. (80–100 cm)
Growth habit: upright
Plants per 11 ft.^{2} (1 m^2): 5 to 6
Use: Garden, singly or in groups, pots, cut flowers, full standards.

'Dirigent'®

Character: The name of this rose means *conductor* in German, and there is music in this variety—and for more than four decades already. Ever-blooming symphony in blood red. For an appropriate appearance, the dense

'Compassion'®

bush needs space, and then the summer garden becomes a rosy concert hall.
Group: shrub
Breeder/Provenance: Tantau
Year of introduction: 1956
Flower color: red
Flowers: semidouble
Flowering pattern: repeat flowering
Growth height: 59–79 in. (150–200 cm)
Growth habit: bushy
Plants per 11 ft.² (1 m²): 1 to 2
Use: Garden, singly or in groups, free-growing, dense hedge rose par excellence, pollen source.
ADR rose: 1958

'Dolly'®

Character: Robust cloud of flowers, which thrive even in less-favorable sites. Its light wild rose scent flatters beds and borders.
Group: bedding
Breeder/Provenance: Poulsen
Year of introduction: 1975
Flower color: dark pink
Flowers: semidouble
Flowering pattern: repeat flowering
Growth height: 24–32 in. (60–80 cm)
Growth habit: bushy
Plants per 11 ft.² (1 m²): 6 to 7
Use: Garden, singly or in groups, tolerates part shade and high altitudes.
ADR rose: 1987

'Dornröschenschloss Sababurg'®

Character: A valuable new introduction, the typical rose fragrance linked with robust foliage. A rose under which one might like to be kissed.
Group: shrub
Breeder/Provenance: Kordes
Year of introduction: 1993
Flower color: pink
Flowers: well doubled
Flowering pattern: repeat flowering
Growth height: 39–59 in. (100–150 cm)

'Dortmund'®

Growth habit: arching
Plants per 11 ft.² (1 m²): 1 to 2
Use: Garden, alone or in groups, free-form hedges

'Dortmund'®

Character: The down-to-earth name fits this robust, winter-hardy, vigorous wall greener. An undemanding hard worker that can always be relied upon.
Group: climber
Breeder/Provenance: Kordes
Year of introduction: 1955
Flower color: red with white eye
Flowers: single
Flowering pattern: repeat flowering
Growth height: 79–118 in. (200–300 cm)
Growth habit: overhanging
Plants per 11 ft.² (1 m²): 2 to 3
Use: Garden, singly or in groups, tolerates frosty situations, will stand part shade, heat tolerant, pollen source, hips.
ADR rose: 1954

'Dortmunder Kaiserhain'®

Character: One of the many robust long-term bloomers from the breeding workshop of Werner Noack. Covers surfaces densely and has fairly won the ADR certificate. A variety with a future, above all for carefree extensive rose plantings.
Group: area
Breeder/Provenance: Noack
Year of introduction: 1994
Flower color: pink
Flowers: double
Flowering pattern: repeat flowering
Growth height: 32–39 in. (80–100 cm)
Growth habit: bushy
Plants per 11 ft.² (1 m²): 3 to 4
Use: Garden, singly or in groups, tolerates part shade.
ADR rose: 1994

'Duftrausch'®

Character: The name says it all—a true rapture of fragrance (*Duftrausch* is a German word that means *fragrant rapture*), awakening memories of sunbathed landscapes. This source of intoxicating rose perfume always invites deep inhaling. For connoisseurs.
Group: hybrid tea with outstanding fragrance
Breeder/Provenance: Tantau
Year of introduction: 1986
Flower color: violet lilac
Flowers: double
Flowering pattern: repeat flowering
Growth height: 32–39 in. (80–100 cm)
Growth habit: upright
Plants per 11 ft.² (1 m²): 6 to 7
Use: Garden, in groups, full standards, flowers for rose recipes and arrrangements.

'Dwarfking'®

Character: A miniature rose that almost grows like a small bedding rose. This vigor provides the necessary robustness to give pleasure in the garden all summer long. Ideal for small, low hedges or for attractive boundary markers.
Group: miniature
Breeder/Provenance: Kordes
Year of introduction: 1978
Flower color: red
Flowers: double
Flowering pattern: repeat flowering
Growth height: 16–24 in. (40–60 cm)

'Dwarfking'®

Growth habit: upright
Plants per 11 ft.² (1 m²): 6 to 7
Use: Garden, in groups, low hedges, pots, troughs, quarter standards.

'Edelweiss'®

Character: One of the most robust white bedding roses, planted often. The variety grows bushy compact and remains low. Ideal color complement for red roses and perennials.
Group: bedding
Breeder/Provenance: Poulsen
Year of introduction: 1969
Flower color: cream white
Flowers: double
Flowering pattern: repeat flowering

'Edelweiss'®

Growth height: 16–24 in. (40–60 cm)
Growth habit: upright
Plants per 11 ft.² (1 m²): 6 to 7
Use: Garden, singly or in groups, very rain-fast, ideal in combination with low-growing perennials.
ADR rose: 1970

'Eden Rose '85'®

Character: A modern cross with the flair of old roses. Dreamily romantic in sprays, each individual flower is pure nostalgia. The foliage is extremely robust and resistant. Needs a place in the sun for optimal flower production.
Group: shrub
Breeder/Provenance: Meilland
Year of introduction: 1985
Flower color: pink
Flowers: double
Flowering pattern: repeat flowering
Growth height: 59–79 in. (150–200 cm)
Growth habit: upright
Plants per 11 ft.² (1 m²): 1 to 2
Use: Garden, singly or in groups, pots, heat tolerant, cut flowers, full standard.

'Elina'®

Character: A secret tip for all connoisseurs who are looking for a robust, yellow-flowered hybrid tea rose. Easy to maintain and resistant, one of the very interesting varieties in this type group. Worth searching for.
Group: hybrid tea with fragrance
Breeder/Provenance: Dickson/Pekmez
Year of introduction: 1983
Flower color: cream yellow
Flowers: well doubled
Flowering pattern: repeat flowering
Growth height: 32–39 in. (80–100 cm)
Growth habit: upright
Plants per 11 ft.² (1 m²): 6 to 7
Use: Garden, in groups, cut flowers, full standard.
ADR rose: 1987

'Elmshorn'

Character: Tried-and-true shrub rose, which bids farewell to summer with an unusually luxuriant fall flowering. A specimen for large garden areas or as a sparkling-red hedge to enclose a property.
Group: shrub
Breeder/Provenance: Kordes
Year of introduction: 1951
Flower color: rose red
Flowers: double
Flowering pattern: repeat flowering
Growth height: 59–79 in. (150–200 cm)
Growth habit: upright
Plants per 11 ft.² (1 m²): 1 to 2
Use: Garden, alone or in groups, free-form hedges, tolerates frosty situations, puts up with part shade.
ADR rose: 1950

'Eroica'®

Character: The velvety red flowers look like heavy brocade, its fragrance is bewitching. For lovers of intense rose fragrance. In addition, the plant grows sturdily and emphatically upright.
Group: hybrid tea rose with fragrance
Breeder/Provenance: Tantau
Year of introduction: 1969
Flower color: red
Flowers: double
Flowering pattern: repeat flowering
Growth height: 32–39 in. (80–100 cm)
Growth habit: upright
Plants per 11 ft.² (1 m²): 6 to 7
Use: Garden, in groups, full standards, flowers for rose recipes, bouquets, and arrangements.
ADR rose: 1969

'Escapade'®

Character: Recommended for a small hedge that is to be very attractive to bees and bumblebees. For this purpose, this pollen source will never run dry since the rose blooms all summer without surcease. For lovers of a rustic rose ambience.
Group: bedding
Breeder/Provenance: Harkness
Year of introduction: 1967
Flower color: lavender/white
Flowers: semidouble
Flowering pattern: repeat flowering
Growth height: 32–39 in. (80–100 cm)
Growth habit: bushy
Plants per 11 ft.² (1 m²): 5 to 6
Use: Garden, singly, in groups, unpruned low hedges, pots, troughs, tolerates part shade, very rain-fast, pollen source, hips.
ADR rose: 1973

'Europas Rosengarten'®

Character: A rose-colored bedding rose that has deserved more attention. Flowers all summer long and remains robust and with tough foliage. A masterpiece from the breeding workshop of Karl Hetzel. Rewarding cut flower.
Group: bedding
Breeder/Provenance: Hetzel
Year of introduction: 1989
Flower color: pink
Flowers: double
Flowering pattern: repeat flowering

'Eden Rose '85'®

'Elina'®

'Escapade'®

Growth height: 24–32 in. (60–80 cm)
Growth habit: bushy
Plants per 11 ft.^{2} (1 m^2): 6 to 7
Use: Garden, singly or in groups, cut flowers.

'Fairy Dance'®

Character: Develops gloriously colored cushions of flowers that cover the ground well and keep down weed growth. Easy-care variety, also for large areas. Do not plant where heat gets trapped.

'Ferdy'®

'Flower Carpet'™

Group: area
Breeder/Provenance: Harkness
Year of introduction: 1979
Flower color: red
Flowers: very double
Flowering pattern: repeat flowering, late
Growth height: 16–24 in. (40–60 cm)
Growth habit: low, bushy
Plants per 11 ft.^{2} (1 m^2): 3 to 4
Use: Garden, singly or in groups, area covering, slopes, embankments.

'Ferdy'®

Character: A very striking shrub rose with an unusual flowering pattern. On the long, arching, several-year-old canes, innumerable flowers line up like pearls on a string. Prune—if at all—only to thin or rejuvenate, needs open space to develop to fullest. Connoisseurs prize the unusual prickles, which ornament the garden in the dull winter months. Should be placed to be seen from a warm place by a window.
Group: shrub
Breeder/Provenance: Keisei
Year of introduction: 1984
Flower color: pink
Flowers: double
Flowering pattern: once flowering
Growth height: 32–39 in. (80–100 cm)
Growth habit: arching
Plants per 11 ft.^{2} (1 m^2): 1 to 2
Use: Garden, singly or in groups, heavily furnished with prickles, free-form hedges, tolerates frosty situations, puts up with part shade, overhanging on walls.

'Flammentanz'®

Character: About the most robust and frost-hardy red climbing rose in this collection. Once flowering but with an intoxicating intensity. Suitable for high altitudes.
Group: rambler
Breeder/Provenance: Kordes
Year of introduction: 1955
Flower color: red
Flowers: double
Flowering pattern: once flowering
Growth height: 118–197 in. (300–500 cm)
Growth habit: flat growing without support
Plants per 11 ft.^{2} (1 m^2): 1 to 2
Use: Garden, singly or in groups, tolerates frosy situations, large hanging baskets, puts up with part shade, weeping standard, overhanging on walls.
ADR rose: 1952

'Flower Carpet'™

Character: A spectacular new cross that has created furor worldwide and today stands like a symbol of a new robust and easily maintained type. The flowers begin late but then keep on without stopping right into frost. The foliage is especially resistant and ornamental as winter green. An important success of hybridization in the direction of the environment-friendly rose.
Group: area
Breeder/Provenance: Noack
Year of introduction: 1988
Flower color: pink
Flowers: semidouble
Flowering pattern: repeat flowering, late
Growth height: 24–32 in. (60–80 cm)
Growth habit: bushy
Plants per 11 ft.^{2} (1 m^2): 2 to 3
Use: Garden, singly or in groups, tolerates frosty situations, pots, troughs, hanging baskets, will stand part shade, very rain-fast, heat tolerant, pollen source, half and full standards.
ADR rose: 1990

'Foxi'®

Character: Lead variety of a new generation of low-growing *rugosa* hybrids. Annual pruning ensures constant rejuvenation and summerlong abundance of flowers. Wonderfully fragrant; the birds find the hips very acceptable.
Group: *rugosa* hybrids with fragrance
Breeder/Provenance: Uhl
Year of introduction: 1989
Flower color: pink
Flowers: double
Flowering pattern: repeat flowering
Growth height: 24–32 in. (60–80 cm)
Growth habit: bushy
Plants per 11 ft.^{2} (1 m^2): 3 to 4
Use: Garden, in groups, free-form hedges, tolerates frosty sites, very rain-fast, pollen source, yellow fall coloring of leaves, hips.
ADR rose: 1993

'Fragrant Cloud'

'Fragrant Cloud'

Character: Uniquely scented rose with worldwide reputation. Although over the last three decades numerous new fragrant roses have been introduced, the scent of this classic always floats like a delicate cloud over countless garden beds. A bedding rose that bestows fragrant delight.
Group: bedding rose with fragrance
Breeder/Provenance: Tantau
Year of introduction: 1963
Flower color: red
Flowers: double
Flowering pattern: repeat flowering
Growth height: 24–32 in. (60–80 cm)
Growth habit: bushy
Plants per 11 ft.² (1 m²): 6 to 7
Use: Garden, singly or in groups, pots, cut flowers, half and full standards, flowers for rose recipes.
ADR rose: 1964

'Fragrant Gold'

Character: The butter-yellow flowers in summer delight the nose with their refreshing perfumed note. A fragrant, first-class, rose relatively impervious to weather.
Group: hybrid tea with fragrance
Breeder/Provenance: Tantau
Year of introduction: 1981
Flower color: yellow
Flowers: double
Flowering pattern: repeat flowering
Growth height: 24–32 in. (60–80 cm)
Growth habit: upright
Plants per 11 ft.² (1 m²): 6 to 7
Use: Garden, in groups, flowers ideal for rose recipes and bouquets.

'Frau Astrid Späth'

Character: A rediscovered bedding rose with proven robustness and abundance of flowers until well into the fall. Problem free and easy care.
Group: bedding
Breeder/Provenance: Spaeth
Year of introduction: 1930
Flower color: rose pink
Flowers: double
Flowering pattern: repeat flowering, early
Growth height: 16–24 in. (40–60 cm)
Growth habit: bushy
Plants per 11 ft.² (1 m²): 6 to 7
Use: Garden, singly or in groups, ideal for combination with perennials and shrubs.

'Freisinger Morgenröte'®

Character: A shrub rose whose flower color is indescribable. The nuances change among orange, yellow, and pink and accustom the eye to ever-new color effects. Robust and resistant.
Group: shrub rose with fragrance
Breeder/Provenance: Kordes
Year of introduction: 1988
Flower color: orange on a yellow ground
Flowers: double
Flowering pattern: repeat flowering
Growth height: 39–59 in. (100–150 cm)
Growth habit: bushy
Plants per 11 ft² (m²): 1 to 2
Use: Garden, singly or in groups, free-form hedges, pots.

'Frau Astrid Späth'

'Frau Dagmar Hartopp'

Character: As a *rugosa* descendent, a bristly companion with robust charm. The old lady among the *rugosa* hybrids does not overage if she is annually pruned back hard.
Group: *rugosa* hybrids with fragrance
Breeder/Provenance: Hastrup
Year of introduction: 1914
Flower color: pastel pink
Flowers: single
Flowering pattern: remontant
Growth height: 24–32 in. (60–80 cm)
Growth habit: upright
Plants per 11 ft.² (1 m²): 4 to 5
Use: Garden, in groups as area cover, tolerates frosty situations, very rain-fast, pollen source, yellow fall coloring of leaves, yellowish red hips.

'Frühlingsgold'

Character: The spring rose for exposed sites in large gardens. As early as May dazzling with a yellow flush of flowers, rings in the summer for the repeat-flowering varieties. Enormously frost hardy, also suitable for high altitudes.
Group: shrub rose with fragrance
Breeder/Provenance: Kordes
Year of introduction: 1937
Flower color: yellow
Flowers: single
Flowering pattern: once flowering, very early

'Freisinger Morgenröte'®

Frühlingsgold

Growth height: 59–79 in. (150–200 cm)
Growth habit: arching
Plants per 11 ft.^{2} (1 m^2): 1 to 2
Use: Garden, singly or in groups, will stand semishade, tolerates frosty sites, pollen source.

'Gartenzauber '84'®

Character: Remains low, compact, and delights with brilliant red flower beauty. A bedding rose to recommend, also for where space is limited.
Group: bedding
Breeder/Provenance: Kordes
Year of introduction: 1984
Flower color: red
Flowers: double
Flowering pattern: repeat flowering
Growth height: 16–24 in. (40–60 cm)
Growth habit: bushy
Plants per 11 ft.^{2} (1 m^2): 6 to 7
Use: Garden, singly or in groups, ideal contrast with perennials and shrubs, annuals, and grasses.

'Ghislaine de Féligonde'

Character: One of the most remarkable rediscoveries of the 1990s. An old variety of the French rose breeder Turbat whose effect never fails, especially in a natural garden. Because of its almost thornless canes, can be considered a child-friendly rose. As a standard (too) a secret tip for connoisseurs and admirers.
Group: shrub rose with fragrance
Breeder/Provenance: Turbat
Year of introduction: 1916
Flower color: salmon pink to yellow
Flowers: double
Flowering pattern: repeat flowering
Growth height: 59–79 in. (150–200 cm)
Growth habit: arching
Plants per 11 ft.^{2} (1 m^2): 1 to 2
Use: Garden, singly or in groups, few prickles, free-form hedges, pots, tolerates semishade, very rain-fast, full standard.

'Ghislaine de Féligonde'

'Golden Medaillon'®

Character: A still young scion from the Kordes breeding workshop. Flowers brilliant golden yellow, contrasting magnificently with the dark foliage. A prize-crowned hybrid tea with a future.
Group: hybrid tea rose
Breeder/Provenance: Kordes
Year of introduction: 1991
Flower color: yellow
Flowers: double
Flowering pattern: repeat flowering
Growth height: 32–39 in. (80–100 cm)
Growth habit: upright
Plants per 11 ft.^{2} (1 m^2): 6 to 7
Use: Garden, in groups, cut flowers, full standard.

'Golden Showers'®

'Golden Showers'®

Character: This climbing rose is one of the earliest among the trellised roses. The dense, dark-green foliage on trellises and pergolas offers privacy from unwanted glances. Tried-and-true yellow variety with which the rose lover cannot do much wrong.
Group: climber
Breeder/Provenance: Lammerts
Year of introduction: 1956
Flower color: yellow
Flowers: double
Flowering pattern: repeat flowering, early
Growth height: 79–118 in. (200–300 cm)
Growth habit: upright
Plants per 11 ft.^{2} (1 m^2): 2 to 3
Use: Garden, singly or in groups, tolerates part shade, good focal point, weeping standard.
AARS winner

'Goldener Sommer '83'®

Character: Attractive yellow contrast in front of dark shrub backdrops. The foliage is tough and resistant, otherwise a variety cannot carry the ADR

rating. The flowers with the color of gold for a golden summer.
Group: bedding
Breeder/Provenance: Noack
Year of introduction: 1983
Flower color: yellow
Flowers: double
Flowering pattern: repeat flowering
Growth height: 16–24 in. (40–60 cm)
Growth habit: bushy
Plants per 11 ft.² (1 m²): 6 to 7
Use: Garden, singly or in groups, in front of dark backrounds.
ADR rose: 1985

'Goldmarie '82'®

'Gruss an Aachen'®

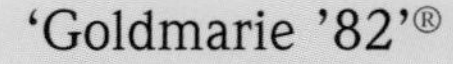

'Goldmarie '82'®

Character: Weather-fast flowers, frost-hardy plants—anyone looking for a robust yellow bedding rose will quickly make eyes at 'Goldmarie'.
Group: bedding
Breeder/Provenance: Kordes
Year of introduction: 1984
Flower color: yellow
Flowers: double
Flowering pattern: repeat flowering
Growth height: 16–24 in. (40–60 cm)
Growth habit: bushy
Plants per 11 ft.² (1 m²): 6 to 7
Use: Garden, singly or in groups, full standards.

'Graham Thomas'®

Character: One of the best of the first English Roses. The dark, full yellow is unusual and is emphasized by the upright growth habit. Ideal sniffing rose for the novice. Named for Graham Stuart Thomas, the famous English garden authority.
Group: shrub rose with fragrance, English Rose
Breeder/Provenance: Austin
Year of introduction: 1983
Flower color: yellow
Flowers: double
Flowering pattern: repeat flowering
Growth height: 39–59 in. (100–150 cm)

'Grand Hotel'®

Growth habit: arching
Plants per 11 ft.² (1 m²): 1 to 2
Use: Garden, singly or in groups, cut flowers, pots, heat tolerant.

'Grand Hotel'®

Character: The flowers of this imposing bush are like those of hybrid tea roses. Their blood-red color glows from afar, the foliage is dense and glossy bright green.
Group: shrub
Breeder/Provenance: McGredy
Year of introduction: 1975
Flower color: red
Flowers: very double
Flowering pattern: repeat flowering
Growth height: 59–79 in. (150–200 cm)
Growth habit: upright
Plants per 11 ft² (1 m²): 1 to 2
Use: Garden, singly or in groups, pots, very rain-fast.
ADR rose: 1977

'Gruss an Aachen'

Character: At the age of 90, this variety is of course still not an old rose but it does not take a backseat in charm. The densely doubled, rosette-like flowers have a yellowish tone at the center that, overcast with pink, changes at the outer edges to a creamy white. For all rose lovers who also do not want to miss out on the flair of grandmother's garden.
Group: bedding
Breeder/Provenance: Geduldig
Year of introduction: 1909
Flower color: cream
Flowers: double
Flowering pattern: repeat flowering
Growth height: 16–24 in. (40–60 cm)
Growth habit: bushy
Plants per 11 ft.² (1 m²): 6 to 7
Use: Garden, singly or in groups.

'Guletta'®

Character: A proven yellow miniature rose, equipped with acceptable robustness—for this rose type. A lemon yellow thumbling, which fits especially well into troughs and boxes.
Group: miniatures
Breeder/Provenance: de Ruiter
Year of introduction: 1976
Flower color: yellow
Flowers: double
Flowering pattern: repeat flowering
Growth height: 12–16 in. (30–40 cm)
Growth habit: upright
Plants per 11 ft.² (1 m²): 8 to 9
Use: Garden, in groups, pots, troughs, boxes, quarter standards.

'Gütersloh'®

Character: This variety has never experienced wide propagation, yet still belongs to the best achievements of modern rose breeding. A problem-free shrub rose that develops a bushy, round habitus.
Group: shrub
Breeder/Provenance: Noack
Year of introduction: 1969
Flower color: red

'Happy Wanderer'®

Flowers: double
Flowering pattern: repeat flowering
Growth height: 39–59 in. (100–150 cm)
Growth habit: upright
Plants per 11 ft.^{2} (1 m^2): 1 to 2
Use: Garden, singly or in groups.

'Happy Wanderer'®

Character: A happy wanderer that really remains low and compact. Good self-cleaning of the flowers, which—compared with other red bedding roses—develop late. This delayed flowering pattern allows skillful combinations with other, early-flowering varieties in designing the private garden.
Group: bedding
Breeder/Provenance: McGredy
Year of introduction: 1974
Flower color: red
Flowers: double
Flowering pattern: repeat flowering, late
Growth height: 16–24 in. (40–60 cm)
Growth habit: very bushy
Plants per 11 ft.^{2} (1 m^2): 6 to 7
Use: Garden, singly or in groups, on the terrace with early-flowering varieties.
ADR rose: 1975

'Heidepark'®

Character: Anyone who has ever been to Heidepark Soltau, Germany, can marvel at this rose in the extensive plantings there. A robust, long-flowering play of color in rose pink, which glows in the large areas of garden of the leisure park even from a distance. Also a leisure-time experience in private gardens.
Group: bedding
Breeder/Provenance: Meilland
Year of introduction: 1988
Flower color: pink
Flowers: semidouble
Flowering pattern: repeat flowering, late
Growth height: 24–32 in. (60–80 cm)
Growth habit: bushy
Plants per 11 ft.^{2} (1 m^2): 6 to 7
Use: Garden, singly or in groups, pots, troughs, very rain-fast, ideal for combination with perennials and annuals.

'Heritage'®

Character: The perfect English Rose, which at the same time develops into a superb round shrub. David Austin himself described it as his most beautiful cross, with which one can only agree. The outer, large petals are whitish rosé and charmingly intersect and surround the inner flower rosette.
Group: shrub rose with fragrance, English Rose
Breeder/Provenance: Austin
Year of introduction: 1984
Flower color: silky pink
Flowers: very double
Flowering pattern: repeat flowering
Growth height: 39–59 in. (100–150 cm)
Growth habit: arching
Plants per 11 ft.^{2} (1 m^2): 2 to 3
Use: Garden, singly or in groups, cutting rose, pots.

'Hidalgo'®

Character: The variety name comes from the Spanish and means *noble*. This captures the class of this robust fragrant rose. Unfortunately, prolonged rain spoils it somewhat. A red, long-stemmed fragrance and cutting rose of the first rank.
Group: hybrid tea with fragrance
Breeder/Provenance: Meilland
Year of introduction: 1979
Flower color: red
Flowers: very double
Flowering pattern: repeat flowering
Growth height: 32–39 in. (80–100 cm)
Growth habit: upright
Plants per 11 ft.^{2} (1 m^2): 6 to 7
Use: Garden, in groups, cutting rose.

'Iceberg'

Character: The most famous and most time-tested white shrub rose in the world. Apparently effortlessly, its elegant flowers appear one after the other, as over the years it becomes a magnificent, self-contained shrub. Ideal beginner variety, harmonizing with many perennials and shrubs.
Group: shrub
Breeder/Provenance: Kordes
Year of introduction: 1958
Flower color: white
Flowers: double
Flowering pattern: repeat flowering, early
Growth height: 39–59 in. (100–150 cm)
Growth habit: bushy
Plants per 11 ft.^{2} (1 m^2): 1 to 2
Use: Garden, singly or in groups, free-form hedges, tolerates frosty situations, pots, will grow in part shade, very rain-fast, tolerates heat, pollen source, sets hips, full and weeping standards.
ADR rose: 1990

'Iceberg'

'IGA '83 München'®

'IGA '83 München'®

Character: One of the most resistant roses in the assortment. After the first flowering it develops an abundance of hips—and thus food for the birds. If the fruits are cut off with pruning shears, however, a second flowering delights the eye in the fall. The decision is thus up to the gardener.
Group: shrub
Breeder/Provenance: Meilland
Year of introduction: 1982
Flower color: rose pink
Flowers: loosely doubled
Flowering pattern: repeat flowering or remontant
Growth height: 32–39 in. (80–100 cm)
Growth habit: bushy
Plants per 11 ft.² (1 m²): 2 to 3
Use: Garden, singly or in groups, free-form hedges, tolerates frosty sites, pots, will stand part shade, very rain-fast, hips.
ADR rose: 1982

'Ilse Haberland'®

Character: Person-high and extraordinarily fragrant. The crimson pink flowers shimmer with many shadings.
Group: shrub rose with fragrance
Breeder/Provenance: Kordes
Year of introduction: 1956
Flower color: rose
Flowers: double
Flowering pattern: repeat flowering
Growth height: 39–59 in. (100–150 cm)
Growth habit: arching
Plants per 11 ft.² (1 m²): 1 to 2
Use: Garden, singly or in groups, free-form hedges.

'Ilse Krohn Superior'®

Character: So far, one of the most interesting of the white climbing roses. Magnificent, intense fragrance, fast growth, yet frost hardy and robust—anyone who is looking for a beautiful ornament for wall or pergola should try it out.
Group: climbing rose with fragrance
Breeder/Provenance: Kordes
Year of introduction: 1964
Flower color: white
Flowers: well doubled
Flowering pattern: repeat flowering
Growth height: 79–118 in. (200–300 cm)
Growth habit: arching
Plants per 11 ft.² (1 m²): 2 to 3
Use: Garden, singly or in groups, pots, fall coloring of leaves, weeping standards.

'Immensee'®

Character: Megavigorous ground cover for larger areas and slopes. Robust hard laborer, develops an impenetrable ground cover. With their strong fragrance, the single, delicately pink flowers invite bees to a flower banquet that lasts for weeks.
Group: area rose with fragrance
Breeder/Provenance: Kordes
Year of introduction: 1982
Flower color: pink
Flowers: single
Flowering pattern: once flowering
Growth height: 12–16 in. (30–40 cm)
Growth habit: flat, vigorous
Plants per 11 ft.² (1 m²): 1 to 2
Use: Larger gardens, singly or in groups, needs space, tolerates part shade, pollen source, full standards.

'Ingrid Bergman'®

Character: Extraordinarily robust hybrid tea rose. Anyone who is ready to cross off fragrance will be repaid with an easy-care, enormously resistant plant. Thus, a hybrid tea rose for sites that are ruled out for the other group colleagues.
Group: hybrid tea with slight scent
Breeder/Provenance: Poulsen
Year of introduction: 1984
Flower color: red
Flowers: well doubled
Flowering pattern: repeat flowering
Growth height: 24–32 in. (60–80 cm)
Growth habit: upright
Plants per 11 ft.² (1 m²): 6 to 7
Use: Garden, in groups, cut flowers, half and full standards.

'Karl Heinz Hanisch'®

Character: An extraordinary rose named for an extraordinary man. Karl Heinz Hanisch was one of the most famous and, at the same time, most imaginative garden writers in Germany. As a great rose lover and expert, he personally selected this rose for sun-warmed locations.
Group: hybrid tea with fragrance
Breeder/Provenance: Meilland
Year of introduction: 1986
Flower color: cream white
Flowers: very double
Flowering pattern: repeat flowering, early
Growth height: 24–32 in. (60–80 cm)
Growth habit: upright
Plants per 11 ft.² (1 m²): 6 to 7
Use: Garden, in groups, cutting rose, ideal with medium-height perennials.

'Kazanlik'

Character: The extraordinarily fragrant oil rose from Bulgaria for one's own garden. If it is given a frost-free place, it will last for many years. Nothing more stands in the way of making one's own rose oil. Also sometimes called 'Trigintipetala'
Group: shrub rose with fragrance

'Kazanlik'

Breeder/Provenance: Bulgaria, *Rosa* x *damascena* offspring
Year of introduction: before 1899
Flower color: rose
Flowers: semidouble
Flowering pattern: once flowering
Growth height: 59–79 in. (150–200 cm)
Growth habit: arching
Plants per 11 ft.^{2} (1 m^2): 1 to 2
Use: Garden, singly or in groups, grows in part shade.

'Kordes' Brilliant'®

Character: Orange counterpart to white 'Iceberg'. Vital, robust, and with unusual flower color. Terrific fall rose.
Group: shrub
Breeder/Provenance: Kordes
Year of introduction: 1983
Flower color: orange to lobster red
Flowers: double
Flowering pattern: repeat flowering
Growth height: 39–59 in. (100–150 cm)
Growth habit: upright
Plants per 11 ft.^{2} (1 m^2): 1 tp 2
Use: Garden, singly or in groups, free-form hedges.

'La Paloma '85'®

Character: This variety belongs to the Floribundas. This means that its pure-white flowers are shaped like hybrid teas but are arranged in large clusters. The color of the flowers makes a pretty contrast to the leathery, dark-green foliage.

'La Paloma '85'®

'Lawinia'®

Group: bedding
Breeder/Provenance: Tantau
Year of introduction: 1985
Flower color: white
Flowers: double
Flowering pattern: repeat flowering
Growth height: 24–32 in. (60–80 cm)
Growth habit: bushy
Plants per 11 ft.^{2} (1 m^2): 6 to 7
Use: Garden, singly or in groups, full standards.

'La Sevillana'®

Character: "Anyone who doesn't know Sevilla, doesn't know what a wonder is," runs a Spanish proverb. Without doubt, that also goes for this rose, which may be counted as one of the healthiest red bedding roses there is. With its loose growth and semidouble flowers, it comes very close to the habitus of wild roses. An ideal, thus a problem-free beginner's variety.
Group: bedding
Breeder/Provenance: Meilland
Year of introduction: 1978
Flower color: red
Flowers: semidouble
Flowering pattern: repeat flowering
Growth height: 24–32 in. (60–80 cm)
Growth habit: bushy
Plants per 11 ft.^{2} (1 m^2): 6 to 7
Use: Garden, singly or in groups, free-form hedges, tolerates frosty sites, pots, troughs, will grow in part shade, very rain-fast, pollen source, hips.
ADR rose: 1979

'Lavender Dream'

Character: Anyone looking for a lavender, fragrant rose that covers the ground well with its trailing growth habit is best served by this variety. Really a dream in a rare color shade reminiscent of the blue of the lavender.
Group: area rose with fragrance
Breeder/Provenance: Interplant
Year of introduction: 1985
Flower color: lavender
Flowers: semidouble
Flowering pattern: repeat flowering, early
Growth height: 24–32 in. (60–80 cm)
Growth habit: low, bushy
Plants per 11 ft.^{2} (1 m^2): 2 to 3
Use: Garden, singly or in groups, heat tolerant, ideal with low perennials.
ADR rose: 1987

'Lawinia'®

Character: Good climbing rose varieties can be counted on both hands—'Lawinia' certainly belongs to this top ten of the colorful climbers. In addition to robust foliage, it has hybrid-tea-shaped flowers, which exude an outstanding fragrance.
Group: climbing roses with fragrance
Breeder/Provenance: Tantau
Year of introduction: 1980
Flower color: pure pink
Flowers: double
Flowering pattern: repeat flowering
Growth height: 79–118 in. (200–300 cm)
Growth habit: overhanging
Plants per 11 ft.^{2} (1 m^2): 2 to 3
Use: Garden, singly or in groups, tubs, weeping standards.

'Leonardo da Vinci'®

Character: A variety that comes up to the brilliance of its name's owner. One of the most interesting 'Romantica' roses of the breeder Meilland. The quartered flower rosettes are stable in color and adorn a plant with a compact growth

'Leonardo da Vinci'®

'Lichtkönigin Lucia'®

habit. The only drop of bitterness is the merely faint scent, but Leonardo, too, experienced the vain search for the perfect creation.
Group: bedding
Breeder/Provenance: Meilland
Year of introduction: 1993
Flower color: rose pink
Flowers: double
Flowering pattern: repeat flowering
Growth height: 24–32 in. (60–80 cm)
Growth habit: bushy
Plants per 11 ft.² (1 m²): 6 to 7
Use: Garden, singly or in groups, ideal for pots, troughs, cut flowers, half standards.

'Lichtkönigin Lucia'®

Character: The classic among the yellow shrub roses. Robust, frost hardy, and developing into a round, densely leaved bush. Used as a single, unpruned plant it is a specimen of the first order.
Group: shrub rose with fragrance
Breeder/Provenance: Kordes
Year of introduction: 1966
Flower color: yellow
Flowers: double
Flowering pattern: repeat flowering, early
Growth height: 39–59 in. (100–150 cm)
Growth habit: upright
Plants per 11 ft.² (1 m²): 1 to 2
Use: Garden, singly or in groups, free-form hedges, pots, very rain-fast.
ADR rose: 1968

'Louise Odier'

Character: An old Bourbon rose that one should give free reign and a prominent place in the garden. Wonderful scented flowers crown a shrub rose with arching-bushy growth. The foliage looks fragile with its light-green color but, nevertheless, is surprisingly disease resistant.
Group: shrub rose with fragrance
Breeder/Provenance: Margottin
Year of introduction: 1851
Flower color: pink
Flowers: double
Flowering pattern: repeat flowering
Growth height: 59–79 in. (150–200 cm)
Growth habit: arching
Plants per 11 ft.² (1 m²): 1 to 2
Use: Garden, singly or in groups, pots, troughs, tolerates part shade.

'Lovely Fairy'®

Character: A sport (mutation) of the well-established variety 'The Fairy'. It is distinguished from the latter by the characteristically stronger rose color of the extremely doubled flowers. Ideal for low hedges in low-maintenance garden areas.
Group: area
Breeder/Provenance: Vurens/Spek
Year of introduction: 1992
Flower color: strong rose
Flowers: densely doubled
Flowering pattern: repeat flowering
Growth height: 24–32 in. (60–80 cm)
Growth habit: bushy
Plants per 11 ft.² (1 m²): 3 to 4
Use: Garden, singly or in groups, pots, troughs, cut flowers, quarter, half, and full standards, low to medium-high hedges.

'Loving Memory'

Character: A robust classic among the hybrid teas. The absolutely elegant flowers are waterproof, the foliage is more than averagely robust for a variety of this group, the plant is very vigorous and winter hardy.
Group: hybrid tea with fragrance
Breeder/Provenance: Kordes
Year of introduction: 1981
Flower color: blood red
Flowers: well doubled
Flowering pattern: repeat flowering
Growth height: 24–32 in. (60–80 cm)
Growth habit: upright
Plants per 11 ft.² (1 m²): 6 to 7
Use: Garden, in groups, pots, cut flowers, full standards.

'Magic Meidiland'®

Character: A true ground cover whose healthy foliage reliably greens the ground. Extraordinarily robust and frost hardy, even suitable for less "rosy" situations. Needs enough space for the unusually wide-spreading growth habit, so two plants per 11 square feet (1 m²) are enough.
Group: area
Breeder/Provenance: Meilland
Year of introduction: 1992
Flower color: rose
Flowers: double
Flowering pattern: repeat flowering
Growth height: 16–24 in. (40–60 cm)
Growth habit: flat, vigorous
Plants per 11 ft.² (1 m²): 1 to 2
Use: Garden, singly or in groups, tolerates frosty sites, troughs, hanging baskets, very rain-fast, heat tolerant, half, full, and weeping standards.
ADR rose: 1995

'Maiden's Blush'

Character: Painter's rose of old Dutch masters, ornamenting numerous still lifes. With its delicate pink color, bursting of full flower pompoms, a worthy and problem-free Old Rose. Hot tip for lovers of kitchen gardens.
Group: shrub rose with sweet scent
Breeder/Provenance: unknown
Year of introduction: before 1500
Flower color: pink
Flowers: double
Flowering pattern: once flowering
Growth height: 39–59 in. (100–150 cm)
Growth habit: arching
Plants per 11 ft.² (1 m²): 2 to 3
Use: Garden, singly or in groups, tolerates frosty sites, kitchen gardens.

'Loving Memory'

'Manou Meilland'®

'Maigold'

Character: This springlike shrub rose is already blooming golden yellow at the end of May and gives off a strong fragrance that awakens the desire for a colorful summer of roses. Enormously frost hardy, therefore can also be planted in high altitudes. Sparser second flowering in the fall.
Group: shrub rose with fragrance
Breeder/Provenance: Kordes
Year of introduction: 1953
Flower color: yellow
Flowers: double
Flowering pattern: very early, remontant
Growth height: 59–79 in. (150–200 cm)
Growth habit: upright
Plants per 11 ft.² (1 m²): 1 to 2
Use: Garden, singly or in groups, will stand part shade, very rain-fast and frost hardy.

'Make Up'®

Character: The right makeup for borders and group plantings. Shallow, splendidly double flower saucers develop salmon-pink rosettes, which adorn a stiffly upright bush. Because of their long stems, best suited for cutting roses. A very few stems produce lush arrangements.
Group: bedding
Breeder/Provenance: Meilland
Year of introduction: 1987
Flower color: rose
Flowers: double
Flowering pattern: repeat flowering
Growth height: 32–39 in. (80–100 cm)
Growth habit: bushy
Plants per 11 ft.² (1 m²): 5 to 6
Use: Garden, singly or in groups, cut flowers, very rain-fast.

'Manou Meilland'®

Character: Fragrant bedding rose with bushy growth. Creates wonderful contrasts for perennials, with the fresh lilac color of its flowers and dark foliage.
Group: bedding rose with fragrance
Breeder/Provenance: Meilland
Year of introduction: 1977
Flower color: rose
Flowers: double
Flowering pattern: repeat flowering
Growth height: 24–32 in. (60–80 cm)
Growth habit: bushy
Plants per 11 ft.² (1 m²): 6 to 7
Use: Garden, singly or in groups.

'Märchenland'

Character: A vital oldie, which can look back over half a century of garden experience. The flowers reach a first climax in July in order, after a short rest in August, to provide a rose-colored preview until fall. Time-tested bedding rose for borders.
Group: bedding
Breeder/Provenance: Tantau
Year of introduction: 1946
Flower color: rose
Flowers: semidouble
Flowering pattern: repeat flowering
Growth height: 24–32 in. (60–80 cm)
Growth habit: upright
Plants per 11 ft.² (1 m²): 6 to 7
Use: Garden, singly or in groups, tolerates frosty sites, pots, will grow in part shade, very rain-fast, pollen source, produces hips.

'Marjorie Fair'®

Character: A Harkness variety with international distribution. Usable in numerous ways, whether as low hedges, enclosures, as red ground-greening cover for large areas. The flowers appear in great clusters; the individual flower is wine red with a white eye and literally winks at the viewer.
Group: area
Breeder/Provenance: Harkness
Year of introduction: 1978
Flower color: red with white eye
Flowers: single
Flowering pattern: repeat flowering
Growth height: 24–32 in. (60–80 cm)
Growth habit: arching
Plants per 11 ft.² (1 m²): 2 to 3
Use: Garden, singly or in groups, free-form hedges, pots, very rain-fast, pollen source.
ADR rose: 1980

'Marjorie Fair'®

'Marguerite Hilling'

Character: A Moyesii variety that puts forth lush, saucer-shaped flowers in the first flowering. In the fall follows a second, somewhat less lavish, flower display. The effect of the very pinnated leaves of this stately shrub is distinctive and fine.
Group: shrub rose
Breeder/Provenance: Hilling
Year of introduction: 1959
Flower color: rose with a lighter center
Flowers: semidouble
Flowering pattern: repeat flowering, remontant
Growth height: 59–79 in. (150–200 cm)
Growth habit: bushy
Plants per 11 ft.² (1 m²): 1 to 2
Use: Garden, singly or in groups, free-form hedges, tolerates frosty situations, pollen source.

'Maria Lisa'

Character: The name suggests a dainty little person, but it is way off. The robust, vigorous variety blooms once lavishly and is emphatically vigorous. Perhaps the delicate name refers to the almost thornless canes—a rose for children's hands.
Group: climber
Breeder/Provenance: Liebau
Year of introduction: 1936
Flower color: rose
Flowers: single
Flowering pattern: once flowering
Growth height: 79–118 in. (200–300 cm)
Growth habit: arching
Plants per 11 ft.² (1 m²): 2 to 3
Use: Garden, singly or in groups, few spines, ideal for pergolas, gateposts.

'Mariandel'®

Character: Many red bedding roses exist. This variety deserves its place in the selection because of its intense color, which neither pales nor goes blue in summer heat. Also, rain has little effect on the weather-fast flowers. The dark foliage is very resistant to powdery mildew.
Group: bedding
Breeder/Provenance: Kordes
Year of introduction: 1984
Flower color: red
Flowers: double
Flowering pattern: repeat flowering
Growth height: 16–24 in. (40–60 cm)
Growth habit: bushy
Plants per 11 ft.² (1 m²): 6 to 7
Use: Garden, singly or in groups, full standards.

'Marguerite Hilling'

'Marondo'®

'Matilda'®

'Marondo'®

Character: An area-greener of the most robust sort. The long, up to 5-feet (1.5-m), canes lie flat against the ground. Rows of rose-pink flowers with bee-attracting golden yellow stamens line up close together. Ideal rose for slopes and banks.
Group: area
Breeder/Provenance: Kordes
Year of introduction: 1991
Flower color: rose
Flowers: semidouble
Flowering pattern: once flowering
Growth height: 24–32 in. (60–80 cm)
Growth habit: flat, vigorous
Plants per 11 ft.² (1 m²): 2 to 3
Use: Garden, singly or in groups, hanging baskets, pollen source, weeping standard, embankments, slopes.
ADR rose: 1989

'Matilda'®

Character: The flowers are delicate pink and look feminine. A lovely bedding rose for sunny locations. Very early flowering, ideal for combination with perennials.
Group: bedding
Breeder/Provenance: Meilland
Year of introduction: 1988
Flower color: delicate pink
Flowers: well doubled

Flowering pattern: repeat flowering, early
Growth height: 16–24 in. (40–60 cm)
Growth habit: bushy
Plants per 11 ft.² (1 m²): 6 to 7
Use: Garden, singly or in groups, pots, troughs, hips, with perennials.

'Matthias Meilland'®

Character: The breeder Alain Meilland calls himself the father of this variety for a special reason, having named it after his son. It is an obligation that this variety, with its brilliant orange-red flowers and its robustness, promises to live up to well. Ideal for the home garden, especially next to lavender.
Group: bedding
Breeder/Provenance: Meilland
Year of introduction: 1988
Flower color: red
Flowers: double
Flowering pattern: repeat flowering
Growth height: 24–32 in. (60–80 cm)
Growth habit: upright
Plants per 11 ft.² (1 m²): 6 to 7
Use: Garden, singly or in groups, pots, troughs.

'Matthias Meilland'®

'Max Graf'

Character: This is a variety that grows very broad and that has shown itself to be absolutely frost hardy during many arctic winters. The canes have striking spines, and the single flowers are attractive to bees and countless other insects. Ideal for flat, impenetrable plantings.
Group: area
Breeder/Provenance: Bowditch
Year of introduction: 1919
Flower color: rose
Flowers: single
Flowering pattern: once flowering
Growth height: 24–32 in. (60–80 cm)
Growth habit: flat, vigorous
Plants per 11 ft.² (1 m²): 1 to 2
Use: Garden, singly or in groups, pollen source.

'Max Graf'

'Mirato'®

Character: A remarkable variety from the breeding workshop of Rosen Tantau. Certainly one of the more interesting ground-cover roses, continuously flowering and of vigorous sturdiness. In appropriate containers on the terrace, a treat to the eye and one that requires no work.
Group: area
Breeder/Provenance: Tantau
Year of introduction: 1990
Flower color: rose
Flowers: double
Flowering pattern: repeat flowering
Growth height: 16–24 in. (40–60 cm)
Growth habit: bushy
Plants per 11 ft.² (1 m²): 3 to 4
Use: Garden, singly or in groups, pots, troughs, hanging baskets, will grow in part shade, half and full standards.
ADR rose: 1993

'Mirato'®

'Montana'®

Character: A bedding rose with many applications, with a long-distance effect—the stiffly upright clusters of flowers glow from far off. Foliage is tough and robust, making a bedding rose even for harsh climates. Safe beginner variety.
Group: bedding
Breeder/Provenance: Tantau
Year of introduction: 1974
Flower color: red
Flowers: double
Flowering pattern: repeat flowering
Growth height: 32–39 in. (80–100 cm)
Growth habit: upright
Plants per 11 ft.² (1 m²): 5 to 6
Use: Garden, singly or in groups, tolerates heat, full standards.
ADR rose: 1974

'Morning Jewel'®

Character: The morning jewels sparkle in an enigmatic rose, surrounded by a lovely fragrance. Flowering lushly and repeatedly, this ADR rose is always worth a visit to wall or pergola.
Group: climbing rose with fragrance
Breeder/Provenance: Cocker
Year of introduction: 1968
Flower color: rose
Flowers: semidouble
Flowering pattern: repeat flowering
Growth height: 79–118 in. (200–300 cm)
Growth habit: arching
Plants per 11 ft.² (1 m²): 2 to 3
Use: Garden, singly or in groups, pollen source, weeping standard.
ADR rose: 1975

'Mountbatten'®

Character: Both a bedding rose and an ornamental shrub rose—the English Rose of the Year in 1982, with its uncommonly large yellow flowers, plays both roles with bravura. The leathery, robust foliage is problem free. The variety honors the British Lord Mountbatten—with the permission of the Royal Family.
Group: beeding/shrub
Breeder/Provenance: Harkness
Year of introduction: 1982
Flower color: mimosa yellow
Flowers: double
Flowering pattern: repeat flowering
Growth height: 32–39 in. (80–100 cm)
Growth habit: bushy
Plants per 11 ft.² (1 m²): 4 to 5

Use: Garden, singly or in groups, heavily spined, free-form hedges, pots, troughs, will grow in part shade.

'Mozart'

Character: Low shrub rose, suitable for many combinations with perennials and shrubs. *Eine kleine Gartenmusik* (a little garden music) with a wild charm. Ideal flower form for freezing in ice cubes. It fits right into the proper space.
Group: area
Breeder/Provenance: Lambert
Year of introduction: 1937
Flower color: rose with white eye
Flowers: single
Flowering pattern: repeat flowering
Growth height: 32–39 in. (80–100 cm)
Growth habit: arching
Plants per 11 ft.^{2} (1 m^2): 3 to 4
Use: Garden, singly or in groups, pollen source, ideal with perennials.

'New Dawn'

Character: Best climbing rose in porcelain pink. An evergreen with winter-green, enormously robust foliage, which scintillates over the summer with countless hybrid tea-like flowers. A sure climbing rose for the rose novice, guaranteed for rosy walls and hillsides.
Group: climbing rose with fragrance, winter green
Breeder/Provenance: Somerset
Year of introduction: 1930
Flower color: mother-of-pearl
Flowers: double
Flowering pattern: repeat flowering
Growth height: 79–118 in. (200–300 cm)
Growth habit: arching
Plants per 11 ft.^{2} (1 m^2): 2 to 3
Use: Garden, singly or in groups, tolerates frosty sites, pots, will grow in part shade, heat tolerant, hips, weeping standards, heights, walls, pergolas, trellises, privacy.

'Mozart'

'New Dawn'

'Nina Weibull'

'Nina Weibull'

Character: The blood-red, nonfading flowers appear lavishly, last a long time, and decorate a compact plant. Durable bedding rose, a border classic. Very winter hardy.
Group: bedding
Breeder/Provenance: Poulsen
Year of introduction: 1992
Flower color: red
Flowers: double
Flowering pattern: repeat flowering
Growth height: 16–24 in. (40–60 cm)
Growth habit: bushy
Plants per 11 ft.^{2} (1 m^2): 5 to 6
Use: Garden, singly or in groups, with perennials, for low hedges.

'Nozomi'

Character: A rewarding ground cover for all garden areas in which there is a romantic porcelain rose. The variety flowers for weeks at a time and decorates embankments and slopes. Robust, natural-looking pot rose.
Group: area
Breeder/Provenance: Onodera
Year of introduction: 1968
Flower color: mother-of-pearl pink
Flowers: single
Flowering pattern: once flowering
Growth height: 16–24 in. (40–60 cm)
Growth habit: flat, not very vigorous grower
Plants per 11 ft.^{2} (1 m^2): 3 to 4
Use: Garden, singly or in groups, will stand part shade, heat tolerant, pollen source, weeping standards.

'Orange Sunblaze'™

Character: An orange-red constant bloomer in miniature. In the garden, they require a lot of care. In troughs and boxes, however, the effect of overwintered fungus spores is very much less than in the bed and the care required can be justified.

Unique luminous power in this rose group, also a hit as a small standard.
Group: miniature
Breeder/Provenance: Meilland
Year of introduction: 1980
Flower color: orange-red
Flowers: double
Flowering pattern: repeat flowering
Growth height: 12–16 in. (30–40 cm)
Growth habit: upright

Group: shrub rose with fragrance, English Rose
Breeder/Provenance: Austin
Year of introduction: 1986
Flower color: crimson
Flowers: enormously doubled
Flowering pattern: repeat flowering, late
Growth height: 39–59 in. (100–150 cm)
Growth habit: upright
Plants per 11 ft.² (1 m²): 1 to 2

Flowering pattern: repeat flowering
Growth height: 24–32 in. (60–80 cm)
Growth habit: bushy
Plants per 11 ft.² (1 m²): 3 to 4
Use: Garden, singly or in groups, will grow in part shade, pots, troughs, with perennials, heat tolerant, full standards.
ADR rose: 1992

Year of introduction: 1963
Flower color: red
Flowers: double
Flowering pattern: repeat flowering
Growth height: 24–32 in. (60–80 cm)
Growth habit: upright
Plants per 11 ft.² (1 m²): 6 to 7
Use: Garden, singly or in small groups, cutting rose.

'Palmengarten Frankfurt'®

'Papa Meilland'®

Plants per 11 ft.² (1 m²): 8 to 9
Use: Singly or in groups in pots, troughs, and boxes, quarter and half standards.

'Othello'®

Character: An English Rose with megalarge rounded flowers—even larger than many other varieties in this group. The shimmering color is hardly describable—one keeps seeing a dark crimson red before one's eyes, then seeing rather a bright crimson purple. Mildew is a problem, but the tendency can be kept within reasonable limits. A powerful fragrance rewards all the effort.

Use: Garden, singly or in groups, cut flowers, pots.

'Palmengarten Frankfurt'®

Character: Decorative coloration for broad areas, clothes the bare soil in a sturdy rose pink. Extraordinarily robust, especially for use together with perennials. Good beginner variety for novices or for less favorable sites, which limit the choice to a robust selection.
Group: area
Breeder/Provenance: Kordes
Year of introduction: 1988
Flower color: strong rose
Flowers: double

'Papa Meilland'®

Character: Particularly, a cutting rose with outstanding, unique scent. However, this miracle of fragrance is fragile. The variety is susceptible to fungal disease, mildew in particular. Its tendency to disease can be reduced by avoiding sites in front of sunny, bright, and glowing-hot south walls without any air circulation. Needs several sprayings during the summer. Only for absolute lovers who do not shy away from any maintenance!
Group: hybrid tea rose with outstanding fragrance
Breeder/Provenance: Meilland

'Pariser Charme'

Character: A sister to the well-known 'Fragrant Cloud', a fragrance rose of world renown. A valuable variety, which brings fragrance together with acceptable foliage robustness—a rare combination in the hybrid tea rose group. The flowers glow in striking pure rose (see jacket) and crown a winter-hardy plant.
Group: hybrid tea with fragrance
Breeder/Provenance: Tantau
Year of introduction: 1965
Flower color: rose
Flowers: double
Flowering pattern: repeat flowering

'Paul Noël'

'Peach Brandy'

Growth height: 24–32 in. (60–80 cm)
Growth habit: upright
Plants per 11 ft.^{2} (1 m^2): 6 to 7
Use: Garden, in groups, full standards.

'Paul Noël'

Character: With some patience, this rambler will grace house and garden walls with a fabulous, fragrant wallpaper of flowers—growing up from underneath as well as hanging down from above. In sites that are airy and breezy, its susceptibility to mildew can be kept within in bounds. Also used as ground cover, but it will tolerate no exposed southern sites or heat traps.
Group: rambler with fragrance
Breeder/Provenance: Tanne
Year of introduction: 1913
Flower color: apricot rose
Flowers: double
Flowering pattern: once flowering, somewhat late
Growth height: 118–197 in. (300–500 cm)
Growth habit: flat without support
Plants per 11 ft.^{2} (1 m^2): 1 to 2
Use: Garden, singly or in groups, will grow in part shade, full and weeping standards, greener for walls, overhanging walls, also ground cover.

'Paul Ricard'®

Character: The unique thing about this fragrant cutting rose is the fresh, very pleasing scent of anise—a fragrance note without parallel in the rose kingdom. A fragrance rose from Provence for lovers of the French art of living.
Group: hybrid tea with anise fragrance
Breeder/Provenance: Meilland
Year of introduction: 1991
Flower color: amber
Flowers: double
Flowering pattern: repeat flowering
Growth height: 24–32 in. (60–80 cm)
Growth habit: upright
Plants per 11 ft.^{2} (1 m^2): 6 to 7
Use: Garden, in groups, pots, will grow in part shade, cut flowers.

'Peace'

Character: With over 100 million plants, the most-sold garden rose of all time. For new varieties in the hybrid tea category, still the measure for robustness, readiness to flower, and vigor.
Group: hybrid tea
Breeder/Provenance: Meilland
Year of introduction: 1945
Flower color: yellow with red border
Flowers: double
Flowering pattern: repeat flowering
Growth height: 32–39 in. (80–100 cm)
Growth habit: upright
Plants per 11 ft.^{2} (1 m^2): 6 to 7
Use: Garden, in groups, tolerates semishade, cut flowers, full standard.
AARS winner

'Peach Brandy'

Character: Interesting rose for troughs and boxes, has a rare, chic apricot color. In containers, the work of maintenance stays within bounds, the inroads of soilborne fungus being very much less, compared with bed culture.
Group: miniature
Breeder/Provenance: ?
Year of introduction: ?
Flower color: apricot
Flowers: double
Flowering pattern: repeat flowering
Growth height: 12–16 in. (30–40 cm)
Growth habit: upright
Plants per 11 ft.^{2} (1 m^2): 8 to 9
Use: Singly or in groups for pots, troughs, and boxes, quarter standards.

'Pheasant'

Character: Careful, the figures for growth height inevitably do not include the more than 6-foot-(2-m) long canes with which this variety covers the ground. For impenetrable garden areas where no uninvited guests are supposed to go. Secret tip: a robust flowering wonder as a weeping standard.
Group: area
Breeder/Provenance: Kordes
Year of introduction: 1985
Flower color: pink
Flowers: double
Flowering pattern: repeat flowering
Growth height: 24–32 in. (60–80 cm)
Growth habit: flat, vigorous
Plants per 11 ft.^{2} (1 m^2): 1 to 2
Use: Garden, singly or in groups, hanging baskets, will stand part shade, very rain-fast, weeping standards.

'Pink Meidiland'®

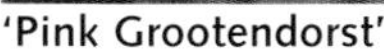

'Pink Grootendorst'

'Pink Symphonie'®

'Pierette'®

Character: A highlight of the new generation within the *rugosa* group. Robust, fragrant (also the leaves), low, compact growing—a problem-free rose for many design variations. Attracts bees and sets hips.
Group: *rugosa* hybrid with fragrance
Breeder/Provenance: Uhl
Year of introduction: 1989
Flower color: rose
Flowers: double
Flowering pattern: repeat flowering
Growth height: 24–32 in. (60–80 cm)
Growth habit: bushy
Plants per 11 ft.^{2} (1 m^2): 3 to 4
Use: Garden, singly or in groups, tolerates frosty sites, very rain-fast, pollen source, hips, yellow fall coloring of leaves.
ADR rose: 1992

'Pink Grootendorst'

Character: This robust variety, a sport of Grootendorst, a Dutch breeder, has unique, carnation-shaped flowers. The rose-pink petals are fringed at the ends like a carnation. The canes are lavishly prickled, predestining it for impenetrable, low hedges. Flowers have slight fragrance. Foliage is bright green with a rather crinkly appearance.
Group: *rugosa* hybrids
Breeder/Provenance: Grootendorst
Year of introduction: 1923
Flower color: rose
Flowers: double
Flowering pattern: once flowering
Growth height: 39–59 in. (100–150 cm)
Growth habit: upright
Plants per 11 ft.^{2} (1 m^2): 1 to 2
Use: Garden, singly or in groups, heavily prickled, unpruned, low hedges, will grow in semishade, very rain-fast, yellow fall coloring of leaves, hips.

'Pink Meidiland'®

Character: A civilized wild rose for the garden, especially when used with other shrubs and perennials. Thanks to prolonged flowering, the salmon-pink flowers with golden yellow stamens provide pollen for bees and bumblebees all summer long—an enormously rich source of bee food that greatly surpasses the pure wild species.
Group: area
Breeder/Provenance: Meilland
Year of introduction: 1984
Flower color: pink with white eye
Flowers: single
Flowering pattern: repeat flowering
Growth height: 24–32 in. (60–80 cm)
Growth habit: arching
Plants per 11 ft.^{2} (1 m^2): 2 to 3
Use: Garden, singly or in groups, free-form hedges, tolerates frosty sites, very rain-fast, heat tolerant, pollen source, hips.
ADR rose: 1987

'Pink Symphonie'®

Character: A miniature rose of the new, robust generation. An adornment for troughs and boxes, a thumbling that tolerates harsh conditions. However, the porcelain pink flowers are also ornamental in the garden or at a sunny grave site. Terrace tip: 'Pink Symphonie' dwarf tree roses in terracotta.
Group: miniature
Breeder/Provenance: Meilland
Year of introduction: 1987
Flower color: pink
Flowers: double
Flowering pattern: repeat flowering
Growth height: 12–16 in. (30–40 cm)
Growth habit: upright
Plants per 11 ft.^{2} (1 m^2): 8 to 9
Use: Garden, singly or in groups, grave sites, pots, troughs, boxes, heat tolerant, quarter and half standards.

'Play Rose'®

Character: A never-say-die fellow with resistant foliage that will also grow in unlikely sites. A sturdy rose color and the setting of hips are further characteristics of this easy-care variety.
Group: bedding
Breeder/Provenance: Meilland
Year of introduction: 1989
Flower color: rose
Flowers: double
Flowering pattern: repeat flowering, early
Growth height: 24–32 in. (60–80 cm)
Growth habit: bushy
Plants per 11 ft.^{2} (1 m^2): 4 to 5
Use: Garden, singly or in groups, tolerates frosty sites, pots, troughs, very rain-fast, pollen source, hips, full standards.
ADR rose: 1989

'Polarstern'®

'Polka '91'®

'Pussta'®

'Polareis'®

Character: The name, Polar Ice, says it all with this *rugosa* hybrid from Russia, although the delicate pink flowers intimate nothing of the enormous winter hardiness of this variety. For high altitudes and plantings in front of shrubs. Particularly for extensive plantings, especially in exposed sites. Annual pruning prevents premature aging of the bush.
Group: *rugosa* hybrid with fragrance
Breeder/Provenance: Strobel
Year of introduction: 1991
Flower color: porcelain pink
Flowers: double
Flowering pattern: repeat flowering
Growth height: 24–32 in. (60–80 cm)
Growth habit: bushy
Plants per 11 ft.^{2} (1 m^2): 3 to 4
Use: Gardens, singly or in groups, tolerates frosty sites best, very rain-fast, pollen source, yellow fall coloring of leaves, hips.

'Polarsonne'®

Character: *Rugosa* selection from Russia, which leaves nothing to be desired as to winter hardiness. Those living in exposed high altitudes or frosty areas who do not want to do without roses will be all right with 'Polarsonne'. Ideal for planting in front of shrub groups.
Group: *rugosa* hybrid with fragrance
Breeder/Provenance: Strobel
Year of introduction: 1991
Flower color: red
Flowers: double
Flowering pattern: repeat flowering
Growth height: 24–32 in. (60–80 cm)
Growth habit: bushy
Plants per 11 ft.^{2} (1 m^2): 3 to 4
Use: Garden, singly or in groups, tolerates frosty sites very well, very rain-fast, pollen source, yellow fall coloring of leaves, hips.

'Polarstern'®

Character: Only a few stems, arranged in a pretty vase, provide a feeling of refreshing coolness on hot summer days. However, the variety has also proven itself in beds and borders thanks to its winter hardiness and robust foliage. All in white—for all occasions that call for contrasts.
Group: hybrid tea with fragrance
Breeder/Provenance: Tantau
Year of introduction: 1982
Flower color: white
Flowers: double
Flowering pattern: repeat flowering
Growth height: 32–39 in. (80–100 cm)
Growth habit: upright
Plants per 11 ft.^{2} (1 m^2): 6 to 7
Use: Garden, in groups, cut flowers, full standards.

'Polka '91'®

Character: A special variation in the wealth of rose flower shapes—very double amber flowers with fringed petals draw the eye. For fragrant hedges or as an elegant star on the garden stage. Robust, vital, and an ideal partner for perennials.
Group: shrub with fragrance
Breeder/Provenance: Meilland
Year of introduction: 1991
Flower color: amber
Flowers: double
Flowering pattern: repeat flowering
Growth height: 39–59 in. (100–150 cm)
Growth habit: upright to arching
Plants per 11 ft.^{2} (1 m^2): 1 to 2
Use: Garden, singly or in groups, free-form hedges, very rain-fast, cut flowers.

'Pussta'®

Character: Dark-red bedding rose with zip. A variety that is very self-cleaning and looks pretty next to low perennials. Flowers very lavishly.
Group: bedding
Breeder/Provenance: Tantau
Year of introduction: 1972
Flower color: red
Flowers: double
Flowering pattern: repeat flowering
Growth height: 24–32 in. (60–80 cm)
Growth habit: bushy
Plants per 11 ft^2 (m^2): 6 to 7
Use: In the garden, singly or in groups.
ADR rose: 1972

'Raubritter'

Character: Fascinating in its flower form and lavish production of them. The arching canes make first-class crowns for rose trees. Prone to mildew, so avoid hot southern exposures with little air movement.
Group: shrub
Breeder/Provenance: Kordes
Year of introduction: 1936
Flower color: rose
Flowers: double
Flowering pattern: once flowering, late
Growth height: 79–118 in. (200–300 cm)
Growth habit: arching
Plants per 11 ft.^{2} (1 m^2): 1 to 2
Use: Garden, singly or in groups, pots, full and weeping standards.

'Raubritter'

'Red Meidiland'®

Character: Forms a low shrub cushion, which—thanks to the dense foliage—covers the ground fast and without much maintenance. In addition, serves as food for the bees and bumblebees. Already proven for years in public gardens, this robust variety also deserves more notice in (natural) home gardens.
Group: area
Breeder/Provenance: Meilland
Year of introduction: 1988
Flower color: red with white center
Flowers: single
Flowering pattern: repeat flowering
Growth height: 24–32 in. (60–80 cm)
Growth habit: low, bushy
Plants per 11 ft.² (1 m²): 2 to 3
Use: Garden, singly or in groups, pots, tolerates semishade, very rain-fast, pollen source, hips.

'Repandia'®

Character: Robust landscape worker that knows how to set to work to cover everything with its canes up to 10 feet (3 m) long, where a rose-covered area is needed fast and without subsequent care required.
Group: area
Breeder/Provenance: Kordes
Year of introduction: 1982
Flower color: rose
Flowers: single
Flowering pattern: once flowering
Growth height: 16–24 in. (40–60 cm)
Growth habit: flat, vigorous
Plants per 11 ft² (m²): 1 to 2
Use: Larger gardens, singly or in groups, will grow in semishade, absolutely rain-fast and resistant, pollen source.
ADR rose: 1986

'Repens Alba' *(Rosa* x *paulii)*

Character: Wherever this rose has established itself, it is a sure thing that no trespasser leaves unharmed. A variety from the rosy watchmen department, ideal for greening of extensively used areas and undesired beaten paths. With canes up to 20 feet (6 m) long and heavily prickled, not suitable for small gardens; typical embankment rose.
Group: *rugosa* hybrid
Breeder/Provenance: G. Paul
Year of introduction: before 1903
Flower color: white
Flowers: single
Flowering pattern: once flowering
Growth height: 12–16 in. (30–40 cm)
Growth habit: flat, vigorous
Plants per 11 ft.² (1 m²): 1 to 2
Use: Embankments, very spiny, tolerates frosty sites and semishade, hips.

'Ricarda'®

Character: An attractive rose by Werner Noack. The salmon-pink flowers set a magnificent scene in front of dark shrub backgrounds. In unfavorable weather conditions, somewhat prone to mildew, otherwise healthy.
Group: bedding
Breeder/Provenance: Noack
Year of introduction: 1989
Flower color: rose
Flowers: semidouble
Flowering pattern: repeat flowering
Growth height: 24–32 in. (60–80 cm)
Growth habit: bushy
Plants per 11 ft.² (1 m²): 6 to 7
Use: Gardens, singly or in groups, hips
ADR rose: 1989

'Robusta'®

Character: A *rugosa* hybrid with upright growth whose name was perfectly chosen for this extremely robust plant. Becomes as tall as an adult, thus needs appropriate space. Ideal boundary protection, excellent for hedges. Like all *rugosas,* wants a regular spring pruning to renew vitality.
Group: *rugosa* hybrid with fragrance
Breeder/Provenance: Kordes
Year of introduction: 1979
Flower color: red
Flowers: single
Flowering pattern: repeat flowering
Growth height: 59–79 in. (150–200 cm)
Growth habit: upright
Plants per 11 ft.² (1 m²): 1 to 2
Use: Garden, singly or in groups, very prickly, free-form hedges, pollen source.
ADR rose: 1980

'Rödinghausen'®

Character: A hearty shrub rose that does not fail even in unfavorable conditions. Forms a hedge of glowing red flowers. The growth remains within limits and is medium vigorous.
Group: shrub
Breeder/Provenance: Noack
Year of introduction: 1987
Flower color: red
Flowers: double
Flowering pattern: repeat flowering
Growth height: 39–59 in. (100–150 cm)
Growth habit: upright
Plants per 11 ft.² (1 m²): 1 to 2
Use: Garden, singly or in groups, loose flower hedges.
ADR rose: 1988

'Romanze'®

Character: The neon pink flowers glow from afar and adorn a compact bush. For loose groups, but can also be used in large borders. One of the more interesting shrub roses with a bushy, tight growth habit.
Group: shrub
Breeder/Provenance: Tantau
Year of introduction: 1984
Flower color: rose pink
Flowers: double
Flowering pattern: repeat flowering
Growth height: 39–59 in. (100–150 cm)
Growth habit: upright
Plants per 11 ft.² (1 m²): 1 to 2
Use: Garden, singly or in groups, pots, will grow in semishade, very rain-fast, heat tolerant, also for large beds and borders.
ADR rose: 1986

'Romanze'®

Rosa arvensis, Field Rose

Character: The common name of this wild rose, field rose, attests to its preference for fields and hedges in its native Europe. Its long, creeping canes gladly clothe emankments and hillsides, even in part shade. It prefers to avoid hot settings, although it tolerates heat. Flowers in July, hips from September.
Group: native wild rose
Breeder/Provenance: Europe
Year of introduction: unknown
Flower color: white
Flowers: single
Flowering pattern: once flowering
Growth height: 32–39 in. (80–100 cm)
Growth habit: flat, vigorous
Plants per 11 ft.^{2} (1 m^2): 1 to 2
Use: Garden, singly or in groups, tolerates semishade, hips.

Rosa centifolia 'Muscosa'

Character: An old cabbage rose from grandmother's garden, which because of its mossy furry sepals—a sport of nature—is termed *moss rose.* That this rose still, after centuries, gains entry into modern gardens lies in its unique Centifolia fragrance, the classic rose fragrance.
Group: shrub rose with Centifolia fragrance, moss rose
Breeder/Provenance: Holland
Year of introduction: 1796
Flower color: pink
Flowers: double
Flowering pattern: once flowering
Growth height: 32–39 in. (80–100 cm)
Growth habit: upright to arching
Plants per 11 ft.^{2} (1 m^2): 1 to 2
Use: Garden, singly or in groups, fragrance and kitchen gardens.

Rosa eglanteria (Syn.: *Rosa rubiginosa*), Sweet Briar

Character: The sweet briar rose is an interesting hedge rose and a bird feeding and protecting shrub to use in wild garden designs. Needs space, unsuitable for small gardens. The leaves often smell of grated apples.
Group: native wild rose
Breeder/Provenance: Europe
Year of introduction: unknown
Flower color: rose
Flowers: single
Flowering pattern: once flowering
Growth height: 79–118 in. (200–300 cm)
Growth habit: arching
Plants per 11 ft.^{2} (1 m^2): 1 to 2
Use: Garden, singly or in groups, very thorny, free-form hedges, pollen source, hips.

Rosa gallica, French Rose

Character: The oldest rose used by humankind. The Latin name indicates France (Gaul), although the species is native over half of Europe. The earlier name, *Rosa rubra,* was more apt, for the flowers have a reddish shimmer. Once used for making vinegar, the mother and foremother of famous old roses like 'Officinalis' and 'Versicolor', with which it has not much in common because of the single, modest flowers. It retains its value today as an important native wild rose. Caution: develops aggressive runners on own-rooted plants.
Group: native wild rose with fragrance
Breeder/Provenance: Europe
Year of introduction: unknown
Flower color: rose
Flowers: single
Flowering pattern: once flowering
Growth height: 32–39 in. (80–100 cm)
Growth habit: upright
Plants per 11 ft.^{2} (1 m^2): 1 to 2
Use: Garden, singly or in groups, extremely prickly, tolerates frosty sites, will grow in part shade but does not thrive in the shadow of large trees, embankments, hips.

Rosa gallica 'Versicolor'

Character: A famous Gallica descendent with light-pink flowers that are striped with crimson and whose loose doubling exposes the golden-yellow stamens to view—a marvelous play of colors. The scent is strangely dominating—not everyone may like it. Also called 'Rosa Mundi'.
Group: shrub rose with fragrance
Breeder/Provenance: Europe
Year of introduction: in cultivation before 1581
Flower color: pink with crimson stripes

Rosa gallica, **'Versicolor'**

Rosa gallica, **'Officinalis'**

Rosa gallica, **French Rose**

Flowers: semidouble
Flowering pattern: once flowering
Growth height: 39–59 in. (100–150 cm)
Growth habit: arching
Plants per 11 ft.^{2} (1 m^2): 1 to 2
Use: Garden, singly or in groups.

Rosa hugonis, Golden Rose of China

Character: This spring rose, already blooming by the middle of May, is also called the Golden Rose of China. Marvelous focal point in front of a dark background of shrubs but also a specimen of the first order. Valuable for its early, lush flowering. The yellow flowers are valuable as a rarity among wild roses.
Group: shrub
Breeder/Provenance: China/Hemsley
Year of introduction: 1899
Flower color: yellow
Flowers: single
Flowering pattern: once flowering, very early
Growth height: 79–118 in. (200–300 cm)
Growth habit: arching
Plants per 11 ft.^{2} (1 m^2): 1 to 2
Use: Garden, singly or in groups, pollen source, hips, May bloomer.

Rosa majalis, Cinnamon Rose

Character: A relatively good (semi) shade-tolerant rose, which is seen in clear areas at the edges of woods and in pasture woodlands. It anchors banks and slopes with its runners. Ideal near water and on the banks of ponds.
Group: native wild rose
Breeder/Provenance: northeastern Europe
Year of introduction: before 1700
Flower color: rose
Flowers: single
Flowering pattern: once flowering, very early

Rosa moyesii, (grafted)

Growth height: 59–79 in. (150–200 cm)
Growth habit: bush
Plants per 11 ft.^{2} (1 m^2): 1 to 2
Use: Banks, hillside gardens, tolerates part shade well, in connection with water.

Rosa moyesii (grafted)

Character: A loose-growing shrub rose, which is very decorative with its large, bottle-shaped hips. A first-class provider of food for the birds. Needs space for full development of its wild beauty.
Group: shrub
Breeder/Provenance: China
Year of introduction: 1890
Flower color: red
Flowers: single
Flowering pattern: once flowering
Growth height: 79–118 in. (200–300 cm)
Growth habit: upright
Plants per 11 ft.^{2} (1 m^2): 1 to 2
Use: Garden, singly or in groups, tolerates frosy sites, will grow in part shade, pollen source, hips.

Rosa nitida

Character: A rose that, with its reddish brown autumn coloring, brings the season to a joyously colorful end. It develops numerous runners, which in toto form a many-caned, knee-high bush. The canes are bristly with prickles.
Group: wild rose
Breeder/Provenance: North America
Year of introduction: 1807
Flower color: rose
Flowers: single
Flowering pattern: once flowering
Growth height: 24–32 in. (60–80 cm)
Growth habit: bushy
Plants per 11 ft.^{2} (1 m^2): 4 to 5
Use: Garden, singly or in groups, heavily prickly, free-form hedges, will grow in semishade, fall coloring of leaves, hips.

Rosa pimpinellifolia (Syn.: *Rosa spinosissima*), The Scotch Rose

Character: The shrub is relatively salt-tolerant and anchors sites needing it with many runners. In its native Europe it commonly grows on sand dunes. A wild rose in whose single, honey-scented flowers the viewer encounters the ancestral form of all roses.
Group: wild rose with fragrance
Breeder/Provenance: Europe
Year of introduction: before 1600
Flower color: cream
Flowers: single

Rosa nitida

Flowering pattern: once flowering, very early
Growth height: 32–39 in. (80–100 cm)
Growth habit: bushy
Plants per 11 ft.^{2} (1 m^2): 1 to 2
Use: Garden, singly or in groups, free-form hedges, anchoring sand dunes, will grow in semishade, pollen source, hips.

Rosa hugonis, **Golden Rose of China**

Rosa repens x *gallica* (Syn.: *Rosa* x *pollineanum*)

Character: A flat ground cover that pleases in June with countless delicate pink flowers. Develops very few hips.
Group: area rose
Breeder/Provenance: unknown
Year of introduction: before 1800
Flower color: rose
Flowers: single
Flowering pattern: once flowering
Growth height: 12–16 in. (30–40 cm)
Growth habit: flat, weak vigor
Plants per 11 ft.^{2} (1 m^2): 2 to 3
Use: Garden, singly or in groups.

Rosa rugosa

Character: A low maintenance, taller shrub rose that is both winter cold hardy and withstands summer heat. Vigorous growing and fragrant, it adds a charming, old-fashioned appeal to a garden.
Group: rambler with fragrance
Breeder/Provenance: Italy
Year of introduction: before 1830
Flower color: rose
Flowers: semidouble
Flowering pattern: once flowering
Growth height: 118–197 in. (300–500 cm)
Growth habit: flat growing without support
Plants per 11 ft.^{2} (1 m^2): 1 to 2
Use: Garden, singly or in groups, will grow in part shade, pergolas, trellises, pillars.

Rosa scabriuscula

Character: A native rose from the sunny forest edges. Ideal for partly sunny garden locations, which duplicate its natural habitat. In June and July attracts numerous insects for pollination. Beautiful hip production.
Group: native wild rose
Breeder/Provenance: Europe
Year of introduction: unknown

Rosa sweginzowii 'Macrocarpa'

Flower color: rose
Flowers: single
Flowering pattern: once flowering
Growth height: 59–79 in. (150–200 cm)
Growth habit: bushy
Plants per 11 ft.^{2} (1 m^2): 1 to 2
Use: Garden, singly or in groups, hips.

Rosa sericea f. *pteracantha*

Character: A shrub rose with many special features. For one, it develops only four petals. For another, very eye-catching winglike, translucent prickles up to 1¼ inches (3 cm) long. A prickery dominating type that needs a great deal of space. Develops no hips.
Group: shrub
Breeder/Provenance: China
Year of introduction: 1890
Flower color: white
Flowers: single
Flowering pattern: once flowering, very early
Growth height: 79–118 in. (200–300 cm)
Growth habit: upright
Plants per 11 ft.^{2} (1 m^2): 1
Use: Garden, singly or in groups, very prickly, tolerates frosty sites, tolerates part shade, no hips.

Rosa sweginzowii 'Macrocarpa'

Character: The uncommonly large hips lavishly adorn this shrub. The old Chinese garden rose grows high and higher—therefore plan enough space for it.
Group: shrub
Breeder/Provenance: northeastern China
Year of introduction: unkown
Flower color: rose
Flowers: single
Flowering pattern: once flowering
Growth height: 79–118 in. (200–300 cm)
Growth habit: upright
Plants per 11 ft.^{2} (1 m^2): 1 to 2
Use: Garden, singly or in groups, very prickly, tolerates frosty sites, will grow in part shade, fall coloring of leaves, decorative hips.

'Rosali '83'®

Character: A fast-growing bedding rose, becoming scarcely more than knee high and yet developing many, many rosette-like flowers. Proven bedding rose with dense foliage.
Group: bedding
Breeder/Provenance: Tantau
Year of introduction: 1983
Flower color: rose
Flowers: double
Flowering pattern: repeat flowering
Growth height: 24–32 in. (60–80 cm)
Growth habit: bushy
Plants per 11 ft.^{2} (1 m^2): 6 to 7
Use: Garden, singly, in groups, or near the terrace.

'Rosarium Uetersen'®

Character: A climbing classic for walls and pergolas, which hardly any rose nursery may be without. The very doubled flowers strewn through its dense, privacy-providing foliage make this climber ideal. Also a sight as a free-form shrub or as a full standard weeping rose.

'Rosarium Uetersen'®

Group: climber, also shrub
Breeder/Provenance: Kordes
Year of introduction: 1977
Flower color: rose
Flowers: double
Flowering pattern: repeat flowering
Growth height: 79–118 in. (200–300 cm)
Growth habit: arching
Plants per 11 ft.² (1 m²): 2 to 3
Use: Garden, singly or in groups, tolerates frosty sites, pots, very rain-fast, tolerates heat, full and weeping standards, walls, pergolas, freestanding.

'Rose de Rescht'

Character: A treat for connoisseurs. The purplish crimson rosette flowers contrast with a foliage that need not hide its face due to robustness. The fragrant rose develops into a round bush that fits harmoniously into many kinds of gardens. Old Rose for beginners.
Group: bedding rose with fragrance
Breeder/Provenance: Persia
Year of introduction: unknown
Flower color: fuchsia red
Flowers: double
Flowering pattern: repeat flowering
Growth height: 32–39 in. (80–100 cm)
Growth habit: upright
Plants per 11 ft.² (1 m²): 3 to 4
Use: Garden, singly or in groups, free-form hedges, pots, troughs, half standards.

'Rosendorf Sparrieshoop'®

Character: This shrub rose is an asset with its wavy petals and its robust foliage. The shrub grows upright-broadly bushy.
Group: shrub
Breeder/Provenance: Kordes
Year of introduction: 1988
Flower color: rose
Flowers: double
Flowering pattern: repeat flowering
Growth height: 39–59 in. (100–150 cm)
Growth habit: upright
Plants per 11 ft.² (1 m²): 1 to 2
Use: Garden, singly or in groups.

'Royal Bassino'®

'Royal Bonica'®

'Rosenresli'®

Character: Shrub rose with very strong fragrance, not taking second place to the classic scented roses in any regard. The flower's changing color play between orange pink and pink orange is fascinating. A fragrant specimen that with its robust canes is also ornamental on small trellises.
Group: shrub with fragrance
Breeder/Provenance: Kordes
Year of introduction: 1986
Flower color: orange pink
Flowers: double
Flowering pattern: repeat flowering
Growth height: 59–79 in. (150–200 cm)
Growth habit: arching
Plants per 11 ft.² (1 m²): 1 to 2
Use: Garden, singly or in groups.
ADR rose: 1984

'Royal Bassino'®

Character: Brilliant fireworks for garden areas that are supposed to attract attention from afar. The robustness of the variety allows planting in large numbers to allow large areas to "burn" bloodred.
Group: area
Breeder/Provenance: Kordes
Year of introduction: 1991
Flower color: red
Flowers: semidouble
Flowering pattern: repeat flowering
Growth height: 16–24 in. (40–60 cm)
Growth habit: bushy
Plants per 11 ft.² (1 m²): 3 to 4
Use: Garden, singly or in groups, pollen source, hips.

'Royal Bonica'®

Character: A royal offspring of the rose-colored variety 'Bonica '82'. The flower clusters are very rain-fast, and the legendary winter hardiness of the mother has been carried on by the daughter. A pretty shrub rose with single to double flowers that keep their color even in great heat and tolerate poor, dry soil near seashores.
Group: *rugosa*
Breeder/Provenance: native to Asia
Year of introduction: ?
Flower color: pink, purple, white, and yellow
Flowers: single to semidouble and double
Flowering pattern: repeat flowering, late
Growth height: 36–72 in. (91–182 cm)
Growth habit: bushy, shrub-like
Plants per 11 ft.² (1 m²): 1 to 2
Use: Garden, singly or in groups, for hedges, borders, and foundation plantings.

'Rugelda'®

Character: Very robust and thoroughly healthy yellow shrub rose. Its descent from *Rosa rugosa* is indicated by its numerous prickles, but the leaf looks like that of the modern shrub rose. As a shrub, it develops a rounded shape and dense foliage. An aesthetically

'Rumba'®

'Salita'®

pleasing, ecologically important provider of breeding area and nest protection for birds.
Group: shrub
Breeder/Provenance: Kordes
Year of introduction: 1989
Flower color: yellow with reddish edge
Flowers: double
Flowering pattern: repeat flowering
Growth height: 59–79 in. (150–200 cm)
Growth habit: upright
Plants per 11 ft.^{2} (1 m^2): 1 to 2
Use: Garden, singly or in groups, very prickly, free-form hedges.
ADR rose: 1992

'Rumba'®

Character: There is rhythm in this variety, the red-yellow parrot-colored flowers radiate verve and good humor. A rewarding cutting rose, especially if only short stems are picked and sufficient foliage is left for another flowering. For good rose situations.
Group: bedding
Breeder/Provenance: Poulsen
Year of introduction: 1960
Flower color: apricot/yellow/copper
Flowers: double
Flowering pattern: repeat flowering
Growth height: 24–32 in. (60–80 cm)
Growth habit: bushy
Plants per 11 ft.^{2} (1 m^2): 6 to 7
Use: Garden, singly or in groups, for cutting.

'Rush'®

Character: This lovely, delicate pink rose has received numerous international prizes and with good reason. It is one of the lushest and longest-flowering shrub roses among the once-flowering varieties. It provides many insects with pollen to feed on for many weeks at a time.
Group: shrub
Breeder/Provenance: Lens
Year of introduction: 1967
Flower color: pink/white
Flowers: single
Flowering pattern: repeat flowering
Growth height: 39–59 in. (100–150 cm)
Growth habit: arching
Plants per 11 ft.^{2} (1 m^2): 1 to 2
Use: Garden, singly or in groups, luxuriant source of pollen.

'Salita'®

Character: A new color shade brought this climber into the circle of the recommendable high rollers. A clear, brilliant orange for designing vertical spaces with roses was missing until the introduction of this variety.
Group: climber
Breeder/Provenance: Kordes
Year of introduction: 1987
Flower color: orange
Flowers: double
Flowering pattern: repeat flowering
Growth height: 79–118 in. (200–300 cm)
Growth habit: trailing
Plants per 11 ft.^{2} (1 m^2): 2 to 3
Use: Garden, singly or in groups, walls, pergolas.

'Santana'®

Character: Covers walls and pergolas in a blood red of full brilliance. The leathery foliage is hard and robust. A red climbing rose with good new growth and abundance of flowers.
Group: climber
Breeder/Provenance: Tantau
Year of introduction: 1984
Flower color: red
Flowers: double
Flowering pattern: repeat flowering
Growth height: 79–118 in. (200–300 cm)
Growth habit: overhanging
Plants per 11 ft.^{2} (1 m^2): 2 to 3
Use: Garden, singly or in groups, walls, pergolas, weeping standards.

'Rush'®

'Santana'®

'Sarabande'®

'Sarabande'®

Character: When this bedding rose came on the market 40 years ago, its geranium red color that lost nothing of its brilliance in hot situations was a sensation. Now as then an indispensable bedding rose, an ideal partner for harmonious combinations with perennials.
Group: bedding
Breeder/Provenance: Meilland
Year of introduction: 1957
Flower color: red with golden-yellow stamens
Flowers: semidouble
Flowering pattern: repeat flowering, early
Growth height: 16–24 in. (40–60 cm)
Growth habit: upright
Plants per 11 ft.² (1 m²): 5 to 6
Use: Garden, singly or in groups, ideal with perennials, pollen source.

'Satina'®

Character: A low area rose whose sturdy rose-colored, very doubled petals cover the bed in dense clouds. Winter hardy, healthy—a tip for small groups and large areas.
Group: area
Breeder/Provenance: Tantau
Year of introduction: 1992
Flower color: rose
Flowers: double
Flowering pattern: repeat flowering
Growth height: 16–24 in. (40–60 cm)
Growth habit: bushy
Plants per 11 ft.² (1 m²): 4 to 5
Use: Garden, singly or in groups, areas.

'Scarlet Meidiland'®

Character: Long-lasting fireworks of flowers in orange red. Develops long, arching canes that are very good for cutting with their doubled flowers. Greens embankments and slopes and when planted on hot, exposed surfaces, does not lose color intensity.
Group: area
Breeder/Provenance: Meilland
Year of introduction: 1986
Flower color: orange red
Flowers: double
Flowering pattern: repeat flowering, late
Growth height: 24–32 in. (60–80 cm)
Growth habit: arching
Plants per 11 ft.² (1 m²): 3 to 4
Use: Garden, singly or in groups, hanging baskets, embankments, slopes, heat tolerant, quarter, half, and full standards.

'Scharlachglut' ('Scarlet Glow')

Character: A Gallica hybrid that has increased the rose red of its mother to a fiery scarlet. Added to this are the glowing golden-yellow stamens, a perfect contrast. A much underappreciated shrub rose that is further adorned with large hips in fall. Enormously frost hardy, thus for high altitudes.
Group: shrub
Breeder/Provenance: Kordes
Year of introduction: 1952
Flower color: red
Flowers: single
Flowering pattern: once flowering
Growth height: 59–79 in. (150–200 cm)
Growth habit: upright
Plants per 11 ft.² (1 m²): 1 to 2
Use: Garden, singly or in groups, tolerates frosty sites, pollen source, develops large hips.

'Scharlachglut' ('Scarlet Glow')

'Schleswig '87'®

Character: A salmon-pink rose for romantics and, at the same time, an indestructable bedding rose for easy-care rose beds. The semidouble flowers look very natural, with golden yellow stamens bejeweling the center of the flowers. Weather-fast.
Group: bedding
Breeder/Provenance: Kordes
Year of introduction: 1987
Flower color: rose
Flowers: semidouble
Flowering pattern: repeat flowering
Growth height: 24–32 in. (60–80 cm)
Growth habit: upright
Plants per 11 ft.² (1 m²): 6 to 7
Use: Garden, singly or in groups, pollen source

'Schneeflocke'®

'Schnee-Eule'®

Character: Pure white flowers, the strongest fragrance, leathery robust foliage, compact growth habit—a variety that apparently without effort brings many good qualities under one roof. A temptation that not only invites use in public areas but also in private gardens.
Group: *rugosa* hybrid with fragrance
Breeder/Provenance: Uhl
Year of introduction: 1989
Flower color: white
Flowers: double
Flowering pattern: repeat flowering
Growth height: 16–24 in. (40–60 cm)
Growth habit: upright
Plants per 11 ft.² (1 m²): 3 to 4
Use: Garden, singly or in groups, tolerates frosty sites, will grow in part shade, very rain-fast, pollen source, yellow fall coloring of leaves, hips.

'Schneeflocke'®

Character: A leading variety of the 'Flowercarpet' series from Werner Noack. And in fact, like a pure white carpet of flowers, it integrates the area and bedding roses into any background. Blooms well into fall and very

'Schöne Dortmunderin'®

'Senator Burda'®

'Silver Jubilee'®

healthy. A dream in white that requires little work.
Group: bedding
Breeder/Provenance: Noack
Year of introduction: 1991
Flower color: white
Flowers: semidouble
Flowering pattern: repeat flowering, early
Growth height: 16–24 in. (40–60 cm)
Growth habit: bushy
Plants per 11 ft.^{2} (1 m^2): 4 to 5
Use: Garden, singly or in groups, area, pollen source, half and full standards.
ADR rose: 1991

'Schöne Dortmunderin'®

Character: One of the varieties that flowers until late fall and rightfully bears the ADR certificate. Despite an enormous flush of flowers and fast and vigorous growth, the rose is not prone to mildew or spot anthracnose. Ideal with perennials and for greening areas.
Group: bedding
Breeder/Provenance: Noack
Year of introduction: 1991
Flower color: rose
Flowers: double
Flowering pattern: repeat flowering
Growth height: 24–32 in. (60–80 cm)
Growth habit: bushy
Plants per 11 ft.^{2} (1 m^2): 4 to 5
Use: Garden, singly or in groups, very rain-fast, heat tolerant.
ADR rose: 1992

'Senator Burda'®

Character: A very robust hybrid tea rose with slight fragrance. The flowers are very doubled and endure even in harsh climate zones. Dedicated to the publishing personality Senator Burda.
Group: hybrid tea
Breeder/Provenance: Meilland
Year of introduction: 1984
Flower color: currant red
Flowers: very double
Flowering pattern: repeat flowering
Growth height: 24–32 in. (60–80 cm)
Growth habit: upright
Plants per 11 ft.^{2} (1 m^2): 6 to 7
Use: Garden, in groups, cut flowers.

'Silver Jubilee'®

Character: Why this uncommonly robust and hardy hybrid tea rose does not have wider distribution remains a mystery. Like scarcely any other fresh-pink-flowered hybrid tea, it is also suitable for nonsunny situations, where the majority of its classmates would certainly fail. Ideal hybrid tea rose for rose newcomers. Many prizes. This renowned classic received its name at the 25th anniversary of the crowning of Queen Elizabeth II.
Group: hybrid tea
Breeder/Provenance: Cocker
Year of introduction: 1978
Flower color: rose
Flowers: double
Flowering pattern: repeat flowering, early
Growth height: 24–32 in. (60–80 cm)
Growth habit: upright
Plants per 11 ft.^{2} (1 m^2): 6 to 7
Use: Gardens, in groups, pots, tolerates part shade, cut flowers, half and full standards.

'Snow Ballet'®

Character: A white ground cover of the modern generation. While suppressing weeds, the dense foliage covers surfaces and beds, especially on slopes. This white rose ballet can be used in versatile ways for planting and for combining.
Group: area
Breeder/Provenance: Clayworth
Year of introduction: 1978
Flower color: white
Flowers: double
Flowering pattern: repeat flowering
Growth height: 16–24 in. (40–60 cm)
Growth habit: flat, vigor weak
Plants per 11 ft.^{2} (1 m^2): 3 to 4
Use: Garden, singly or in groups, hanging baskets, slopes, embankments, quarter, half, full, and weeping standards.

'Sommermärchen'®

Character: An interesting area and bedding rose of the 1990s. A strong pink that glows from afar and, even with strong sun beating on it, does not lose its intensity is the mark of this robust, very healthy rose for many garden areas—the stuff of dreams.

'Sommermärchen'®

Group: area
Breeder/Provenance: Kordes
Year of introduction: 1992
Flower color: pink
Flowers: semidouble
Flowering pattern: repeat flowering
Growth height: 16–24 in. (40–60 cm)
Growth habit: bushy
Plants per 11 ft.² (1 m²): 3 to 4
Use: Garden, singly or in groups, troughs, pollen source, quarter, half, and full standards, ideal with perennials and shrubs.

'Sommermorgen'®

Character: Rose-colored bedding rose that never disappoints and enriches the summer in the garden in a problem-free fashion. The variety grows broadly bushy and also tolerates less rosy situations. Great standard rose.
Group: bedding
Breeder/Provenance: Kordes
Year of introduction: 1991
Flower color: rose
Flowers: double
Flowering pattern: repeat flowering
Growth height: 24–32 in. (60–80 cm)
Growth habit: bushy
Plants per 11 ft.² (1 m²): 4 to 5
Use: Garden, singly or in groups, will grow in part shade, full standards.

'Sonnenkind'®

Character: Because of their proneness to disease, many miniature rose varieties are only very limitedly recommended as garden roses. Thus with the dwarf roses, choosing miniature varieties must be undertaken very carefully. 'Sonnenkind' falls comfortably within the limits, for this vigorous variety grows astonishingly well despite fungus attacks. Above all, in containers—less at risk from fungus spores because of reduced soil contact—it is a problem-free as well as decorative ornament for terrace and balcony.
Group: miniature
Breeder/Provenance: Kordes
Year of introduction: 1986
Flower color: yellow
Flowers: double
Flowering pattern: repeat flowering
Growth height: 12–16 in. (30–40 cm)
Growth habit: upright
Plants per 11 ft.² (1 m²): 8 to 9
Use: Garden, singly or in groups, pots, troughs, quarter standards.

'Sonnenkind'®

'Sunsprite'

'Sunblest'

Character: Very well proven and rewarding hybrid tea rose in yellow. Its winter hardiness is worth mentioning since it can be called very good for this color and rose type. Pure yellow ornament for a vase with utter elegance.
Group: hybrid tea
Breeder/Provenance: Tantau
Year of introduction: 1970
Flower color: yellow
Flowers: very double
Flowering pattern: repeat flowering
Growth height: 24–32 in. (60–80 cm)
Growth habit: upright
Plants per 11 ft.² (1 m²): 6 to 7
Use: Garden, in groups, cutting rose, full standards.

'Sunblest'

'Surrey'

'Sunsprite'

Character: The classic among the yellow bedding roses. Combines sturdiness, compact growth, long flowering, and fragrance in an aesthetic manner. Popular color miracle, yellow symphony for many garden situations.
Group: bedding rose with fragrance
Breeder/Provenance: Kordes
Year of introduction: 1973
Flower color: yellow
Flowers: double
Flowering pattern: repeat flowering, early
Growth height: 24–32 in. (60–80 cm)
Growth habit: upright
Plants per 11 ft.² (1 m²): 6 to 7
Use: Garden, singly or in groups, low hedges, pots, troughs, heat tolerant, full standards.
ADR rose: 1973

'Surrey'

Character: A firm element in modern commercial selections, suitable both for private gardens and for planning public green spaces. The breeder Kordes describes it as a large-flowered 'The Fairy'—a good comparison, which emphasizes the versatility of this all-purpose rose. Easy care and good for use with perennials.
Group: area
Breeder/Provenance: Kordes
Year of introduction: 1985
Flower color: rose
Flowers: semidouble
Flowering pattern: repeat flowering
Growth height: 16–24 in. (40–60 cm)
Growth habit: bushy

Plants per 11 ft.² (1 m²): 4 to 5
Use: Garden, singly or in groups, with perennials, tolerates frosty sites, pots, troughs, grows in part shade, very rain-fast, heat tolerant, pollen source, quarter, half, and full standards.
ADR rose: 1987

'Souvenir de la Malmaison'

Character: The darling of experienced rose lovers with a preference for Old Roses. Why? In the first place, the growth of this old Bourbon rose stays within bounds, which also makes it good for small gardens. In the second place, it is outstandingly fragrant. In the third place, it blooms constantly into the fall. An Old Rose for beginners.
Group: shrub rose with fragrance
Breeder/Provenance: Béluz
Year of introduction: 1843
Flower color: rose
Flowers: double
Flowering pattern: remontant
Growth height: 32–39 in. (80–100 cm)
Growth habit: bushy
Plants per 11 ft.² (1 m²): 4 to 5
Use: Garden, singly or in groups, pots, troughs.

'Stadt Eltville'®

Character: Anyone who, on an excursion to the Rheingau, takes a stroll through the rose gardens of the Eltville castle moat will be attracted by the glowing power of the fiery red rose roundel as if by magic. The fuel of this circle of fire is—its name, city of Eltville, says it all—the rose variety 'Stadt Eltville'. This bedding rose may with justice be regarded as one of the most robust red bedding roses. Its healthy foliage is reddish when it first appears in the spring, becoming dark green for the rest of the summer. A problem-free rose for the beginner.
Group: bedding
Breeder/Provenance: Tantau

'Suaveolens'

'Super Dorothy'®

Year of introduction: 1990
Flower color: red
Flowers: very double
Flowering pattern: repeat flowering
Growth height: 24–32 in. (60–80 cm)
Growth habit: upright
Plants per 11 ft.² (1 m²): 6 to 7
Use: Garden, in groups, tolerates frosty sites, grows in part shade.

'Suaveolens'

Character: This old shrub rose offers a sweet fragrance, which it exudes lavishly in the summer. However, it needs some space; smaller areas are not suitable for this grower. The arching overhanging shrub bears longish hips in the fall.
Group: shrub rose with fragrance
Breeder/Provenance: unknown
Year of introduction: in culture before 1750
Flower color: white
Flowers: double
Flowering pattern: once flowering
Growth height: 24–32 in. (60–80 cm)
Growth habit: upright
Plants per 11 ft.² (1 m²): 6 to 7
Use: Garden, singly or in groups, numerous elongated hips.

'Super Dorothy'®

Character: A repeat-flowering rambler that has taken over the many good qualities of 'Dorothy Perkins' but does not, however, have its susceptibility to mildew. The robust variety is good for beautifying pergolas and trellises as well for using as a ground cover.
Group: rambler
Breeder/Provenance: Hetzel
Year of introduction: 1986
Flower color: rose
Flowers: double
Flowering pattern: repeat flowering
Growth height: 118–197 in. (300–500 cm)
Growth habit: flat without support
Plants per 11 ft.² (1 m²): 1 to 2
Use: Garden, singly or in groups, pots, troughs, hanging baskets, grows in part shade, tolerates heat, fall coloring of leaves, full and weeping standards.

'Super Excelsa'®

Character: Like its sister variety 'Super Dorothy', 'Super Excelsa' is also the hybridized improvement of an old variety, in this case 'Excelsa'. It is suitable for linked pyramids and columns, walls, pergolas, and as a ground cover when flat. A rosy ornamental, versatile design material.
Group: rambler
Breeder/Provenance: Hetzel
Year of introduction: 1986
Flower color: crimson pink
Flowers: double
Flowering pattern: repeat flowering
Growth height: 118–197 in. (300–500 cm)
Growth habit: grows flat without support
Plants per 11 ft.² (1 m²): 1 to 2
Use: Garden, singly or in groups, pots, troughs, hanging baskets, grows in part shade, heat tolerant, fall coloring of leaves, full and weeping standards.
ADR rose: 1991

'Swany'®

Character: The name fits the delicate swan white of the shallow, densely doubled flowers. A rose that satisfactorily spreads its flower cloak over walls and embankments. It loves breezy situations. In sites in which the air is trapped as in a kettle—such as in the inner city—it

exhibits a tendency toward spot anthracnose. Tree rose classic.
Group: area
Breeder/Provenance: Meilland
Year of introduction: 1977
Flower color: white
Flowers: double
Flowering pattern: repeat flowering
Growth height: 16–24 in. (40–60 cm)
Growth habit: flat, weak vigor
Plants per 11 ft.² (1 m²): 3 to 4
Use: Garden, singly or in groups, pots, troughs, hanging baskets, very rain-fast, quarter, half, and full standards.

'Sympathie'

Character: Well-proven red super climber, which swoops over rose arches and pergolas without any difficulty. No wall is too high for it, no trellis too steep. A climbing rose with a worldwide reputation, with its velvety dark-red and wild rose fragrance, unsurpassed.
Group: climber with fragrance
Breeder/Provenance: Kordes
Year of introduction: 1964
Flower color: red
Flowers: double
Flowering pattern: repeat flowering
Growth height: 79–118 in. (200–300 cm)
Growth habit: arching
Plants per 11 ft.² (1 m²): 2 to 3
Use: Garden, singly or in groups, weeping standards.
ADR rose: 1966

'The Fairy'

Character: A roly-poly. Displays countless little flowers all summer long and thus closely covers the ground. Even semishady sites are hidden, though less densely. Also widely used as a standard for pots and gardens.
Group: area
Breeder/Provenance: Bentall
Year of introduction: 1932
Flower color: rose
Flowers: densely doubled

'The Fairy'

Flowering pattern: repeat flowering, late
Growth height: 24–32 in. (60–80 cm)
Growth habit: bushy
Plants per 11 ft.² (1 m²): 4 to 5
Use: Garden, singly or in groups, grows in frosty sites, pots, troughs, hanging baskets, cut flowers, yellow fall coloring of leaves, quarter, half, full, and weeping standards.

'The McCartney Rose'™

Character: Named after Paul McCartney but not just a collector's variety for Beatles fans. A rose with an outstanding fragrance. Caution must be taken against mildew, so never water the leaves or plant in a corner where there is no air circulation. Best suited for cut flowers; just a few flowers fill a room with intoxicating rose perfume. For all lovers of fragrant roses. Note that 'The McCartney Rose' should not be confused with the 'Macartney Rose' *(Rosa bracteata)*.
Group: hybrid tea with outstanding fragrance
Breeder/Provenance: Meilland
Year of introduction: 1991

'The Queen Elizabeth Rose'®

Flower color: rose
Flowers: double
Flowering pattern: repeat flowering
Growth height: 24–32 in. (60–80 cm)
Growth habit: upright
Plants per 11 ft.² (1 m²): 6 to 7
Use: Garden, in groups, cutting rose.

'The Queen Elizabeth Rose'®

Character: The growth habit of this old-timer has a gawky, long-legged look—like a long-legged flamingo it stalks through gardens and plantings. The variety is considered indestructible, the large foliage healthy. Predestined because of its tallness to provide a joyous background before or behind fences and low walls.
Group: bedding
Breeder/Provenance: Lammerts
Year of introduction: 1954
Flower color: rose
Flowers: double
Flowering pattern: repeat flowering
Growth height: 39–59 in. (100–150 cm)
Growth habit: upright
Plants per 11 ft.² (1 m²): 5 to 6
Use: Garden, singly or in groups, free-form hedges, tolerates frosty sites, part shade, and heat, sets hips.
AARS winner

'The Squire'

Character: David Austen has said that he knows of no dark crimson rose that has such magnificent flowers as 'The Squire'. Unfortunately, its growth is only thin and it is very disease prone. This—from David Austin—about his own hybrid. Nevertheless, the flower color and the fragrance of the variety attract not only experts to choose it—but some care has to be figured in when choosing.
Group: shrub rose with fragrance, English Rose
Breeder/Provenance: Austin
Year of introduction: 1977

'The Squire'

'Träumerei'®

'Westerland'®

'Vogelpark Walsrode'®

Flower color: red
Flowers: double
Flowering pattern: repeat flowering
Growth height: 39–59 in. (100–150 cm)
Growth habit: arching
Plants per 11 ft.² (1 m²): 1 to 2
Use: Garden, singly or in groups, cut flowers.

'Träumerei'®

Character: The salmon orange of the tea rose-shaped flowers is reminiscent of lobster, the fragrance of perfume; the variety flowers lavishly and is robust. A particularly elegant bedding rose, whose stems last a long time as cut flowers.
Group: bedding rose with fragrance
Breeder/Provenance: Kordes
Year of introduction: 1974
Flower color: salmon orange
Flowers: double
Flowering pattern: repeat flowering
Growth height: 24–32 in. (60–80 cm)
Growth habit: bushy
Plants per 11 ft.² (1 m²): 6 to 7
Use: Garden, singly or in groups, cut flowers, full standards.

'Venusta Pendula'

Character: Secret tip for gateposts, pergolas, light trees—a megagrower with lavish flower production. Once flowering—every year the gardener waits impatiently for this flower fantasy. Very winter hardy.
Group: rambler
Breeder/Provenance: unknown
Year of introduction: introduced 1928 by Kordes
Flower color: rose/white
Flowers: semidouble
Flowering pattern: once flowering
Growth height: 118–197 in. (300–500 cm)
Growth habit: flat without support
Plants per 11 ft.² (1 m²): 1 to 2
Use: Garden, singly or in groups, tolerates frosty sites, grows in part shade.

'Vogelpark Walsrode'®

Character: A shrub rose that does not disappoint. Whether solo or in groups, the delicate pink flowers of this long bloomer beautify every planting. Problem-free and versatile, for beginners.
Group: shrub, fall flowering
Breeder/Provenance: Kordes
Year of introduction: 1988
Flower color: rose
Flowers: double
Flowering pattern: repeat flowering, early
Growth height: 39–59 in. (100–150 cm)
Growth habit: bushy
Plants per 11 ft.² (1 m²): 1 to 2
Use: Garden, singly or in groups, pots, heat tolerant.
ADR rose: 1989

'Warwick Castle'

Character: A compact English Rose that fits into small gardens. Typical Austin flower doubling, which makes romantic hearts beat harder. Named for the restored Victorian rose gardens at Warwick Castle, which are well worth seeing.
Group: bedding rose with fragrance, English Rose
Breeder/Provenance: Austin
Year of introduction: 1986
Flower color: rose red
Flowers: double
Flowering pattern: repeat flowering
Growth height: 24–32 in. (60–80 cm)
Growth habit: bushy
Plants per 11 ft.² (1 m²): 6 to 7
Use: Garden, singly or in groups, very good as cut flowers.

'Westerland'®

Character: Lushly flowering, fragrant shrub rose with unique copper tone to blossoms. Develops into a compactly tight, self-contained shrub that leaves designers nothing to be desired.
Group: shrub rose with fragrance
Breeder/Provenance: Kordes
Year of introduction: 1969
Flower color: apricot
Flowers: semidouble
Flowering pattern: repeat flowering
Growth height: 59–79 in. (150–200 cm)
Growth habit: bushy
Plants per 11 ft.² (1 m²): 1 to 2
Use: Garden, singly or in groups, free-form hedges, pots, heat tolerant.
ADR rose: 1974

'White Meidiland'®

Character: Ideal white area rose with enormous flower fullness and size, very resistant to mildew. Already proven its worth in public gardens for

many years, also a feast for the eye in the home garden.
Group: area
Breeder/Provenance: Meilland
Year of introduction: 1985
Flower color: white
Flowers: very double
Flowering pattern: repeat flowering
Growth height: 16–24 in. (40–60 cm)
Growth habit: low, spreading
Plants per 11 ft.² (1 m²): 4 to 5
Use: Garden, singly or in groups, pots, grows in part shade. For sites with mildew-favorable conditions.

'Wife of Bath'

Character: An enchanting small shrub rose that develops a dense, pretty bush. The delicate pink of the flowers is deceptive, for the 'Wife of Bath' is a robust lady who knows how to keep disease at a distance. The fragrance of the flowers is reminiscent of myrrh and is—obligatory for an English Rose—outstandingly strong.
Group: shrub rose with fragrance, English Rose
Breeder/Provenance: Austin
Year of introduction: 1969
Flower color: pink
Flowers: double
Flowering pattern: repeat flowering
Growth height: 32–39 in. (80–100 cm)
Growth habit: arching
Plants per 11 ft.² (1 m²): 1 to 2
Use: Garden, singly or in groups, cutting rose, free-form hedges, pots.

'Wildfang'®

Character: The name, which means *wild catch,* indicates the wild robustness of this ADR rose. A variety for the 1990s, the result of the breeder's best selections. Indestructible variety for embankments, hillsides, but also for garden areas.
Group: area
Breeder/Provenance: Noack
Year of introduction: 1989
Flower color: rose
Flowers: semidouble
Flowering pattern: repeat flowering
Growth height: 24–32 in. (60–80 cm)
Growth habit: bushy
Plants per 11 ft.² (1 m²): 2 to 3
Use: Garden, singly or in groups
ADR rose: 1991

'Yellow Dagmar Hastrup'

Character: Important yellow *rugosa* variety for beds and surfaces. Strong fragrance. Wants annual rejuvenation through pruning. Fantastic color variety for planting in front of shrubs.
Group: *rugosa* hybrid with fragrance
Breeder/Provenance: Moore
Year of introduction: 1989
Flower color: yellow
Flowers: semidouble
Flowering pattern: repeat flowering
Growth height: 24–32 in. (60–80 cm)
Growth habit: upright
Plants per 11 ft.² (1 m²): 3 to 4
Use: Garden, singly or in groups, tolerates frosty stiuations, will stand semishade, pollen source, yellow fall coloring of leaves.

ROSE KNOW-HOW FROM A TO Z

A

AARS trials: This acronym stands for All-American Rose Selections. With the intention of testing all new hybrids outdoors, the All-American Rose Competition was started in the United States in 1938. The roses are tested in 25 test gardens over a period of 2 years. Only American growers are allowed to compete. The winning variety receives the very desirable AARS certificate, an important advantage in trade and sales promotion.
Abiotic: Has a general sense of not living. Is used to denote abiotic injury, i.e., injuries that are not caused by animal pests, fungi, bacteria, or viruses but by weather, physical damage, and faulty maintenance.
ADR roses: ADR stands for Allgemeine Deutsche Rosenneuheiten-Prüfung (Universal German New-Rose Trials). The ADR trials are conducted by a team of rose breeders and coworkers from independent trial gardens. Once a year, the screening information is collected, and the ADR certificate is given to those rose varieties that score high enough. The trial roses are not treated with any chemicals during the several years of the trial period; the varieties' powers of resistance are of the greatest importance in the evaluation.
Aged/Permanent humus: See Humus.
Alba roses: In cultivation by the Romans and Greeks, the oldest known garden rose. Forms and varieties with very good winter hardiness, robust growth, and outstanding scent.
All-American Rose Selections (AARS): A nonprofit organization of U.S. producers and introducers who organize rose test gardens throughout the United States. Roses that receive the AARS award are considered exceptional new varieties.
Alpenrose: German name for *Rhododendron* genus; has only the name in common with roses.
Alpine Rose: *Rosa pendulina,* native wild rose with rose-colored flowers and very few prickles.
American Rose Society (ARS): Primary national society (with regional chapters) devoted to promoting rose growing and education about roses and rose growing.
Anther: The pollen-bearing portion of the stamen.
Aphid-parasitic hymenopteran: Beneficial insect. Lays its eggs in the egg, larvae, or pupa of another insect, especially of the aphid.
Aphids: Small, green insects that attack young, still soft canes. In warm, dry weather massively reproduce, starting in April.
Apothecary Rose: *Rosa gallica* 'Officinalis' ("Red Rose of Lancaster"). Semidouble, crimson-red rose, has been cultivated since the thirteenth century in Provins near Paris. Fragrant and medicinal rose. On both sides of the main street of the town are numerous drugstores and pharmacies that send medicines produced from the roses to all parts of the world.
Apple Rose: *Rosa villosa;* native wild rose with rose-pink flowers, up to 79 inches (2 m) tall. Hips with unusually high vitamin C content.
Arboretum: From the Latin *arbor* meaning tree. Parklike garden for woody plants, often attached to botanical gardens.
Arctic rose: *Rosa acicularis.* Dark-red flowers borne in May, from September pear-shaped hips. Few real prickles but numerous needlelike bristles.
Area rose: Umbrella term for the group of ground-cover and small shrub roses; also variously called landscape or prostrate roses. All varieties of this group act as weed suppressors owing to the way their dense leaf placement covers the surfaces. Ground-covering varieties grow flat and ground hugging. Increasingly offered commercially as own-root stock, which cannot develop tiresome suckers.
Austrian Briar rose: *Rosa foetida,* wild rose with shrublike habit, which blooms deep yellow in June and then gives off a strong flower scent that many find unpleasant. The familiar variety is *R. foetida bicolor,* the orange-scarlet 'Austrian Copper' that is the ancestor of all the orange and yellow garden roses. See also Lutea hybrids.

B

Balling: Tendency of rose flowers to stick together and develop so-called flower mummies after heavy rain, often seen in *rugosa* hybrids, for example.
Bamboo: Fascinating plant group belonging to the grasses. Identifying characteristics are woody, perennial culms. About 100 genera with more than 1000 species. Species and varieties are classified as clump- and runner-forming.
Bare-root roses: Form in which many roses are sold, especially by mail order; rosebush without soil on the roots.
Bark mulch: Chopped bark that is applied as ground covering.
Bark spot disease: Fungal disease; causes brownish red spots on the canes, which can easily be mistaken for frost damage.
Bedding roses: Collective term for polyantha and floribunda roses (Polyantha hybrids). Usually for the purpose of area effects, low roses planted with less space between them for beds and borders. Develop well-branched canes, which are striking, with numerous bunches of more or less doubled individual flowers in red, pink, white, and yellow.
Bee food: Plants that serve as an important source of nutrition for bees and other insects because of their flower nectar and pollen. In return, the visitors to the flowers facilitate extensive pollination.
Belgian stone: Very fine whetstone with which pruning tools can be sharpened to razor sharpness.
Beneficial insects: They combat pests in the garden, e.g., lacewings, ladybugs, earwigs, parasitical hymenopters, and syrphid flies.
Bengal roses: See China roses
Berries: Fleshy, juicy fruit.
Bird-food shrub: Shrub whose fruits are used as a source of food by birds of various species.
Bird-protecting shrub: Shrub that provides safe nesting and brooding with its dense branches or armored canes.
Black rose: Truly black roses have hitherto belonged to the realm of fantasy. There are very dark-red rose varieties that are repeatedly being advertised in the catalogs. One of the darkest so far was the very weakly growing variety 'Nigrette' introduced in 1933 by the German breeder Max Krause. With a great deal of goodwill it can be called black. As a rule, these varieties are not very vigorous and are prone to disease.
Black spot: Fungus that forms black spots on sugar-containing excretions of aphids and scale on rose leaves (See Honeydew).
Blind cane: Canes without flower buds, often occurring as a result of insufficient light. They use up a plant's nutrients, water, and energy and are cut back.
Blue rose: The varieties sold as blue are lavander to lilac-colored; a true marine blue has so far remained a breeder's dream. The anthocyanin (a type of pigment) delphinidin is responsible for the pure blue in other plant species. To date, breeders have not succeeded in transferring it to roses. Until this occurs, truly blue roses are only a reality in prospectus pictures so colored.

Bone meal: High-phosphate organic fertilizer of animal origin.
Botrytis: Fungal disease; the fungus causes rotten spots on petals and buds.
Bourbon roses: The first Bourbon rose and later offspring varieties very probably arose as crossings of Damask roses with China roses on the Île de Bourbon (today La Réunion) in the Indian Ocean at the beginning of the nineteenth century. Later, they were brought to France, where they were hybridized to develop the class of Bourbon roses. The important variety in this class is 'Souvenir de la Malmaison'. Further crossings of Bourbon roses with, e.g., tea roses, led to the development of the lushly flowering hybrid perpetuals.
Breeding: Process for achieving new rose varieties through crossing of mother and father varieties. Two days before the actual crossing, the yellow stamens of the father variety are cut off with scissors. The pollen grains are allowed to dry in a little dish with light excluded. One day before the complete opening of the mother variety, all its petals and stamens are removed by hand in the early hours of the morning and thus the possibility of self-pollination is excluded. The breeder dips the castrated flowers of the mother variety into the pollen of the father variety. Immediately after crossing, the flowers are covered with paper bags. After development of hips, the sowing of seeds and selection of seedlings takes place. It takes at least seven years before a new variety reaches the marketplace.
Broad-leaf evergreens: Evergreen foliage shrubs such as Buxus, Ilex, firethorn, rhododendron; keeping their foliage at least two summers.
Budding: See Grafting
Bud union: Place at which the eye of the desired rose plant is joined by grafting to grow on the understock. Is found either at the root neck or—in standard roses—at the crown level, recognized as a knotlike thickening.

C

Cabbage rose: *Rosa centifolia.*
Calcium (Ca): Nutrient contained in lime; also regulates the pH value of soil.
Cane borer: Insect; causes wavy canes through passageways eaten into the canes. Downward-workng rose cane borers hollow canes from top to bottom, upward-working ones from bottom to top.
Cane death: See Valsa disease.
Cane strength: Thickness of a cane.
Capsule: Many-seeded fruit, formed of several carpellary leaves, various opening mechanisms.
Cascade standards (weeping standards): Form of tree rose in which usually climbers or overhanging area roses are grafted at a height of 55 inches (140 cm).
Centifolia roses: The Centifolias—also called cabbage or Provence roses—are considered the absolute epitome of the Old Roses. Lush doubled flowers and fantastic fragrance distinguish this group. They grow a good $6^1/_2$ feet (2 m) high. See also Moss roses.
Chestnut Rose: *Rosa roxburghii.* Species that develops unusual hips that are covered with prickles and look like chestnuts.
China roses (Bengal roses): Rose group stemming from China of the species *Rosa chinensis,* which revolutionized rose breeding with their appearance at the end of the eighteenth century. China roses introduced repeat flowering and low growth to already existing roses; results were ultimately the modern hybrid tea and bedding roses. In addition, the China roses decisively expanded the color spectrum, for besides pink, white, and orange yellow they introduced a pure red. In our latitudes, China roses usually are low growing; their disadvantage is their often insufficient winter hardiness, and winter protection is thus urgently recommended.
Cinnamon Rose: *Rosa majalis.* In May, lavender flowers appear on this native wild rose for banks and slopes.
Claw: Tool for unhilling roses in spring and for loosening soil.
Climbing roses: They developed from a mutation, a rose form with whiplike canes that are in culture as climbing descendants of existing bedding and hybrid tea rose varieties, especially in southern European countries, e.g., 'Climbing Peace'. Many of these climbing varieties need warmer climates to develop fully into impressive climbing roses. The climbing growth habit arises from the missing end bud on the cane.
Compost: Organic fertilizer, obtained by returning organic materials to the biocycle. Compost production occurs through collection of organic house and garden debris. Compost management with the earthworms and the microorganisms as important helpers.
'Conditorum': *Rosa gallica,* the Hungarian Rose. The semidouble ruby-red flowers were once used in the production of confectioners' products and Hungarian rose water.
Conifers: Woody plants that, with very few exceptions (gingko), bear needles and grow as trees or shrubs.
Container roses: In the gardening trade, *container* refers to a pot of plastic, foil, or recycled paper in which, among other things, roses are cultivated. Container roses permit summer planting of flowering rosebushes. Minimum size 2 quarts (2 l).
Corymb: Inflorescence; clusters in which the flowers with stems of different lengths end at one level.
Crown drip line: Outermost boundary of the crown umbrella area of a tree. At the edge of the crown along the so-called drip line is where water from rain and condensation is shed. It especially affects any roses placed under it and makes difficult the drying off of their leaves. No site for roses.
Cut roses: See Field-grown cut roses, Hothouse cut roses.
Cyme: Panicle whose branching stems end in a flat head, umbrella-like.

D

Damask roses: Damask roses are classified into two types: the summer-flowering forms and those that flower in summer and again in the fall, known as Autumn Damasks (see Portland Roses). Typical is the gray-green, soft foliage and the heavy, luxurious fragrance.
Delphinidin: See Blue roses.
Dethorner: Useful implement for preparation of cut roses. The spines at the lower end of the rose stem can be easily stripped with it.
Disbudding flowers: Removal of certain, just-developing flower buds influences the development of the remaining buds of cutting roses. With hybrid teas the accessory buds are removed; with bedding and area roses, the uppermost main buds are taken.
Discounter: Supermarkets sporadically offer packaged roses at low prices, usually nameless and often utterly unusable varieties for the garden, which often flower entirely differently from the description. Anyone who wishes to realize his or her design plan successfully with the help of a careful selection of varieties should keep at a distance from these roses.
Distillation: Method of obtaining rose oil. Hot steam removes the essential oils from fresh flowers. After cooling, the lighter, absolutely pure essential oil is floating on the water and is siphoned off.
Dog rose: *Rosa canina,* a native wild rose with delicate pink flowers appearing in July, producing an abundance of hips. The more-than-600-years-old rose at Hildesheim Cathedral is an *R. canina* form. Its rather unattractive common name probably goes back to its medicinal use of the production of extracts from its roots. This

medicine was prescribed well into the sixteenth century as a certain remedy for the bite of a rabid dog.
Dormant eye: See Eye.
Downward-working rose cane borer: See Cane borer.
Downy mildew: See False mildew.
Dwarf Bengal roses: See Miniature roses.
Dwarf roses: See Miniature roses.
Dwarf standard: See Quarter standards.

E

Ear, spike/false ear: Form of inflorescence with unstemmed flowers situated in an axis in the axilla of small supporting leaves.
Earwig: Beneficial insect. Hunts aphids at night. Rose balls filled with excelsior keep them warm inside and are ideal nesting places for earwigs.
Ecology: Study of the changing relationships between, among others, plants and the environment.
Enfleurage: Method of obtaining rose oil. Rose petals are mixed with tallow. The flowers are repeatedly sprinkled on tablets of beef fat. The fat fixes the essential oils, which are later extracted with alcohol.
English Roses: New rose class with very fragrant, usually repeat-flowering bedding and shrub roses, which have been developed by Englishman David Austin in more than 30 years of hybridizing work. The large, very doubled rosette or round flowers are reminiscent of Old Roses.
Extraction: Method of obtaining rose oil. Dried flowers are moistened and layered in stainless steel vats on perforated sieve plates. With the addition of hexane, the plant cells break open; after heating and cooling the so-called concrete, a waxlike mass with the fragrant essential oil is left. From this absolute, the pure essence is obtained.
Eye: Growth point (bud) protected by leaf stem in the leaf axil or, if dormant, as a dormant eye (accessory bud) situated on old canes.

F

False mildew (downy mildew): Fungal disease; evidenced by whitish gray field of mold on the undersides of leaves and dark spots on tops of leaves. Affected leaves wither and fall off; appears especially in late summer and fall.
Ferns: Perennials, which either drop their leaves in fall or survive the winter as wintergreens or evergreens. Love damp, humus soil in shady spots. Spread by rhizomes usually; the feathery leaves appear in spring as fronds.
Fertilizing: Additional application of nutrients for increasing the yield in agriculture and in commercial and home gardening. A soil test gives information about the nutrients already present in the soil and in a form available for the plants.
Field-grown cut roses: In the growing trend toward the summer sale of cut roses, the so-called field-grown roses are playing an increasingly larger role. In contrast with the hothouse roses, these are for the most part of native production. The so-called spray roses, that is, very cluster-flowered varieties with many blossoms per stem, are especially a runaway seller for use in arrangements.
Field rose: *Rosa arvensis,* native wild rose; its long canes cover embankments and slopes, also in semishady sites, flowers in July, hips begin in September.
Flavonoids: Cellular elements that are responsible for the yellow color in roses.
Floribunda-Grandiflora roses: Very large-flowered floribunda roses, used the same way. Example: 'The Queen Elizabeth Rose'.
Floribunda roses (Polyantha hybrids): Large-flowered, winter-hardy, low-bedding roses, whose elegant flowers resemble the shape of the hybrid teas. The first rose in this group is considered to be 'Gruss an Aachen'. Transitions to the Polyantha roses are fluid. See also Floribunda-Grandiflora roses.
Flower gatherer shears: Special pruners developed for clipping cut roses, which hold the stem of the rose by a thick blade after it is cut. Some brands are designed with a prickle remover.
Foetida hybrids: See Lutea hybrids.
Foliage: Leaf organs with a high content of chlorophyll. Photosynthesis takes place in the leaf. In this process, carbon dioxide and water are changed through the energy of the sunlight into sugar compounds, and oxygen is released.
French Rose: See *Rosa gallica.* Garden rose of our ancestors, extremely fragrant. Many varieties were developed in the nineteenth century. Also called Provins Rose, see also Apothecary Rose.
Full standard: Tree rose with the bud union at a height of 36 inches (90 cm), where bedding, hybrid tea, or area roses are grafted. Ideal for the home garden.
Full-sun site: The rose is completely in the sun from sunup to sundown.
Fungal diseases: In roses, designated as powdery mildew, false mildew, rose rust, spot anthracnose.
Fungicide: From the Latin *fungus* and *caedere,* "to kill"; chemical toxic substances that can be used in gardening against injurious fungal diseases.

G

Gallica roses: Group of varieties going back to the oldest rose species cultivated by humans, *Rosa gallica. Rosa gallica* is considered the ancestor of all modern rose varieties and was already being widely planted in the thirteenth century. Its rose oil was used as a medicine and as a cooking ingredient. Later, the variety 'Officinalis', the Apothecary Rose, gained very great importance. Most Gallicas are accidental seedlings, arising through cross-pollination by insects.
Garden center: Place of sale for everything concerning plants in the house and garden. As a rule, do not produce their own plants.
Garden nurseries: Garden nurseries are recognized quality nurseries that provide private customers with, among other things, a wide assortment of woody plants and perennials throughout the year. They offer high professional expertise, service, and advice about all garden matters. For growers with their own plantings, see also Nursery.
Golden Rose of China: *Rosa hugonis,* yellow spring rose, valuable for early, lavish flowering, starting mid-May.
Grafting: Most commonly used method of propagation for roses, in which an eye is inserted into prepared understock furnished with a T-cut.
Grafting knife: Special knife for grafting work, either with the bark stripper on the blade or with a stripper of plastic at the end of the knife.
Grandiflora roses: Transitional form between floribunda and hybrid tea roses. Repeated crossing of both forms has developed cluster-flowered hybrid tea roses.
Green manuring: Summer sowing from April of *Tagetes erecta* (marigold) before roses are planted improves the soil structure and—through increasing the humus—the fertility of the soil. In addition green manure plants are important food plants for bees and other beneficial insects. Used to combat nematodes in nurseries.
Green Rose, The: *Rosa chinensis viridiflora,* rose that flowers all summer long with light-green "petals," arising through backcrossing of petals and stamens to leaves. Sport of nature, produced from a China rose. Introduced commercially by an English nursery in 1856.
Ground-cover roses: Old term for area roses that indicates the ground-covering, very low growth some varieties in this group display. Examples: 'Immensee', 'Swany'.

H

Habitus: External look of a tree or a shrub, depends on the kind of branching; direction of growth of the leader, degree of woodiness of branches, and leaf form.
Half standard: Tree roses with the graft at 24 inches (60 cm); usually bedding or area rose varieties are used as grafts. Ideal standard roses for pots.
Hardwood cuttings: Propagation method. Woody cane sections of about 8 inches (20 cm) in length (shrub and climbing roses) are stuck into the soil up to the last eye to root.
Healing plants: Plants that can be used for production of medications because of the substances they contain.
Herbicide: From Latin *herba* meaning *herb* and *caedere* meaning *to kill;* chemical substance for combatting weeds.
Hilling: Covering the base of the rosebush to a depth of up to 8 inches (20 cm) with leaf compost or garden compost as a protection from winter winds and freezing.
Hip: Orange, yellow, brown, black, or red syncarp or pseudocarp of roses with high vitamin C content; see also Nut.
Home-builders' supply centers: Many builder's supply centers have discovered the green world for themselves and include garden departments. As a rule, the standard selections are available, and this goes for the rose department. In the search for possible varieties, the price comparison may pay in certain cases.
Honeydew: Sugar-containing plant juice that is excreted by sucking aphids and, among other places, deposited onto leaves of roses planted under affected trees (such as linden). Spot anthracnose finds ideal breeding conditions on the sticky leaves.
Horn chips: Organic plant and soil-protecting nitrogen fertilizer of animal origin.
Hothouse cut roses: In commercial horticulture, rose plants grown in hothouses for the production of cut flowers. The targets of modern cut-rose breeding are above all high productivity, highest mildew resistance, good keeping power, good cane behavior, good transportability of the roses, and, recently again, fragrance. See also Field-grown cut roses.
Hudson Bay rose: *Rosa blanda,* wild rose with single rose-pink flowers in May and June.
Humidity: Air with an extremely high percentage of moisture within a container or greenhouse for providing, say, cuttings with enough leaf moisture until they have rooted.
Humus: Totality of dead organic substances of animal or vegetable origin in the soil. Dark color typical. Humus serves to feed the soil organisms, improves the soil structure, contains nutrients, holds in soil moisture, and reduces erosion from wind and rain.
Hungarian rose: See 'Conditorum'.
Husk: Outer covering of a seed or fruit.
Hybridization: Produced by means of the wind, animals, or the human hand, sexual union of different parent varieties resulting in the occurrence of a new type of offspring.
Hybrid perpetuals: Sometimes called remontant roses. An important link between the old and modern roses, developed at the beginning of the nineteenth century from crossings of, among others, Bourbon, tea, and Portland roses. The varieties in this group are remontant, that is, they have the ability to develop a second flowering after the main flowering. There is a definite resting spell between the first and the repeat flowering. Most hybrid perpetual roses have a shrub habit.
Hybrid tea roses: Characteristics of this group are long stems and elegantly shaped flowers occurring one to a stem. The hybrid teas were produced from crosses with Chinese tea roses, but they surpass them in winter hardiness. The variety 'La France', introduced in 1867 by the Frenchman Jean Baptiste Guillot, is considered the first hybrid tea. Very many varieties available.

I

Inorganic fertilizers (short-term fertilizers): High-powered nutrient concentrates that are easily water soluble and are constantly activated with appropriate soil moisture. Inorganic fertilizers have none of the soil-improving characteristics but serve as fast remedies for the appearance of deficiency symptoms.
Insecticide: From Latin *insectum* meaning *notched* (see Insects) and *caedere* meaning *to kill;* chemical substances for killing insects.
Insects: From Latin *insectum* meaning *notched;* refers to the sharp indentations in the bodies of these animals that separate their heads, chests, and lower bodies.
Integrated pest control: Integrates the natural resistance of plants into the pest control and employs application of chemical pesticides only after damage thresholds are exceeded.
Internodes: Intervals between the eyes (buds) on a cane.
In vitro culture: See Meristem propagation.
Iron (Fe): Nutrient. Iron deficiency becomes evident through chlorosis, a yellow coloration of leaves and fruit.

K

Kordesii roses: Variety group created by Wilhelm Kordes of especially winter-hardy, long-flowering, robust varieties.

L

Lacewings: Also called aphid lions. Beneficial insects. The larvae suck on aphids.
Ladybug: Beneficial insect. Larvae, pupae, and beetles can destroy large quantities of aphids and scale insects.
Lambertiana roses: Named for Peter Lambert, well-known rose breeder. He developed the repeat-flowering shrub rose 'Trier', which became the starting variety of the Lambertiana roses. See also Musk roses.
Largest rose nursery in the world: Jackson & Perkins, United States. Fifteen million rose plants are grafted annually.
Leafhoppers: See Rose leafhoppers.
Leaflet: The rose leaf is odd-pinnate; depending on the species, the rose leaf consists of 5 to 15 leaflets (pinnules).
Lettuce-leaved Rose: *Rosa centifolia* 'Bullata', Centifolia rose with unique, very large, crinkly leaves.
Lime: See Calcium.
Loppers: Tool for cutting of branches up to 1½ inches (4 cm) thick on old rosebushes.
Lutea hybrids (Pernetiana roses, Foetida hybrids): A rose class arising from crossings of *Rosa foetida* 'Persian Yellow' with the hybrid perpetual rose 'Antoine Ducher' by the Frenchman Pernet-Ducher at the end of the nineteenth century, with many wonderul colors, among others pinks, oranges, and apricots. Highly susceptible to spot anthracnose; see also Austrian Briar Rose.

M

Magnesium (Mg): Nutrient and a constituent of chlorophyll (see Foliage). Magnesium deficiency is evidenced by mosaiclike yellowing of the leaves.
Meristem: Undifferentiated tissue capable of division, e.g., at the growing tip of a rose. Used in meristem propagation.
Meristem propagation: A still new method of propagating roses. The principle involved is to develop a viable rose identical to the parent variety from an isolated plant cell under laboratory conditions (in vitro).
Minature roses (dwarf roses, dwarf China roses): Probably,

these varieties have their origins in China. The miniature growth makes possible many different kinds of use on balconies, in gardens and rock gardens, or as pot roses. Many varieties are subject to fungal diseases and need several sprayings with fungicides during the summer months.
Monoculture: Massive, monotonous use of only one rose species or variety in a certain area. Promotes species impoverishment and with it increased vulnerability to damage from species-specific insects and diseases.
Moss: See Moss roses.
Moss roses: Arose from a spontaneous change (see Mutation) of the flower bud of the Centifolia roses; characteristic of the moss roses are inflorescences, ovaries, and calyxes covered with mosslike glands.
Mulching: The term derives from Low German *mölsch,* which means something like *soft, at the beginning of decay.* Mulching denotes covering the soil with organic debris. Effect: weed suppression, promotion of soil moisture, and water conservation. Depending on the mulch used, addition of a nitrogen fertlizer may be advisable.
Multinutrient fertilizer: See Rose fertilizer.
Musk hybrids: See Musk roses.
Musk roses: Crosses, which go back to *Rosa moschata,* of a wild rose from India and China. Even in antiquity writers were describing their musklike flower fragrance. The first hybrid Musk was 'The Garland', a breeding success of the Englishman Wells in the year 1835. Around 1900, Peter Lambert continued the crossing with *Rosa moschata* and created the famous variety 'Trier'. The Lambertiana roses were named in his honor.
Mutation: Spontaneous alteration of genes; for example, a different type of cane develops on a rose (see Climbing roses). If a piece of such a cane is propagated, a new variety is thus developed with altered characteristics, such as a change in color of flowers or of growth habit. Such a new variety is also called a *sport.*
Mycorrhiza: Symbiotic relationship consisting of roots and fungi colonizing on them.

N

Native shrubs: Shrubs that are distributed throughout North America.
Near-humus: See Humus.
Nematodes: Tiny threadworms that suck on the roots of roses and other plants. Through the wounds thus caused, fungi enter the root hairs and damage these vital organs. May be a cause of soil exhaustion.
Nitrogen (N): Nutrient that, along with others, the rose needs for long growth. Nitrogen deficiency causes light-green leaves.
Noisette roses: Class of repeat-flowering roses that go back to the Frenchman Louis Noisette. By using seedlings from his brother Philippe, who was living in the United States, Louis made crosses at the beginning of the nineteenth century that resulted in low and climbing varieties.
Nomenclature: Procedure for the correct naming of living things. Every organism is named with a genus (first name) and species (second name). A variety name (third name) can be added to this so-called binary nomenclature: e.g., *Rosa foetida bicolor.*
Nonrecurrent: Roses that produce flowers only once a year.
Nonsunny situation: The plants are situated either in the heavy shadow of buildings for a large portion of the day or are in sloping sites that face more than 30 percent toward north.
Nursery: A nursery is a commercial enterprise devoted to growing and nurturing young plants until they are large and/or strong enough to be sold either to the public directly, to landscaping companies, or to other nurseries that will grow and sell the plant material. Many nurseries also propagate the plants they grow either from seed or by vegetative cuttings until they reach sellable sizes. Nurseries can grow plants outdoors or in sheltered structures such as greenhouses, lathhouses, or cold frames. Retail nurseries sell directly to the public by catalog or on-site sales. Other nurseries, known as wholesale nurseries, sell only larger quantities of plant material to landscaping companies, retail garden centers or nurseries, and other wholesale nurseries who grow the plants further to sell. In the home garden, a nursery area is a place where young plants grow until they are transplanted to a permanent site in the garden.
Nut: Usually single-seeded indehiscent fruit. Botanically, rose seeds are regarded as small nuts and are collected in a fruit, the hip.

O

Oil rose: Term that most often refers to *Rosa* x *damascena* 'Trigintipetala'. Almost half the world's production of rose oil comes from Ukraine, a quarter from Bulgaria, the rest from Turkey, Morocco, and other places. The semidouble flowers are harvested in the early morning hours. About 6,640 pounds (3,000 kg) of flowers are necessary for 2.2 pounds (1 kg) of rose oil.
Old Roses: The so-called Old Roses are classified as old according to the year of their introduction. The American Rose Society established the definition in 1966: "A rose is an old rose if it belongs to a class that was in existence before 1867—the year of the introduction of the variety 'La France' as the first hybrid tea."
Old wood: Describes the wood of canes that are several years old.
Organic fertilizers: After microbial breakdown by the soil organisms, these fertilizers release their nutrients in a slow flow. Examples of organic fertilizers are horn chips, compost, guano, and horse manure.
Ornamental shrub roses: Repeat-flowering shrub roses.
Ovary: The basal or lower portion of a pistil that becomes a fruit.

P

Panicle: Inflorescence, several branched clusters with an end flower.
Park roses: Once-flowering shrub roses.
Patio roses: English term for Miniature roses, especially planted in balcony boxes.
Pedicel: The stalk of an individual flower in an inflorescence.
Perennials: Herbaceous plants that live several years, surviving the winters by means of their undergound organs.
Pernetiana roses: See Lutea hybrids.
Petals: Botanical term for flower parts (from the Greek *petalon*). One unit, usually colored and showy, of a whole flower unit.
Phosphorus (P): Nutrient, among other things necessary for protein synthesis, essential substance in cell nucleus. Phosphorous deficiency shows itself through undersized, bluish green leaves that display a purplish bronze coloration along the edges.
pH value: Expresses the concentration of soil acidity. Very acidic soils (suitable for plants such as heaths and heathers) have a very low pH value around 4. For roses, a pH value in the neutral range, around 6.5, is ideal. An increase in pH is possible by adding lime, see Calcium.
Pinching back: Shortening of new growth. Hybrid tea roses, especially, tend to develop very few new canes from the base and to grow sparsely and loose. Pinching back new growth beginning in May stimulates the plant to further branching. This promotes bushiness and multiple canes in the varieties with an awkward growth habit.
Pistil: The female part of the flower consisting of the stigma, style, and ovary.

Pod: Dry, many-seeded dehiscent fruit; fruit with many seeds formed from a single elongated carpellary leaf, which opens on the underside.
Polyantha hybrids: See Floribunda roses.
Polyantha roses: Small-flowered, low-growing bedding roses with large, many-flowered inflorescences. Produced from crossing *Rosa multiflora* (Syn.: *Rosa* x *polyantha*) and *Rosa chinensis.* Transitions to Floribunda roses are fluid.
Portland roses: Probably a form arising from crosses of the repeat-flowering Autumn Damask rose (see Damask roses) with *Rosa gallica* 'Officinalis', often with shrub habit, fragrant, robust.
Potash: See Potassium.
Potassium (K): Important nutrient for regulation of water balance of roses. Fertilizing with potassium or potash promotes maturing of wood and thus lowers the rose cane's vulnerability to winterkill.
Potted roses: Small-growing miniature roses for the house and balcony, usually offered in pots 4 to 5 inches (10 to 12 cm) in diameter. The main producing country for cutting-propagated pot roses in Europe is Denmark, which exports almost all the potted roses produced in that country. The potted rose production is highly specialized and automated; after 12 to 14 weeks, the roses are ready for sale. Production takes place all year long. In Denmark, there are over 30 million, in all of Europe about 54 million potted roses per year.
Powdery mildew: Fungal disease; a floury white deposit that can be wiped away appears on the tops of young leaves as well as on flower calyxes and canes. With severe infection, the affected leaves crinkle and turn red; appears in early summer.
Predatory mites: Beneficial insects. Suck out the eggs of the red spider.
Prickles: Roses have prickles, botanically regarded as outgrowths of the upper bark layer that are easily removed— in contrast with thorns, which are grown as one with the bark.
Proliferation: Process in which a flower grows through, i.e., a new flower stem grows from the inside of a blooming rose. Often these double-decker flowers can be observed in old roses and bedding roses.
Provence Rose: See Centifolia roses.
Provins Rose: See French rose.
Pruning knife: Curved garden knife that is very well suited for removing suckers on roses.

Q

Quality standards: Properly, "FLL Quality Standards for Nursery Plants," a set of rules to regulate plant quality within the nursery trade. Recognized Europe-wide.
Quarter standard (minature rose trees): Tree roses with the bud union at a height of 16 inches (40 cm); usually the miniature rose varieties are used for grafting.

R

Raceme: Inflorescence, stemmed individual flowers on an axis sitting in the axil of small supporting leaves.
Rambler: Climbing rose that develops especially long, soft, thin canes. Grows into light trees, greens pergolas, and arches.
Recurrent: Roses that produce another flowering after the first, main one; also refers to roses that bloom continuously.
Red-leaved rose: See *Rosa glauca.*
"Red Rose of Lancaster": See Apothecary Rose
Red spider: See Spider mite.
Remontant: Reflowering; recurrent; used here to mean roses that have a second flowering later in the season.
Remontant roses: See Hybrid perpetuals.
Replant disease: See Specific replant disease.
Romantica® roses: New product line from French breeder Meilland with rose varieties that are reminiscent of Old Roses with their very double rosettes and round flowers.
Root balling: The roots of the rose are placed into a ball of earth, which is held together by a net or a carton to avoid drying out of the roots during the selling period.
Root neck: Transition between the root and the stem. In roses is often at the level of the bud union.
Rosa glauca: The red-leaved rose, a wild rose that blooms in June with dark-pink flowers and has frosted, bluish red leaves. Formerly called *R. rubrifolia.*
Rosa moyesii: A loosely growing shrub rose that is decorated with large, bottle-shaped bristled hips; bird-food shrub.
Rosa multiflora* (Syn.: *Rosa* x *polyantha): Formerly an important understock for the grafting of roses, its importance is declining today. Parent rose in breeding of many climbing varieties. White flowers from June to July, growing in dense clusters.
Rosa nitida: Wild rose that develops numerous runners, which ultimately develop into a many-caned knee-high bush. Canes are bristling with prickles, foliage has fall coloring.
Rosa pimpinellifolia: Native wild rose very common on sand dunes and slopes, relatively salt tolerant, puts out many runners, flowers have a honey scent.
Rosa rugosa: Wild rose, which came to Europe from Japan in the middle of the eighteenth century and today is considered native there. Has characteristically wrinkled leaves that turn golden in fall. *Rosa rugosa* and its varieties are robust, considered salt tolerant, and bear an abundance of hips.
Rosa scabriuscula: Native wild rose of sunny forest edges, in June and July attracts numerous insects for pollination, good hip production.
Rosa sericea* f. *pteracantha: Native wild rose that develops only four petals and is notable for winglike, up to 1¼ inches (3 cm) spines, especially on young canes. This form originated in central Asia.
Rosa tomentosa: Native wild rose, flowering rose pink in June.
Rose ball: Decorative shelter for earwigs.
Rose fertilizer: Multinutrient fertilizer, contains the necessary organic and mineral nutrient components for roses in the right quantities and proportions.
Rose gall wasp: Mosslike, clearly visible swellings on the canes; noticeable by the hairlike outgrowth, the so-called "sleep apple," which can reach a diameter of 2 inches (5 cm). Sleep apples are thought to induce sleep.
Rose leafhoppers: The leaves are whitish spotted on the upper sides. On the undersides are greenish white, aphidlike insects, which spring away.
Rose leaf roller wasp: Attack can be recognized by rolled up leaves of roses, starting in May. The larvae develop in the little leaf rolls.
Rose of Jericho: As Easter plants, are sometimes for sale at fairs. *Selaginella lepidophylla,* a club moss from the dry regions of Mexico and Central America, the so-called Rose of Jericho. The plants form rolled rosettes the size of tennis balls, which can live all year long without water. When one throws a rosette into hot water, it opens, "blooms," to the astonishment of viewers.
Rose rust: Fungal disease; after the new growth in the spring, orange spore deposits are seen on the undersides of the leaves. In fall, the pustules are dark brown.
Rose sickness: See Specific replant disease.
Rose spading fork: Spading fork with only two tines of about 10 inches (25 cm) in length and 1½ in. (4 cm) apart. Ideal for protecting roots while loosening soil.
Rose test gardens: See All-American Rose Selections.
Rust: see Rose rust

S

Scale: Brown insect, protected by a hard shell, which excretes a sticky juice and sits on canes and leaves.
Selection: Process in rose breeding in which the seedlings are evaluated according to certain criteria, e.g., robustness and readiness to flower, and selected for further propagation. See also Breeding.
Self-cleaning: Ability of a rose blossom to clean itself after fading. Some varieties, on the other hand, have the characteristic of keeping their flowers on the stem until they turn brown.
Semishady site: The plant is shadowed by light moving or changing shadow for more than half the day but is always shaded during the middle of the day.
Sepals: Technical term in botany for a modified leaf, one of several that make up the calyx.
Shady site: The plant receives no direct sunlight, e.g., in situations directly under trees or in the darkest shadow of buildings and walls.
Shallow rooted: Woody plants whose main roots run shallowly and close to the surface, e.g., birches.
Short-term fertilizer: See Inorganic fertilizers.
Shrub: Woody plant that, in contrast with trees, branches at the base or undergound.
Shrub roses: Umbrella term for once-flowering park roses and ever-blooming ornamental shrub roses. All roses are "shrubs"; the term shrub roses, however, denotes species and varieties that are striking because of greater growth vigor in height and width. The flowers can be single, semidouble, but also doubled and like a hybrid tea rose.
Sleep apple: See Rose gall wasp.
Slow-release fertilizer: Inorganic fertilizers consisting of grains of fertilizer that are surrounded by a semipermeable resinous shell and that give up their nutrients according to the temperature. Washing out of the nutrients is avoided.
Snout weevils: About 4-inch (10-cm) long, black weevils that leave chewed holes in the buds, leaves, and canes where they have settled.
Soft cuttings: Propagation method. Herbaceous cane sections with two to three eyes are made to root in the summer under plastic or glass.
Specific replant disease: Complex, so-far unexplained phenomenon. Recognizable by decreased vigor after replanting of roses in an area where roses have been grown previously. So-called rose sickness.
Spider mites (e.g., red spider): Very tiny, orange-red creatures that settle on the undersides of leaves, sucking, and can be seen only with a magnifying glass. The upper sides of the leaves are sprinkled with brownish yellow until the leaves drop. With the common spider mite, a fine webbing occurs between leaves, leaf stems, and canes.
Sport: See Mutation.
Spot anthracnose: Fungal disease; star-shaped, violet-brown to black spots are visible on upper sides of leaves, with rays running out from them. Usually appears in late summer and fall, but also can be seen starting in June in very wet summers.
Spray roses: Especially cluster-forming varieties with numerous flowers per stem, in particular among the bedding roses. See also Field-grown cut roses.
Spreading climbers: Climbing roses, especially ramblers, are among the spreading climbers. They grow upward, in so doing always searching for a hold with their hooked spines. If the search is successful, they hook fast.
Standard: Also called tree roses, consisting of a bud grafting of a rose variety onto the top of a straight, leafless rootstock stem.
Steinfurther Rosefest: It takes place every two years in July, community show of all rose firms in Hessian Bad Nauheim-Steinfurth. Among other things, there is a rose show with many hundreds of varieties exhibited; Sundays: rose parade.
Stigma: The top part of a pistil that receives pollen.
Stipule: An outgrowth from the base of a leaf stalk.
Stone fruit: Juicy fruit with woody seed casing (episperm).
Stratification: Layers of seeds are placed into damp sand until the resting phase of the seed is overcome. Increases the germination rate markedly and ensures a uniform swelling of the seed. Is used with seeds of wild roses.
Style: The narrow section of the pistil that bears the stigma.
Substrate: Premixed potting soil for tubs, troughs, hanging baskets, and propagation trays.
Sucker: Canes from the root or the rootstock of grafted roses below the bud union or from the trunk of standard roses. As a rule, the finer, smaller foliage is easily differentiated from the larger foliage of the grafted variety.
Sunny site: Between sunrise and sunset, the plants are almost continuously irradiated by direct sunlight.
Sweet Briar rose: *Rosa eglanteria,* also Eglantine Rose (formerly called *R. rubiginosa*), native wild rose, interesting hedge rose, bird protection and bird-food shrub, leaves with apple fragrance. The English Lord Penzance bred a number of healthy, very robust Eglanteria varieties at the end of the nineteenth century.
Syrphid flies: Beneficial insects. The larvae spear the aphids with their mouthparts and suck the juice out of them. *Rugosa* hybrids are favorite landing targets.
Systemic pesticides: Preparations that penetrate to the interior of the plant and protect plants from within since sucking insects take up the pesticide with the stream of juice.

T

Taproot: Arrowlike, very deep-growing root. Roses usually develop an arrowlike root.
T-cut: See Grafting.
Tea roses: Old rose group with very double flowers. Scarcely winter hardy in colder climates. Arise from red, rose-pink, and light-yellow roses that sailors brought back from China (not to be confused with China roses) on tea ships. At the beginning of the nineteenth century there began intensive breeding work on the tea roses.
Thorns: Thorns are outgrowths of the outer bark layer and are firmly attached to the branch. Roses do not have thorns, botanically speaking, but have prickles.
Thorn stripper: See Dethorner.
Threadworms: See Nematodes.
Thrips: Insects; primarily found in buds about to break. The petals deform, brown flecks appear on the edges of the petals.
Topping: Removal of wild crown of grafted roses in spring.
Trademark: Identifying mark on words and pictures. Only the trademark owner may use his or her trademark. Almost all new rose varieties are legally protected by trademark. The user can tell whether a rose variety enjoys trademark and/or variety protection by the label that is attached to the plant.
Tree saws: Saws with foxtail-like teeth with a narrow, replaceable, sometimes folding blade.
Trenching: Deep digging of the garden soil with the goal of mixing the different soil layers.

U

Umbel/false umbel: Inflorescence; compressed cluster whose stemmed flowers extend raylike from a single point.
Understock: Rose seedling onto which a rose is grafted.
Unhilling: Springtime removal of the soil heaped up around the base of a rosebush as a winter protection.
Upward-climbing rose cane borer: See Cane borer.

V

Valsa disease: Fungal disease, also called cane death; causes withering of the canes, sometimes even the death of the whole rose plant.

Variety and protection: The U.S. Patent and Trademark Office grants protection of a variety after reporting of a new plant variety that is different from others already existing. The protection secures the right of the breeder to his or her product and through the granting of licenses enables the recovery of the costs of breeding. The user knows whether a rose variety is protected by the labeled tags fastened to the roses.

W

Weeping standard: See Cascade standards.
Wetted sulfur: Inexpensive means of combatting powdery mildew; however, leaves spray spots on leaves.
"White Rose of York": Synonym for 'Semiplena', a *Rosa* x *alba* hybrid. Grown in Bulgaria for production of rose oil.
Winter hand grafting: Method of propagation. In the winter, runners of cutting-rose varieties are grafted onto sturdy understocks in the hothouse.
Woods stock: Wild rose plant harvested in woods for use as grafting understock for standard roses; usually used in earlier times in contrast with the modern practice of growing plants for the purpose.

Y

Yellow Rose of Texas: 'Harison's Yellow', *Rosa* x *harisonii.* Species hybrid with bright-yellow, semidouble flowers, which is the subject of the Texas folk song, the "Yellow Rose of Texas."
"York-and-Lancaster Roses": Synonym for *Rosa* x *damascena* 'Versicolor'. Very old variety with bicolored, pink-and-white, fragrant flowers, which are supposed to be the roses in the coats of arms of the English noble houses of York (white rose) and Lancaster (red rose).
Young wood: Denotes the wood of one-year-old canes.

APPENDIX

U.S. Rose Gardens to Visit

American Rose Society
8877 Jefferson-Paige Road
Shreveport, LA 71119
318-938-5402
750 plants

Bellingrath Gardens
12401 Bellingrath Gardens Road
Theodore, AL 36582
334-973-2217
3,500 plants

Boerner Botanical Gardens
5879 South 92nd Street
Hales Corner, WI 53130
414-425-1131
3,500 plants

Boys Town AARS Constitution Rose Garden
Father Flanagan's Boys Home
Boys Town, NE 68010
402-498-1104
2,000 plants

Brooklyn Botanic Garden
Cranford Rose Garden
1000 Washington Avenue
Brooklyn, NY 11225
718-622-4433
more than 5,000 plants

Chicago Botanic Garden
Bruce Krasberg Rose Garden
Lake Cook Road
Glencoe, IL 60022
847-835-8325
5,000 plants

Columbus Park of Roses
3923 North High Street
Columbus, OH 43214
614-645-6648
more than 11,000 plants

Descanso Gardens International Rose Garden
1418 Descanso Drive
La Canada, CA 91011
818-952-4396
5,000 plants

Elizabeth Park Rose Garden
Prospect & Asylum Avenues
West Hartford, CT 06119
860-722-4321
15,000 plants

Exposition Park Rose Garden
701 State Drive
Los Angeles, CA 09937
213-748-4772
15,000 plants

International Rose Test Garden
400 Southwest Kingston Avenue
Portland, OR 97201
503-823-3636
10,000 plants

Lyndale Park Rose Garden
4125 East Lake Harriet Parkway
Minneapolis, MN 55409
612-661-4875
3,000 plants

Michigan State University
Horticultural Demonstration Gardens
East Lansing, MI 48823
517-353-4800
1,000 plants

Missouri Botanic Garden
Gladney & Lehmann Rose Gardens
4344 Shaw Boulevard
St. Louis, MO 63110
314-577-5189
3,500 plants

Robert Pyle Memorial Rose Garden
Rts. 1 & 796
Jennersville, PA 19390
800-458-6559
2,100 plants

The Biltmore Estate
1 Lodge St.
Asheville, NC 28803
704-274-6246
3,000 plants

Tyler Municipal Rose Garden
420 South Rose Park Drive
Tyler, TX 75702
903-531-1200
30,000 plants

Woodland Park Rose Garden
5500 Phinney Avenue North
Seattle, WA 98103
206-684-4863
5,000 plants

Mail-order Nursery Sources for Roses

Antique Rose Emporium
Rt. 5, Box 143
Brenham, TX 77833
800-441-0002
Catalog $5

Edmond's Roses
6235 S.W. Kahle Road
Wilsonville, OR 97979
503-682-1476

Heirloom Old Garden Roses
24062 Riverside Drive Northeast
Saint Paul, OR 97137
503-538-1576
Catalog $5

Heritage Rose Gardens
Tanglewood Farms
16831 Mitchell Creek Drive
Fort Bragg, CA 95437
707-964-3748
Catalog $1.50

Hortico, Inc.
723 Robson Road
R.R. 1
Waterdown, Ontario L0R 2H1
Canada
416-689-6984

Jackson & Perkins
1 Rose Lane
Medford, OR 97501
800-292-4769

Milaeger's Gardens
4838 Douglas Avenue
Racine, WI 53402-2498
414-639-1855

Nor'East Miniature Roses, Inc.
58 Hammond Street
P.O. Box 307
Rowley, MA 01969
800-426-6485

Roses of Yesterday & Today
802 Brown's Valley Road
Watsonville, CA 95076
408-724-2755
Catalog $3.00

Spring Valley Roses
N7637-330th Street
Spring Valley, WI 54767
715-778-4481

W. Atlee Burpee & Co.
Warminter, PA 18974
800-888-1447

Wayside Gardens
Hodges, SC 29695-0001
800-845-1124

White Flower Farm
Litchfield, CT 06759-0050
203-496-9600

Literature

American Rose Society, *American Rose Annual.* Published annually by the ARS.

Austin, David, *The Heritage of the Rose,* Woodbridge, England, Antique Collector's Club, 1988.

Bales, Suzanne Frutig, *Burpee American Gardening Series: Roses,* New York, MacMillan, 1994.

Beales, Peter, *Classic Roses,* New York, Henry Holt, 1985.

Christopher, Thomas, *In Search of Lost Roses,* New York, Summit Books, 1989.

Harkness, Jack, *How to Grow Roses,* St. Alban's England: The Royal National Rose Society, 1988.

LeRougetel, Hazel, *A Heritage of Roses,* Owing Mills, MD, Stemmer House. 1988.

MacCaskey, Michael, and Richard, Ray, *Roses: How to Select, Grow, and Enjoy,* Tucson, AZ, HP Books, 1985.

Macoboy, Stirling, *The Ultimate Rose Book,* New York, Harry Abrams, 1993.

Martin, Clair G., *100 English Roses for the American Garden,* New York, Workman Publishing, 1997.

McCann, Sean, *Miniature Roses: Their Care and Cultivation,* New York, Prentice Hall, 1991.

Oster, Maggie, *10 Steps to Beautiful Roses,* Pownal, VT, Storey Communications, 1989.

Oster, Maggie, *The Rose Book,* Emmaus, PA., Rodale Books, 1994.

Pesch, Barbara B., ed., *Roses,* Brooklyn, NY, Brooklyn Botanic Garden, 1990.

Phillips, Roger, and Martyn Rix, *The Random House Guide to Roses,* New York, Random House, 1988.

Reddell, Rayford Clayton, *Growing Good Roses,* New York, Harper & Row, 1988.

Reddell, Rayford Clayton, *The Rose Bible,* New York, Harmony Books, 1994.

Scanniello, Stephen, *A Year of Roses,* New York, Henry Holt & Co., 1997.

Scanniello, Stephen, and Tania Bayard, *Roses of America,* New York, Henry Holt & Co., 1990.

Schneider, Peter, *Taylor's Guide to Roses,* New York, Houghton Mifflin Co., 1995.

Thomas, Graham Stuart, *Climbing Roses Old & New,* London, J.M. Dent & Sons, 1983.

Thomas, Graham Stuart, *Shrub Roses of Today,* London, J.M. Dent & Sons, 1980.

Thomas, Graham Stuart, *The Old Shrub Roses,* London, J.M. Dent & Sons, 1955. Revised 1978.

Walheim, Lance, and The Editors of The National Gardening Association, *Roses for Dummies,* Foster City, CA, IDG Books, 1997.

Welch, William C., *Antique Roses for the South,* Dallas, TX, Taylor Publishing, 1990.

Periodicals

American Rose Rambler, a bimonthly newsletter edited by Peter Schneider. Contact: Peter Schneider, P.O. Box 677, Mantua, OH 44255.

Bev Dobson's Rose Letter, bimonthly newsletter. Contact: Beverly R. Dobson, 215 Harriman Road, Irvington, NY 10533

Rose Societies and Organizations

American Rose Society
8877 Jefferson Paige Road
P.O. Box 30000
Shreveport, LA 71119-8817
Tel. 318-938-5402
FAX: 318-938-5405

Canadian Rose Society
c/o Anne Graber
10 Fairfax Crescent, Scarborough, Ontario M1L 1Z8, Canada

The Heritage Roses Foundation
c/o Mr. Charles Walker, Jr.
1512 Gorman Street
Raleigh, NC 27606

CD-ROM

Botanica's Roses—from Macmillan & Co., New York.

Internet Sources for Rose Information

Web sites devoted soley to roses:

Rose Resource—www.rose.org

Santa Clara County Rose Society Links Page—this site offers dozens of links to rose societies, related home pages, commericial nurseries, extension services, colleges, and government agricultural information—http://mejac.palo-alto.ca.us/orgs/sccrs/other.html

The Rosarian - www.rosarian.com

American Rose Society—www.ars.org

Yesterday's Rose: A Tribute to Old and Old-Fashioned Roses—www.Country Lane.com

These sites provide general gardening information and links to information on roses:

The Garden Gate—www. prairienet.org/garden-gate

PLANTAMERICA—www. plantamerica.com

National Gardening Association—www. garden.org

Garden Net—www. gardennet.com

The Virtual Garden—www. vg.com

Garden Escape—www.garden.com

Index

Picture Credits

Archiv für Kunst und Geschichte: 10, 11, 12 right, 13, 14, 15 left, 15 right, 16, 17 bottom
Bieker: 8-9, 15 top right, 105, 121 bottom, 122 left, 184-185
Borstell: 1, 4-5, 28-29, 34 top left, 34 bottom left, 35 center right, 39, 46-47, 47, 48 top, 52, 54, 61 left, 61 top right, 61 bottom right, 69 bottom left, 70 bottom, 75 bottom, 80-81 top, 81 top right, 82 top, 84 top, 84 bottom, 84-85, 88, 91, 97, 98-99, 99 right, 100 top, 100 bottom, 103 bottom, 109, 125, 130 bottom, 185, 191 top, 196 top left, 197, 198 top right, 201 top, 202 top right, 204 bottom, 207 right, 209 left, 210 right, 213 top, 214 top, 214 bottom, 216 bottom right, 217 center, 220 right, 222 top right, 223 bottom, 224-225
Fischer: 75 top, 85, 106
Hagen: 90 right
Hoppe: 71 bottom right
Kögel: 24 top left, 24-25 top, 53, 69 right, 89 top, 123, 188 bottom right, 208 right, 218 left
Kordes/Klein Offenseth-Sparrieshoop: 59 top left, 103 top, 190 top center, 196 bottom right, 198 left, 198 bottom right, 199 top, 204, 205 center, 215 top, 216 top right, 219 center, 221 right, 223 top
Meilland-BKN Strobel/Pinneberg: 21 bottom, 60, 102, 107 bottom left, 107 top right, 115, 163 top, 183 bottom, 190 bottom center, 207 center, 208 center, 210 top left, 215 bottom
Morell: 12 left, 40 top, 69 center left, 141, 201 bottom, 212 bottom left
Pforr: Title, 20 right, 21 top left, 29, 36 right, 70 top right, 71 center left, 71 center right, 76-77, 103 center, 121 top, 136 top right, 138, 148 bottom, 163 bottom, 164 bottom, 165 left, 165 right, 166 right, 167, 168 top left, 168 center, 168 top right, 169 bottom left, 196 top right, 200, 202 bottom, 206 bottom, 212 center, 216 top left, 218 top center, 223 center
Redeleit: 59 bottom left, 71 bottom left, 75 center, 157 right, 183 top
Reinhard: 2-3, 20 left, 24-25 bottom, 27, 30 top, 30 bottom, 37, 38, 40 center, 40 bottom, 41, 43 top, 48 bottom, 57, 66 top, 68, 90 left, 92 right, 95 center, 101, 107 top left, 107 center left, 112, 117, 120, 121 second from top, 122 bottom right, 122-123, 140-141, 157 left, 158, 161, 169, 186, 188 left, 189 top left, 189 top right, 189 bottom, 190 left, 190 top right, 191 bottom, 193 top, 193 bottom, 194 left, 194 center, 195 left, 195 center, 196 bottom left, 199 center, 203, 204 top, 205 left, 206 left, 206 center, 207 left, 208 left, 209 right, 210 bottom left, 211 left, 212 top, 213 bottom, 216 bottom left, 217 top, 218 bottom center, 218 right, 219 bottom, 220 left, 221 left, 222 top left, 222 bottom
Romeis: Endpapers, 6-7, 9, 36-37, 82 bottom, 219 top
Rosen Tantau/Uetersen: 107 center right, 107 bottom right, 179, 181 top, 181 bottom
Sammer: 34 bottom right, 55 top, 80 top, 81 top left, 110-111, 111, 114, 122 top right, 127 top, 127 bottom, 128 top left, 128 top right, 128 bottom, 129, 130 top, 132, 134, 136 top left, 136 bottom, 137, 139, 145, 159 top, 160 top, 166 left, 194 right, 212 bottom right
Seidl: 34 top right, 89 center, 89 bottom, 94, 95 top, 95 bottom, 121 second from bottom, 192 bottom, 199 bottom, 205 right, 217 bottom
Stangl: 148 top, 176 bottom
Stein: 92 left, 108
Timmermann/Rottenburg: 71 top left
All other photographs by the author

The illustration on page 17 top is used with kind permission from the book *The Little Prince* by Antoine de Saint Exupéry (Karl Rauch Publishers, Düsseldorf, Germany).

Jacket picture: 'Pariser Charme'
Page 1: 'Donau' and 'Bobbie James'
Pages 2–3: 'Graham Thomas'
Pages 4–5: 'Direktor Benschop'
Pages 6–7: 'Peace'

Author's and Publisher's Acknowledgements
. . . to three people who have the letters of the rose in their names: Klaus-Jürgen Strobel, Ute and Wilhelm Kordes, as well as to Birgit Markley, who corrected the proofs, and to the firms of BKN Strobel, W. Kordes' Söhne, Noack Rosen, Rosen Tantau, and Rosen-Union for making their picture material available.

Title of original German edition: *DIE BLV ROSEN-ENZYKLOPÄDIE*

All inquiries should be addressed to:
Barron's Educational Series, Inc.
250 Wireless Boulevard
Hauppauge, New York 11788
http://www.barronseduc.com

Library of Congress Catalog Card No. 99-64236

International Standard Book No. 0-7641-5193-2

PRINTED IN HONG KONG
9 8 7 6 5 4 3 2 1

Plant Hardiness Zone Map

The United States Department of Agriculture (USDA) Zone Map divides the country into 11 major climatic zones. A zone is an area of the country that has roughly the same average annual minimum temperature.

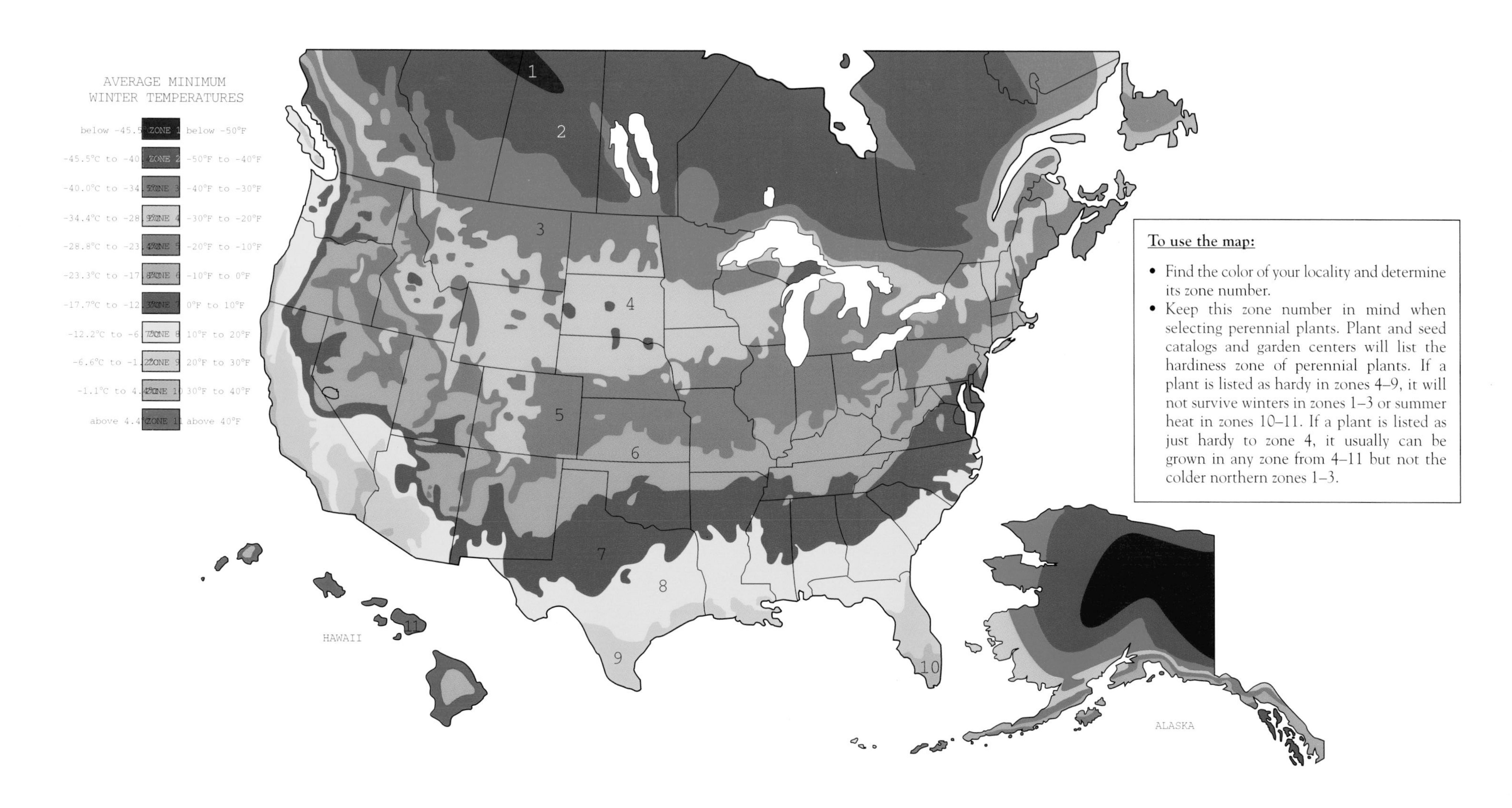

To use the map:

- Find the color of your locality and determine its zone number.
- Keep this zone number in mind when selecting perennial plants. Plant and seed catalogs and garden centers will list the hardiness zone of perennial plants. If a plant is listed as hardy in zones 4–9, it will not survive winters in zones 1–3 or summer heat in zones 10–11. If a plant is listed as just hardy to zone 4, it usually can be grown in any zone from 4–11 but not the colder northern zones 1–3.